Studies in Classification, Data Analysis, and Knowledge Organization

Titles in the Series

E. Diday, Y. Lechevallier, M. Schader, P. Bertrand, and B. Burtschy (Eds.)
New Approaches in Classification and Data Analysis. 1994 (out of print)

W. Gaul and D. Pfeifer (Eds.)
From Data to Knowledge. 1995

H.-H. Bock and W. Polasek (Eds.)
Data Analysis and Information Systems. 1996

E. Diday, Y. Lechevallier, and O. Opitz (Eds.)
Ordinal and Symbolic Data Analysis. 1996

R. Klar and O. Opitz (Eds.)
Classification and Knowledge Organization. 1997

C. Hayashi, N. Ohsumi, K. Yajima, Y. Tanaka, H.-H. Bock, and Y. Baba (Eds.)
Data Science, Classification, and Related Methods. 1998

I. Balderjahn, R. Mathar, and M. Schader (Eds.)
Classification, Data Analysis, and Data Highways. 1998

A. Rizzi, M. Vichi, and H.-H. Bock (Eds.)
Advances in Data Science and Classification. 1998

M. Vichi and O. Opitz (Eds.)
Classification and Data Analysis. 1999

W. Gaul and H. Locarek-Junge (Eds.)
Classification in the Information Age. 1999

H.-H. Bock and E. Diday (Eds.)
Analysis of Symbolic Data. 2000

H. A. L. Kiers, J.-P. Rasson, P. J. F. Groenen, and M. Schader (Eds.)
Data Analysis, Classification, and Related Methods. 2000

W. Gaul, O. Opitz, and M. Schader (Eds.)
Data Analysis. 2000

R. Decker and W. Gaul (Eds.)
Classification and Information Processing at the Turn of the Millenium. 2000

S. Borra, R. Rocci, M. Vichi, and M. Schader (Eds.)
Advances in Classification and Data Analysis. 2001

W. Gaul and G. Ritter (Eds.)
Classification, Automation, and New Media. 2002

K. Jajuga, A. Sokołowski, and H.-H. Bock (Eds.)
Classification, Clustering and Data Analysis. 2002

M. Schwaiger and O. Opitz (Eds.)
Exploratory Data Analysis in Empirical Research. 2003

M. Schader, W. Gaul, and M. Vichi (Eds.)
Between Data Science and Applied Data Analysis. 2003

H.-H. Bock, M. Chiodi, and A. Mineo (Eds.)
Advances in Multivariate Data Analysis. 2004

D. Banks, L. House, F. R. McMorris, P. Arabie, and W. Gaul (Eds.)
Classification, Clustering, and Data Mining Applications. 2004

D. Baier and K.-D. Wernecke (Eds.)
Innovations in Classification, Data Science, and Information Systems. 2005

M. Vichi, P. Monari, S. Mignani, and A. Montanari (Eds.)
New Developments in Classification and Data Analysis. 2005

D. Baier, R. Decker, and L. Schmidt-Thieme (Eds.)
Data Analysis and Decision Support. 2005

C. Weihs and W. Gaul (Eds.)
Classification – the Ubiquitous Challenge. 2005

M. Spiliopoulou, R. Kruse, C. Borgelt, A. Nürnberger, and W. Gaul (Eds.)
From Data and Information Analysis to Knowledge Engineering. 2006

V. Batagelj, H.-H. Bock, A. Ferligoj, and A. Žiberna (Eds.)
Data Science and Classification. 2006

Sergio Zani · Andrea Cerioli
Marco Riani · Maurizio Vichi
Editors

Data Analysis, Classification and the Forward Search

Proceedings of the Meeting of the Classification and Data Analysis Group (CLADAG) of the Italian Statistical Society, University of Parma, June 6–8, 2005

With 118 Figures and 50 Tables

Springer

Prof. Sergio Zani
Department of Economics
Section of Statistics
and Computing
University of Parma
Via Kennedy 6
43100 Parma, Italy
sergio.zani@unipr.it

Prof. Marco Riani
Department of Economics
Section of Statistics
and Computing
University of Parma
Via Kennedy 6
43100 Parma, Italy
mriani@unipr.it

Prof. Andrea Cerioli
Department of Economics
Section of Statistics
and Computing
University of Parma
Via Kennedy 6
43100 Parma, Italy
andrea.cerioli@unipr.it

Prof. Maurizio Vichi
Department of Statistics,
Probability and Applied
Statistics
University of Rome
"La Sapienza"
Piazzale Aldo Moro 5
00185 Roma, Italy
maurizio.vichi@uniroma1.it

ISSN 1431-8814
ISBN 10 3-540-35977-X Springer Berlin Heidelberg New York
ISBN 13 978-3-540-35977-7 Springer Berlin Heidelberg New York

Springer-Verlag is a part of Springer Science+Business Media

springer.com

Cover: Erich Kirchner, Heidelberg
Production: LE-TEX, Jelonek, Schmidt & Vöckler GbR, Leipzig

SPIN 11789703 Printed on acid-free paper – 43/3100 – 5 4 3 2 1 0

Preface

This volume contains revised versions of selected papers presented at the biennial meeting of the Classification and Data Analysis Group (CLADAG) of the Italian Statistical Society, which was held in Parma, June 6-8, 2005. Sergio Zani chaired the Scientific Programme Committee and Andrea Cerioli chaired the Local Organizing Committee.

The scientific programme of the conference included 127 papers, 42 in specialized sessions, 68 in contributed paper sessions and 17 in poster sessions. Moreover, it was possible to recruit five notable and internationally renowned invited speakers (including the 2004–2005 President of the International Federation of Classification Societies) for plenary talks on their current research work. Among the specialized sessions, two were organized by Wolfgang Gaul with five talks by members of the GfKl (German Classification Society), and one by Jacqueline J. Meulman (Dutch/Flemish Classification Society). Thus, the conference provided a large number of scientists and experts from home and abroad with an attractive forum for discussion and mutual exchange of knowledge. The topics of all plenary and specialized sessions were chosen to fit, in the broadest possible sense, the mission of CLADAG, the aim of which is "to further methodological, computational and applied research within the fields of Classification, Data Analysis and Multivariate Statistics".

A peer-review refereeing process led to the selection of 46 extended papers, which are contained in this book. The more methodologically oriented papers focus on developments in clustering and discrimination, multidimensional data analysis, data mining, and robust statistics with a special emphasis on the novel Forward Search approach. Many papers also provide significant contributions in a wide range of fields of application. Customer satisfaction and service evaluation are two examples of such emerging fields. This suggested the presentation of the 46 selected papers in six parts as follows:

1. Clustering and Discrimination
2. Multidimensional Data Analysis and Multivariate Statistics
3. Robust Methods and the Forward Search
4. Data Mining Methods and Software
5. Multivariate Methods for Customer Satisfaction and Service Evaluation
6. Multivariate Methods in Applied Science

We wish to express our gratitude to the other members of the Scientific Programme Committee

B. Chiandotto, N.C. Lauro, P. Monari, A. Montanari, C. Provasi, G. Vittadini

and to the specialized session organizers

F. Camillo, M. Chiodi, W. Gaul, S. Ingrassia, J.J. Meulman

for their ability to attract interesting contributions, and to the authors, whose enthusiastic participation made the meeting possible. We would also like to extend our thanks to the chairpersons and discussants of the sessions for their stimulating comments and suggestions. We are very grateful to the referees for their careful reviews of all submitted papers and for the time spent in this professional activity.

We gratefully acknowledge the University of Parma and its Department of Economics for financial support and hospitality. We are also indebted to *Istat - Istituto Nazionale di Statistica* and *SAS* for their support.

We thank all the members of the Local Organizing Committee

A. Corbellini, G. Gozzi, L. Grossi, F. Laurini, M.A. Milioli, G. Morelli, I. Morlini

for their excellent work in managing the organization of the CLADAG-2005 conference. Special thanks go to Prof. Isabella Morlini, for her skilful accomplishment of the duties of Scientific Secretary of CLADAG-2005, and to Dr. Fabrizio Laurini for his assistance in producing this volume.

Finally, we would like to thank Dr. Martina Bihn of Springer-Verlag, Heidelberg, for her support and dedication to the production of this volume.

Parma and Rome,
June 2006

Sergio Zani
Andrea Cerioli
Marco Riani
Maurizio Vichi

Contents

Part I

Clustering and Discrimination

Genetic Algorithms-based Approaches for Clustering Time Series

Roberto Baragona[1] and Salvatore Vitrano[2]

[1] Department of Sociology and Communication, University of Rome "La Sapienza", Italy roberto.baragona@uniroma1.it
[2] Statistical Office, Ministry for Cultural Heritage, Italy svitrano@beniculturali.it

Abstract. Cluster analysis is to be included among the favorite data mining techniques. Cluster analysis of time series has received great attention only recently mainly because of the several difficult issues involved. Among several available methods, genetic algorithms proved to be able to handle efficiently this topic. Several partitions are considered and iteratively selected according to some adequacy criterion. In this artificial "struggle for survival" partitions are allowed to interact and mutate to improve and produce a "high quality" solution. Given a set of time series two genetic algorithms are considered for clustering (the number of clusters is assumed unknown). Both algorithms require a model to be fitted to each time series to obtain model parameters and residuals. These methods are applied to a real data set concerned with the visitors flow recorded, in state owned museums with paid admission, in the Lazio region of Italy.

1 Introduction

Clustering time series, that is division of time series into homogeneous subgroups, is composed of several steps. First, pre-processing is almost always needed for removing or softening unwanted characteristics that may bias the analysis (for instance, outliers, missing observations and Easter and trading day effects). Moreover, adjustment for seasonality and trend could possibly be required to allow some methods to run properly. Among many available methods X-12 ARIMA (Findley et al. (1998)) and Tramo-Seats (Gómez and Maravall (1996)) are currently adopted by Statistical Authorities in many Countries. They are well founded on theoretical grounds and supported by computer programs that make easier their application.

Then, the extraction of measurements may take place so that either the usual matrix units (time series) per variables (the measurements) or a matrix of distances between each pair of time series is available. Liao (2005) in a comprehensive survey distinguishes whether methods are based on the raw data directly, on features extracted from the data or on models built on the data.

Third step, choosing the cluster method, is closely related to the preceding step as the method largely depends on the available data structure. Four main classes may be distinguished, that is partitioning methods, hierarchical methods, density-based clustering and grid-based clustering (see, for instance, Berkhin (2002) for a comprehensive survey).

The fourth step is concerned with the choice of the algorithm. As clustering problems arise in so many fields (these range from sociology and psychology to commerce, biology and computer science) the implementation and design of algorithms continue to be the subject of active research. Over the last two decades a new class of algorithms has been developed, namely the optimization heuristics. Examples are evolutionary algorithms (simulated annealing, threshold accepting), neural networks, genetic algorithms, tabu search, ant colony optimization (see, for instance, Winker and Gilli (2004)). Optimization heuristics may cope with problems of high complexity whose potential solutions are a large discrete set. This is the case of the admissible partitions of a set of time series. In addition, assumptions on the form of the final partition (either hard or fuzzy assignments), for instance, or on the number of clusters (either known or unknown) are easily handled by optimization heuristics and require only slight modifications of the basic algorithm. Clustering time series by meta heuristic methods was investigated by Baragona (2001) while Pattarin et al. (2004) examined genetic algorithms-based approaches.

In this paper genetic algorithms (GAs) are used for implementing two model-based-methods, the first one based on the cross correlations (Zani (1983)), the second one based on the autoregressive distance (Piccolo (1990)). Other optimization heuristics may be of use, but GAs seem to ensure most flexibility and vast choice to meet the special requirements involved in clustering time series. Both algorithms are tested on the data set of the visitors of museums, monuments and archaeological sites in the Lazio region of Italy.

The rest of the paper is organized as follows. The next Section includes a description of the two clustering time series methods and the GAs are described. Results of the application to the real data set are displayed in Section 3. Section 4 concludes.

2 Clustering Methods and Genetic Algorithms

Given a set of time series two methods are considered for clustering. Both methods require a model to be fitted to each time series to obtain model parameters and residuals. The number of clusters g is assumed unknown. The fitted models are autoregressive integrated moving-average (ARIMA) models (Box et al. (1994)). The first method is aimed at grouping together time series according to the residuals cross correlations. The second method is aimed at grouping together time series that share a similar model's structure.

Let the time series $\{x_t\}$ be generated by the ARIMA$(p,d,q)(P,D,Q)_s$ model

$$\phi(B)\Phi(B^s)\nabla^d\nabla_s^d x_t = \theta(B)\Theta(B^s)a_t, \tag{1}$$

where $\{a_t\}$ is a white noise process with finite variance σ^2. The polynomials in (1) are defined

$$\phi(B)\Phi(B^s) = (1-\phi_1 B - \ldots - \phi_p B^p)(1-\Phi_1 B^s - \ldots - \Phi_P B^{Ps}) \tag{2}$$

$$\theta(B)\Theta(B^s) = (1-\theta_1 B - \ldots - \theta_q B^q)(1-\Theta_1 B^s - \ldots - \Theta_Q B^{Qs}), \tag{3}$$

and have not common factors. B is the back-shift operator, that is $B^k x_t{=}x_{t-k}$. Model (1) is stationary if the polynomial (2) has all roots outside the unit circle and is invertible if the polynomial (3) has all roots outside the unit circle. Under this latter assumption, the time series $\{x_t\}$ admits the infinite autoregressive representation $x_t = \Sigma\pi_k x_{t-k}$. The coefficients π_k are called π-weights.

For the first method a cluster is required to satisfy the following condition (Zani (1983)). Given a set of k time series $\{\mathbf{x}_1,\ldots,\mathbf{x}_k\}$, $\mathbf{x}_i = (x_{i,1},\ldots,x_{i,n})$, $i=1,\ldots,k$, a subset C which includes k' series $(k'<k)$ is said to form a group if, for each of the $k'(k'-1)/2$ residuals cross-correlations $\rho_{i,j}(\tau)$, we have

$$|\rho_{i,j}(\tau)| > c(\alpha) \tag{4}$$

for at least a lag τ between $-m$ and m, and $i,j \in C, i \neq j$. A positive integer m has to be pre-specified which denotes the maximum lag. If all time series have n as a common number of observations, then choosing the significance level $\alpha = 0.05$, say, gives the figure $c(\alpha) = 1.96/\sqrt{n}$ in (4). The previously stated definition does not exclude that a time series may belong to more than a single group. Then there are possibly several allowable partitions to consider, and their number may be very large. Meta heuristic methods, in particular GAs, were proposed by Baragona (2001) to find the best feasible partition. As an overall objective function a modification of the k-min cluster criterion (Sahni and Gonzalez (1976)) was assumed

$$f^+(C_1,C_2,\ldots,C_g;g) = \sum_{\omega=1}^{g}\sum_{i,j\in C_\omega, i\neq j} d^+_{i,j}, \tag{5}$$

where (5) has to be maximized and

$$d^+_{i,j} = \max\{|\rho_{i,j}(\tau)|\}, \qquad \tau = -m,\ldots,m. \tag{6}$$

When using (5), it is crucial that each cluster be a group, according to (4), for, otherwise, any algorithm, unless prematurely ended, will put together all time series into a single cluster.

The second method is new and has been developed along the same guidelines, except that the autoregressive distance proposed by Piccolo (1990) is adopted instead of the cross-correlations-based dissimilarity index. Each time series $\{x_t, t = 1, 2, \ldots n\}$ is associated to the first m π-weights of the autoregressive representation of the ARIMA model. The first m π-weights may be computed from the coefficients of the ARIMA model (1) using the equations given in Box et al. (1994). The positive integer m is a truncation point that has to be pre-specified. For each time series there is a set of measurements $\pi_{v,1}, \pi_{v,2}, \ldots \pi_{v,m}, v = 1, 2, \ldots k$ that allows clusters to be determined. The π-weights define the autoregressive distance

$$d_{i,j} = \sqrt{\sum_{h=1}^{m} (\pi_{i,h} - \pi_{j,h})^2}. \tag{7}$$

The distributional properties of the autoregressive distance (7) were studied by Piccolo (1990) under the assumption that the time series are uncorrelated. The presence of correlation between time series was shown (Corduas (1992)) to modify the distribution of (7). In Corduas (2000) an approximation is provided that allows a formal test to be established. The squared autoregressive distance $d_{i,j}^2$ is approximately distributed as a random variable $a\chi_\nu^2 + b$ where χ_ν^2 is the chi-squared random variable with ν degrees of freedom. The constants a, b and ν depend on the common (under the null hypothesis that the time series $\{x_{i,t}\}$ and $\{x_{j,t}\}$ are generated by the same ARIMA model) variance-covariance matrix of the estimated π-weights. The condition for a set of time series to form a group has to be re-formulated in terms of the approximate critical values that may be computed for the squared distance (7). Two time series are allowed to be included in the same cluster if their squared distance is less than $a\chi_\nu^2(\alpha) + b$, where $\chi_\nu^2(\alpha)$ is the $(1-\alpha)$-quantile of the chi-squared distribution with ν degrees of freedom and α is the significance level.

It may be argued, extending the results reported by Corduas (1992), that if the cross correlations $\rho_{i,j}$ are close to unity the two methods are likely to yield similar results. Note that Tong and Dabas (1990) include the index (6), with $\tau = 0$, among the measures of dissimilarity for a set of time series models though no threshold-based constraints were introduced.

GAs were introduced by Holland (1975) to provide evolutionary models for the adaptation process to the environment of individuals belonging to a given population. If the evolution of the best fit individual is recorded, GAs may be viewed as optimization tools. In this case, a numerical measure is used to evaluate the adaptation to the environment. This measure is called fitness function and it is the objective function which has to be maximized. The fitness function is not required to possess special mathematical properties but to be positive and non-decreasing function of the adaptation to the environment. Each potential solution to the optimization problem has to be coded as a string of ℓ characters, for instance a binary string of length ℓ (this string is usually called chromosome). There is no need to enumerate all solutions, which has to be considered fairly impossible, but only a set of rules by which any string may be decoded in a meaningful way and assigned a positive real number. In practice GAs are useful because they allow very large spaces to be searched for solutions and very mild assumptions are required for the objective function. For a detailed description of GAs see, for instance, Goldberg (1989) and Haupt and Haupt (2004). Convergence properties have been discussed by Reeves and Rowe (2003), among others.

GAs start with an initial population of candidate solutions (individuals) that are said to form a population though they are actually a sample from the set of potential solutions. The population evolves through a iterative procedure. Each iteration usually includes three steps, that is selection, crossover and mutation. Selection is aimed at choosing the individuals with high fitness. Selection is done with replacement, so that many copies of the same individual may enter the next generation. The chromosomes of the selected individuals are possibly combined, by means of the crossover operator, to produce new individuals. Mutation of some characters within the genetic pool may take place with usually small probability, 0.001 for instance. The generational gap is the fraction of the old population that is replaced by new individuals. If the new population entirely replaces the past one the generational gap is equal to unity. Iterations continue either after a pre-specified number of generations or when a stopping criterion is met. Often the elitist strategy is used, that is the best fit individual found in a iteration is inserted in the next generation unless a better individual is found (see, for instance, Jennison and Sheehan (1995)).

Both methods are implemented by GAs with permutation encoding, often called ordered GAs (Jones and Beltramo (1991)). Time series are assigned labels $1, 2, \ldots k$, and several random permutations are considered. For each one, a random number of cluster g is generated and the g time series at the top of the list are assumed as cluster centers. Then, aggregation of the remaining time series to the nearest center is performed. To improve the solutions, the permutations are evolved through some iterations by the GAs operators selection, crossover and mutation. These have to be specially designed according to chosen encoding method and distance measure. The ordered GAs proved to be very effective in practice as far as clustering procedures are concerned.

3 Application to Real Data

The number of visitors to cultural sites and state owned museums (according to the definition given by the International Council of Museums (ICOM)), with admission fees, in the Lazio region of Italy is collected monthly by the Statistical Office at the Ministry for Cultural Heritage. We considered 37 time series from January 1996 to December 2003. Adjustment for outliers and missing data, and ARIMA model (1) building were performed by the computer program Tramo-Seats (Gómez and Maravall (1996)). This program may be downloaded from the web site http://www.istat.it. All series were reduced to have common number of observations 78. Two time series were discarded because the sites happened to be closed many times during the observation period. For all remaining 35 time series the ARIMA coefficients were used to compute the π-weights. Also, residuals from ARIMA models were recorded and the cross correlations were computed. All estimated ARIMA models used for cluster analysis passed the Ljung-Box test (based on the first 24 residuals autocorrelations) at the 5% significance level (at the 1% significance level for series 10).

The first method groups series with cross correlations absolute values greater than 0.2219 (1.96 divided by the square root of the number of observations). The clusters that have been formed are reported in Table 1 (series are numbered from 1 to 35).

Cluster	Time series
(1)	2 5 10 11 14 15 21 22
(2)	19 23 24 25 26 29 32 35
(3)	6 7 27 30 31
(4)	8 9 16 17 18 20
(5)	4 12 13 28 33
(6)	1 34
(7)	3

Table 1. Clusters of cultural sites visitors in Lazio, Italy (first method)

The second method groups the time series according to the π-weights. Time series may be grouped together only if their pairwise squared autoregressive distance is less than a threshold whose values for this data set have been found to vary in the interval $[0.12, 0.17]$ (for this method thresholds for acceptance vary through time series pairs). The two methods are likely to yield similar results if the cross correlations are large. This is not the present case, however, as most cross correlations are less than 0.3, some are between 0.3 and 0.5 and only one cross correlation exceeds 0.5 (it is about 0.6). We obtained 9 clusters that are reported in Table 2. First and second clusters contain 23 out of 35 series. Clusters 3 and 6 are very small. The first one includes series that are similar to series in clusters 1 and 2, while the second one contains series that are considerably different. As far as time series 26, 34, 12, 17 and 23 are concerned, each one forms a single cluster.

The first method seems to cluster together time series according to spatial and typological closeness. Cluster 1 includes museums located in Rome. Cluster 2 include archaeological areas in Rome too. Cluster 3 includes archaeological areas as well, but out of Rome. The sites that belong to the historic / artistic local authority are in cluster 4. Evidence of typological clustering is provided by cluster 5, which includes four sites mostly archae-

Cluster	Time series
(1)	3 4 6 7 11 13 15 28 30 32 35
(2)	1 2 9 10 16 18 19 20 21 24 25 33
(3)	5 27 31
(4), (5)	26, 34
(6)	8 14 22 29
(7), (8), (9)	12, 17, 23

Table 2. Clusters of cultural sites visitors in Lazio, Italy (second method)

ological museums. The second method seems to group time series according to sites easy to access as perceived by visitors. This is a new interesting issue that emerges from cluster analysis. Foreign tourists, for instance, are unlikely to visit remote sites while it is easy for a school to visit sites nearby. Also, scholars may consider important to visit sites though difficult to access. These circumstances produce different time series dynamic behaviors. The first cluster includes most archaeological sites out of Rome that are likely to attract mostly local public. Cluster 3 is very similar. Cluster 2 includes many museums in Rome that foreign tourists, for instance, hardly miss to visit. Clusters with single time series are explained by some special characteristics of the site. Cluster 4, for example, includes only Villa d'Este - Tivoli. Unlike most sites out of Rome, this site attracts a rather regular visitors flow. Another example is cluster 5 that includes the archaeological tour: Colosseum, Palatine Hill and Forum. This site too is peculiar as it sells combined tickets, that is tickets that allow the tourist to visit these and other sites.

4 Concluding Remarks

Genetic algorithms were designed to implement two methods for clustering time series. The first one is based on the residuals cross correlations, the second one on the time series models structures. As an example, the two methods are applied for clustering the time series of the visitors to the museums, monuments and archaeological areas of the Lazio region of Italy. The first method seems to group time series according to the spatial locations of the sites. The second method seems to group the time series together according to both visitors typology and sites characteristics. Cross correlations are rather small (though larger than the critical values) and the two methods are expected to produce different results in this case. Performance of genetic algorithms may be considered quite good. Computations were fast and the results seem to be reliable and accurate.

References

BARAGONA, R. (2001): A Simulation Study on Clustering Time Series with Meta-Heuristic Methods. *Quaderni di Statistica, 3, 1–26.*

BERKHIN, P. (2002): *Survey of Clustering Data Mining Techniques.* Technical Report, Accrue Software, San José, California, http://citeseer.ist.psu.edu/berkhin02survey.html.

BOX, G.E.P., JENKINS, G.M. and REINSEL, G.C. (1994): *Time Series Analysis. Forecasting and Control (3rd Edition).* Prentice Hall, San Francisco.

CORDUAS, M. (1992): Una Nota sulla Distanza tra Modelli ARIMA per Serie Storiche Correlate. *Statistica, 52, 515–520.*

CORDUAS, M. (2000): La Metrica Autoregressiva tra Modelli ARIMA: una Procedura in Linguaggio GAUSS. *Quaderni di Statistica, 2, 1–37.*

FINDLEY, D.F., MONSELL,B.C., BELL, W.R., OTTO, M.C. and CHEN, B.-C. (1998): New Capabilities and Methods of the X-12-ARIMA Seasonal Adjustment Program (with discussion). *Journal of Business and Economic Statistics, 16, 127–176.*

GOLDBERG, D.E. (1989): *Genetic Algorithms in Search, Optimization and Machine Learning.* Addison-Wesley, Reading, Massachusetts.

GÓMEZ, V. and MARAVALL, A. (1996): *Programs Tramo and Seats: Instructions for Users.* Technical Report 9628, The Banco de España, Servicios de Estudios.

HAUPT, R.L. and HAUPT, S.E. (2004): *Practical Genetic Algorithms (2nd Edition).* John Wiley & Sons, Hoboken, New Jersey.

HOLLAND, J.H. (1975): *Adaptation in Natural and Artificial Systems.* University of Michigan Press, Ann Arbor.

JENNISON, C. and SHEEHAN, N. (1995): Theoretical and empirical properties of the genetic algorithm as a numerical optimizer, *Journal of Computational and Graphical Statistics, 4, 296-318.*

JONES, D.R. and BELTRAMO, M.A. (1991): Solving Partitioning Problems with Genetic Algorithms. In: R.K. Belew and L.B. Booker (Eds.): *Proceedings of the Fourth International Conference on Genetic Algorithms.* Morgan Kaufmann, San Diego, California, 442–449.

LIAO, T.W. (2005): Clustering of Time Series Data–A Survey. *Pattern Recognition, 38, 1857–1874.*

PATTARIN, F., PATERLINI, S. and MINERVA, T. (2004): Clustering Financial Time Series: an Application to Mutual Funds Style Analysis. *Computational Statistics & Data Analysis, 47, 353–372.*

PICCOLO, D. (1990): A Distance Measure for Classifying ARIMA Models. *Journal of Time Series Analysis, 11, 153–164.*

REEVES, C.R. and ROWE, J.E. (2003): *Genetic Algorithms - Principles and Perspective: a Guide to GA Theory.* Kluwer Academic Publishers, London.

SAHNI, S. and GONZALEZ, T. (1976): P-Complete Approximation Problems. *Journal of the Association for Computing Machinery, 23, 555–565.*

TONG, H. and DABAS, P. (1990): Cluster of Time Series Models: an Example. *Journal of Applied Statistics, 17, 187–198.*

WINKER, P. and GILLI, M. (2004): Application of Optimization Heuristics to Estimation and Modelling Problems. *Computational Statistics & Data Analysis, 47, 211–223.*

ZANI, S. (1983): Osservazioni sulle Serie Storiche Multiple e l'Analisi dei Gruppi. In: D. Piccolo (Ed.): *Analisi Moderna delle Serie Storiche.* Franco Angeli, Milano, 263–274.

On the Choice of the Kernel Function in Kernel Discriminant Analysis Using Information Complexity

Hamparsum Bozdogan[1], Furio Camillo[2], and Caterina Liberati[2]

[1] Department of Statistics, Operations, and Management Science
The University of Tennessee
Knoxville, TN 37996-0532 U.S.A.
bozdogan@utk.edu

[2] Dipartimento di Scienze Statistiche
Università di Bologna, Italy
{fcamillo, liberati}@stat.unibo.it

Abstract. In this short paper we shall consider the Kernel Fisher Discriminant Analysis (KFDA) and extend the idea of Linear Discriminant Analysis (LDA) to nonlinear feature space. We shall present a new method of choosing the optimal kernel function and its effect on the KDA classifier using information-theoretic complexity measure.

1 Introduction and the Problem

Discriminant analysis (DA) is one of the popular multivariate methods which has a long history. DA is a classification problem that consists of assigning or classifying an individual or object to one of several known or unknown alternative classes (or groups) on the basis of many measurements on the individuals, objects, or cases. The goal of discriminant analysis is: given a data set with two or more than two classes (or groups), say, find the best feature or feature set either linear or non-linear to discriminate between the classes and maximize average class separation. Equivalently, we attempt to minimize the probability of missclassification.

Recently in statistical data mining and knowledge discovery, kernel-based methods have attracted attention from many researchers. As a result, many kernel-based methods have been developed. They have become popular tools for classification, clustering, and regression analysis in the machine learning community since the introduction of support vector machines (SVMs) during the early 1990s. The popularity of the method stems from the fact that kernel methods almost always outperform traditional multivariate statistical techniques. Now, we can carry out kernel based approaches to all the classical multivariate procedures. Examples of these include, kernel principal component analysis (KPCA), kernel logistic regression (KLR), kernel Fisher discriminant analysis (KFDA), or in short kernel discriminant analysis (KDA), kernel canonical correlations (KCC), etc., to mention a few. These methods are characterized by transformation of the input data to a high dimensional feature space, followed by application of the technique in question to the transformed data.

In this paper we shall consider Kernel Fisher Discriminant Analysis (KFDA) and extend the idea of Linear Discriminant Analysis (LDA) to nonlinear feature space. We shall present a new method of choosing the optimal kernel function and explore its effect on the KDA classifier. In general the problem of which is the most appropriate kernel for a particular real application or problem is still an open problem in the literature. In this short paper, we will introduce a new special form of the information-theoretic measure of complexity of Bozdogan (1988, 1990, 1994, 2000, 2004) to choose the optimal kernel function.

We will illustrate our result using a toy example on a benchmark data set of Ripley (1994) and discuss future work on model selection in kernel methods.

2 Kernel Discriminant Analysis (KDA)

Reproducing Kernel Hilbert Space (RKHS) were developed by Aronszajn in 1950. A *RKHS* is defined by a positive definite kernel function

$$K : \mathcal{R}^d \times \mathcal{R}^d \rightarrow \mathcal{R} \tag{1}$$

on pairs of points in data space.

If these kernel functions satisfy the Mercer's condition (Mercer, 1909, Cristianini and Shawe-Taylor, 2000), they correspond to non-linearly mapping the data to a higher dimensional *feature* space $\mathcal{F}$ by a map

$$\Phi : \mathcal{R}^d \rightarrow \mathcal{F} \tag{2}$$

and taking the dot product in this space (Vapnik, 1995):

$$K(x, y) = \Phi(x) \cdot \Phi(y). \tag{3}$$

This means that any linear algorithm in which the data only appears in the form of dot products $< \mathbf{x}_i, \mathbf{x}_j >$ can be made nonlinear by replacing the dot product by the kernel function $K(\mathbf{x}_i, \mathbf{x}_j)$ and doing all the other calculations as before. In other words, each data point is mapped nonlinearly to a higher dimensional feature space.

As is well-known the Fisher Linear Discriminant analysis (FLDA) or in short LDA, is one of the most frequently used classification techniques. In order to make LDA applicable to nonlinear data in a feature space induced by a Mercer kernel, we need to develop and utilize kernel methods also referred to *"kernel machines"*. This approach gives rise to a nonlinear pattern recognition method which has very impressive performance on real data sets.

Assume that we are given the input data set $\mathcal{I}_{XY} = \{(x_1, y_1), ..., (x_n, y_n)\}$ of training vectors $\mathbf{x}_i \in \chi \subseteq \mathcal{R}^d$ and the corresponding values of $y_i \in \mathcal{Y} = \{1, 2\}$. The y_i are sets of indices of training vectors belonging to the first $y = 1$ and the second $y = 2$ class, respectively. The class separability in a direction of the weights $\alpha = [\alpha_1, \ldots, \alpha_n]'$ in the *feature space* $\mathcal{F}$ is defined such that the Fisher criteria:

$$J_F(w) = \frac{\alpha' S_B^{\Phi} \alpha}{\alpha' S_W^{\Phi} \alpha}, \tag{4}$$

is maximized, where S_B^{Φ}, S_W^{Φ} are respectively the *between and within covariance matrices* in the future space. That is,

$$\begin{aligned} S_B^{\Phi} &= (m_1^{\Phi} - m_2^{\Phi})(m_1^{\Phi} - m_2^{\Phi})^T \\ &= (\overline{\kappa}_1 - \overline{\kappa}_2)(\overline{\kappa}_1 - \overline{\kappa}_2')', \end{aligned} \tag{5}$$

$$\begin{aligned} S_W^{\Phi} &= \sum_{i=1,2} \sum_{x \in X_i} (\Phi(x) - m_i^{\Phi})(\Phi(x) - m_i^{\Phi})^T \\ &= KK' - \sum_{k=1}^{2} n_k \overline{\kappa}_k \overline{\kappa}_k' \end{aligned} \tag{6}$$

with

$$K = [\kappa(x_i, x_j)]_{(n \times n)}, \text{ and}$$

$$\overline{\kappa}_k = \frac{1}{n_k} \sum_{j \in I_k} K_j,$$

where K_j is the j-th column of K and I_k the index set of group k.

The kernel discriminant function $f(x)$ of the binary classifier

$$q(\mathbf{x}) = \begin{cases} 1 & for\ f(x) \geq 0, \\ 2 & for\ f(x) < 0 \end{cases} \tag{7}$$

can be written as

$$\begin{aligned} f_y(x) &= \sum_{i=1}^{n} \alpha_i \kappa(x_i, x) + b_y \\ &= < \alpha_y, \kappa(\mathbf{x}) > + b_y, \quad y \in \mathcal{Y}. \end{aligned} \tag{8}$$

With α being solved from (4), the intercept (or the bias) b of the discriminant hyperplane (8) is determined by forcing the hyperplane to pass through the mid point of the two group means. That is,

$$b = -\alpha' \frac{(\overline{\kappa}_1 + \overline{\kappa}_2)}{2}. \tag{9}$$

3 Regularized Kernel Discriminant Analysis (RKDA)

Without loss of generality, dropping the superscript of S_W^{Φ} and S_B^{Φ}, the coefficients α are given by the leading eigenvector of $S_W^{-1} S_B$.

Since the matrix S_W is at most of rank $n-1$, it is not strictly positive and can even fail to be positive semi-definite due to numerical problems.

Therefore, we regularize it by adding a penalty function μI to overcome the numerical problem caused by singular within-group covariance S_W. In this case, the criterion

$$J_F(w) = \frac{\alpha' S_B \alpha}{\alpha'(S_W + \mu I)\alpha} \tag{10}$$

is maximized, where the diagonal matrix μI in the denominator of the criterion (10) serves as the regularization term. If μ is sufficiently large, then S_W is numerically more stable and becomes positive definite. Another possible regularization would be to add a multiple of the kernel matrix K to S_W as suggested by Mika (2002, p. 46). That is,

$$S_W(\mu) = S_W + \mu K, \quad \mu \geq 0, \tag{11}$$

but this does not work well in practice.

If we let the covariance matrix $\hat{\Sigma}_W$

$$\hat{\Sigma}_W = \frac{1}{n} S_W, \tag{12}$$

then $\hat{\Sigma}_W$ degenerates when the data dimension p increases. In cases when the number of variables p is much larger than the number of observations n, and in general, it makes sense to utilize improved methods of estimating the covariance matrix Σ_W. We call these estimators *smoothed*, *robust*, or *stoyki* covariance estimators. These are given as follows.

- The *stipulated diagonal covariance estimator (SDE)*:

$$\hat{\Sigma}_{SDE} = (1-\pi)\hat{\Sigma}_W + \pi Diag(\hat{\Sigma}_W), \tag{13}$$

where $\pi = p(p-1)\left[2n\left(trR^{-1} - p\right)\right]^{-1}$ and where

$$R = (Diag(\hat{\Sigma}_W))^{-1/2}\hat{\Sigma}_W(Diag(\hat{\Sigma}_W))^{-1/2} \tag{14}$$

is the correlation matrix.

The SDE estimator is due to Shurygin (1983). SDE avoids scale dependence of the units of measurement of the variables.

- The *convex sum covariance estimator (CSE)*:

Based on the quadratic loss function used by Press (1975), Chen (1976) proposed a *convex sum covariance matrix estimator (CSE)* given by

$$\hat{\Sigma}_{CSE} = \frac{n}{n+m}\hat{\Sigma}_W + (1 - \frac{n}{n+m})\hat{D}_W, \quad (15)$$

where $\hat{D}_W = (\frac{1}{p}tr\hat{\Sigma}_W)\mathbf{I}_p$. For $p \geq 2$, m is chosen to be

$$0 < m < \frac{2[p(1+\beta) - 2]}{p - \beta}, \quad (16)$$

where

$$\beta = \frac{(tr\hat{\Sigma}_W)^2}{tr(\hat{\Sigma}_W^2)}. \quad (17)$$

This estimator improves upon $\hat{\Sigma}_W$ by shrinking all the estimated eigenvalues of $\hat{\Sigma}_W$ toward their common mean. Note that there are other smoothed covariance estimators. For space considerations, we will not discuss them in this paper. For more on these, see Bozdogan (2006).

4 Choice of Kernel Functions

One of the important advantages of kernel methods, including KDA, is that the optimal model parameters are given by the solution of a convex optimization problem with a single, global optimum. However, optimal generalization still depends on the selection of a suitable kernel function and the values of regularization and kernel parameters. See, e.g., Cawley and Talbot (2003, p. 2).

There are many kernel functions to choose from. The most common kernel functions are *Gaussian RBF* ($c \in \mathcal{R}$), *polynomial* ($d \in \mathcal{N}, c \in \mathcal{R}$), *sigmodial* ($a, b \in \mathcal{R}$), *PE kernel* ($r \in \mathcal{R}, \beta \in \mathcal{R}_+$), *Cauchy kernel* ($c \in \mathcal{R}_+$), and *inverse multi-quadric* ($c \in \mathcal{R}_+$) kernel functions are among the most common ones.

The main idea is that kernel functions enables us to work in the feature space without having to map the data into it.

Name of Kernel	$K(\mathbf{x}_i, \mathbf{x}_j) =$
Gaussian RBF	$\exp[-\frac{\lVert\mathbf{x}_i - \mathbf{x}_j\rVert^2}{c}]$
Polynomial	$((x_i.x_j) + c)^d$
Power Exponential (PE)	$\exp[-(\frac{\lVert\mathbf{x}_i - \mathbf{x}_j\rVert^2}{r^2})^\beta]$
Hyperbolic tangent or *Sigmoidal*	$\tanh[a(\mathbf{x}_i \cdot \mathbf{x}_j) + b]$
Cauchy	$\frac{1}{1 + \frac{\lVert\mathbf{x}_i - \mathbf{x}_j\rVert^2}{c}}$
Inverse multi-quadric	$\frac{1}{\sqrt{\lVert\mathbf{x}_i - \mathbf{x}_j\rVert^2 + c^2}}$

There are many other kernel functions, such as spline functions that are used for support vector machines. In addition, many kernels have been developed for specific applications. However, in general it is very difficult to decide which kernel function is best suited for a particular application. This is an open problem among others.

5 Information Complexity

The choice of the best mapping function is not so simple and automatic. Presently a valid method for selecting the appropriate kernel function does not exist in the literature. Here, we propose to use the information complexity criterion of Bozdogan (1988, 1990, 1994, 2000, 2004) as our model selection index as well as our criterion for feature variable selection.

Since in the kernel methods make use of orthogonal and highly sparse matrices, in this paper we propose the modified entropic complexity of a covariance matrix.

Under a multivariate normal model, the maximal information-based complexity of a covariance matrix $\hat{\Sigma}$ is defined by

$$\begin{aligned} C_1(\hat{\Sigma}) &= \frac{s}{2}\ln(\frac{tr(\hat{\Sigma})}{s}) - \frac{1}{2}\ln\left|\hat{\Sigma}\right| \\ &= -\frac{1}{2}\ln(\prod_{j=1}^{s}(\frac{\lambda_j}{\overline{\lambda}_a}) \\ &= \frac{s}{2}\ln(\frac{\overline{\lambda}_a}{\overline{\lambda}_g}) \end{aligned} \tag{18}$$

where $\overline{\lambda}_a = 1/s\sum_{j=1}^{s}\lambda_j \equiv tr(\hat{\Sigma})/s$ is the arithmetic mean of the eigenvalues (or singular values) of $\hat{\Sigma}$,and $\left|\hat{\Sigma}\right|^{1/s} \equiv \overline{\lambda}_g = \left(\prod_{j=1}^{s}\lambda_j\right)^{1/s}$ is the geometric mean of the eigenvalues of $\hat{\Sigma}$, and $s = rank(\hat{\Sigma})$.

- Note that $C_1(\hat{\Sigma}) = 0$ only when all $\lambda_j = \overline{\lambda}_a$.
- $C_1(\cdot)$ is scale invariant with $C_1(c\hat{\Sigma}) = C_1(\hat{\Sigma})$, $c > 0$. See, Bozdogan (1990).

Under the orthogonal transformation T, the maximal complexity in (18) can be written as

$$C_1^*(\hat{\Sigma}) = -\frac{1}{2}\sum_{j=1}^{s}\ln(s\lambda_j) \tag{19}$$
$$\cong \frac{1}{4}\sum_{j=1}^{q}(s\lambda_j - 1)(s\lambda_j - 3) - \frac{1}{6}\sum_{j=1}^{q} O\left(s\lambda_j - 1\right)^3,$$
$$0 < \lambda_j < \frac{2}{s},\ j = 1, 2, \ldots, s$$

where $O(\cdot)$ denotes the order of the argument and the Taylor series expansion of $\ln(s\lambda_j)$ used in (19) is about the neighborhood of the point

$$\lambda_1 = \lambda_2 = \cdots = \lambda_s = \frac{1}{s}. \tag{20}$$

At the point of eigenvalue equality $C_1^*(\cdot) = 0$ with $C_1^*(\cdot) > 0$ otherwise. See, Morgera (1985, p.610).

We note that (19) is only one possible measure of covariance complexity. Any convex function $\phi(\cdot)$, like $-\ln(\cdot)$, whose second derivative exists and is positive, may be used as a complexity measure, i.e.,

$$C_\phi^*(\cdot) = c\sum_{j=1}^{q}[\phi(\lambda_j) - \phi(\frac{1}{q})] \tag{21}$$

leads to an entire family of complexity measures, where c is a constant.

With this in mind, van Emden (1971, p. 63, eq. 311) suggested a second measure of complexity of a covariance matrix based on the Frobenius norm given by

$$C_F(\hat{\Sigma}) = \frac{1}{s}\left\|\hat{\Sigma}\right\|^2 - \left(\frac{tr\hat{\Sigma}}{s}\right)^2 \tag{22}$$

where $\left\|\hat{\Sigma}\right\|^2 = tr(\hat{\Sigma}'\hat{\Sigma})$, the square of the Frobenius norm of $\hat{\Sigma}$. In terms of the eigenvalues (or singular values), $C_F(\hat{\Sigma})$ reduces to

$$C_F(\hat{\Sigma}) = \frac{1}{s}\sum_{j=1}^{s}(\lambda_j - \overline{\lambda}_a)^2. \tag{23}$$

Note that $C_F(\cdot) \geq 0$ with $C_F(\cdot) = 0$ only when all $\lambda_j = \overline{\lambda}$. Hence $C_F(\cdot)$ measures the absolute variation in the eigenvalues and it is translation invariant. That is, $C_F(\hat{\Sigma} + kI)) = C_F(\hat{\Sigma})$.

Since we can approximate $C_1(\hat{\Sigma})$ as

$$C_1(\hat{\Sigma}) \cong \frac{1}{4}\sum_{j=1}^{s}(\frac{\lambda_j - \overline{\lambda}_a}{\overline{\lambda}_a})^2, \tag{24}$$

in terms of the eigenvalues (or singular values) λ_j, $j = 1, 2, \ldots, s$, we can relate $C_1(\hat{\Sigma})$ to the Frobenius norm characterization of complexity $C_F(\hat{\Sigma})$ of $\hat{\Sigma}$ (Bozdogan, 1988) by introducing $C_{1F}(\hat{\Sigma})$ given by

$$C_{1F}(\hat{\Sigma}) = \frac{s}{4}\frac{C_F(\hat{\Sigma})}{(\frac{tr(\hat{\Sigma})}{s})^2} = \frac{s}{4}\frac{\frac{1}{s}\left\|\hat{\Sigma}\right\|^2 - \left(\frac{tr\hat{\Sigma}}{s}\right)^2}{(\frac{tr(\hat{\mathcal{F}}^{-1})}{s})^2} \tag{25}$$

$$= \frac{s}{4}\frac{\frac{1}{s}tr(\hat{\Sigma}'\hat{\Sigma}) - \left(\frac{tr\hat{\Sigma}}{s}\right)^2}{(\frac{tr(\hat{\Sigma})}{s})^2}$$

$$= \frac{s}{4}\frac{1}{s\overline{\lambda}_a^2}\sum_{j=1}^{s}(\lambda_j - \overline{\lambda}_a)^2$$

$$= \frac{1}{4\overline{\lambda}_a^2}\sum_{j=1}^{s}(\lambda_j - \overline{\lambda}_a)^2.$$

We note that $C_{1F}(\cdot)$ is a second order equivalent measure of complexity to the original $C_1(\cdot)$ measure. Also, we note that $C_{1F}(\cdot)$ is scale-invariant and $C_{1F}(\cdot) \geq 0$ with $C_{1F}(\cdot) = 0$ only when all $\lambda_j = \overline{\lambda}$.Also, $C_{1F}(\cdot)$ measures the relative variation in the eigenvalues.

When it is assumed that the covariances are common between the classes or groups, we define ICOMP in KDA given by

$$ICOMP(\hat{\Sigma}_W) = np\log 2\pi + n\log|\hat{\Sigma}_W| + np + 2C_{1F}(\hat{\Sigma}_W) \tag{26}$$

$$= np\log 2\pi + n\log|\frac{1}{n}\hat{S}_W^{\Phi}| + np + 2\left[\frac{1}{4\overline{\lambda}_a^2}\sum_{j=1}^{s}(\lambda_j - \overline{\lambda}_a)^2\right]$$

In our numerical results, however, it suffices just to use and score $C_{1F}(\hat{\Sigma}_W)$ by itself to choose the optimal kernel function in KDA in the next section. In the literature cross-validation based criteria have been used. These type of criteria due to the high dimensionality of the feature space are too time-consuming. Our approach shortens the model selection time.

6 A Numerical Example

In this section we illustrate our results using the binary classifier trained by the KDA on Ripley's (1994) two dimensional toy data of $n_{trn} = 250$ training observations. Then, the classifier that is found is evaluated on the testing data of Ripley which has $n_{test} = 1000$ observations using support vector classifier (svmclass) to classify the input vector $\mathbf{x}$. We obtain both the training and test error. Our results are based on the modified version of the STPRTool Matlab Modules of Franc and Hlaváč (2004). In our computation, we use *svds* function in Matlab to find the singular values of the large sparse within-group covariance matrix $\hat{\Sigma}_W$. We experimented with our results by retaining $k = 4$, and 15 largest singular values of $\hat{\Sigma}_W$ and scored the information-theoretic complexity $C_{1F}(\hat{\Sigma}_W)$ for each of the alternative kernel functions. The results from this experiment are summarized in Table 1 below. We report the results for $k = 4$ largest singular values only. The results up to the 15 largest singular values of $\hat{\Sigma}_W$ are the same in terms of the ordering of the complexity $C_{1F}(\hat{\Sigma}_W)$.

Kernel Function	Training Error	Test Error	$C_{1F}(\hat{\Sigma}_W)$
Linear	17.20%	14.20%	5.0276
RBF	13.60%	9.60%	4.5207
Polynomial [2 1]	15.60%	9.20%	4.6401
Sigmoid [2 1]	12.00%	9.00%	4.0747*

Table 1. Results KDA using different kernels functions and SVM Classifier.

Note that the regularization parameter μ was set to a small value 0.001, and the regularization constant C was set to 10. Looking at the above table we see that sigmoid kernel function seems to be a better choice based on the minimum value of the complexity measure $C_{1F}(\hat{\Sigma}_W)$ for this data set with better training and test error percentages. The visualization of the classifiers as SVM classifiers are shown in Figures 1, and 2.

7 Conclusion and Future Work

In this sort paper, we introduced the information-theoretic complexity $C_{1F}(\hat{\Sigma}_W)$ as a new method for model selection in choosing the optimal kernel function in kernel discriminant analysis (KDA). We showed our results on a toy benchmark data set of Ripley to evaluate the performance of the optimal classifier based on the choice of the kernel function. Our method shortens the model selection time over the more time-consuming cross-validation method.

The future work in this direction will involve several important problems in automating the choice of the regularization constant C, the regularization parameter μ and to study their effect on the classifier across different benchmark data sets and show the generalization ability of this new method. Our results, will be applied to real micro data mining data sets and the results will be reported elsewhere.

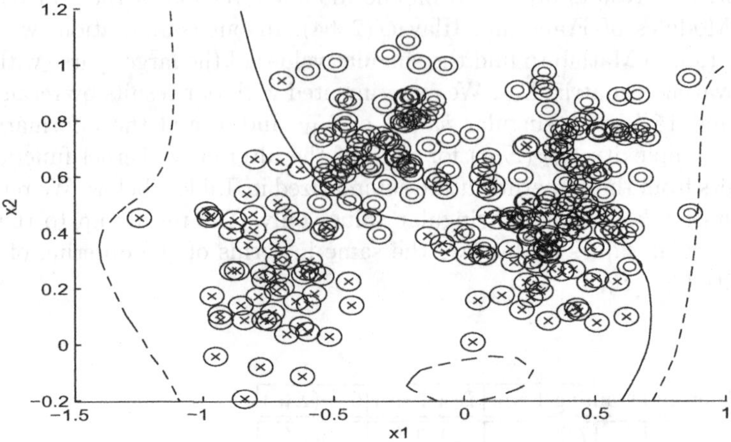

Fig. 1. Pattern of Ripley Training Data in the Feature Space.

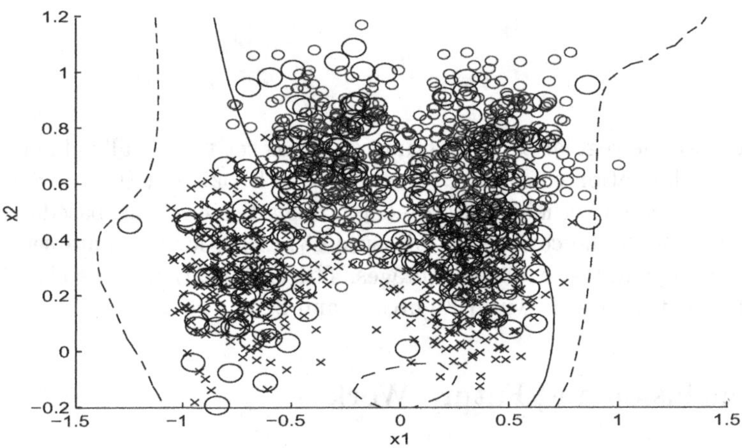

Fig. 2. Pattern of Ripley Test Data in the Feature Space.

Acknowledgements

The authors would like to express their thanks to Professor Isabella Morlini of the Università di Modena e Reggio Emilia, Italy for kindly formatting the paper, and to Dr. Russ Zaretzki of the Department of Statistics, Operations, and Management Science at the University of Tennessee, Knoxville, Tennessee, for reading and commenting on this paper.

References

ARONSZAJN, N. (1950): Theory of reproducing kernels. *Transactions of the American Mathematical Society, 68, 337-404.*

BOZDOGAN, H. (1988): ICOMP: A New Model-Selection Criterion. In Hans H. Bock (Ed.): *Classification and Related Methods of Data Analysis*, Elsevier Science Publishers B. V. (North Holland), Amsterdam, 599-608.

BOZDOGAN, H. (1990): On the information-based measure of covariance complexity and its application to the evaluation of multivariate linear models. *Communications in Statistics Theory and Methods, 19(1), 221-278.*

BOZDOGAN, H. (1994): Mixture-model cluster analysis using model selection criteria and a new informational measure of complexity. In H. Bozdogan (Ed.), *Multivariate Statistical Modeling*, Vol. 2. Kluwer Academic Publishers, Dordrecht, the Netherlands, 69-113.

BOZDOGAN, H. (2000): Akaike's information criterion and recent developments in informational complexity. *Journal of Mathematical Psychology, 44, 62-91.*

BOZDOGAN, H. (2004). Intelligent statistical data mining with information complexity and genetic algorithms. In H. Bozdogan (Ed.) *Statistical Data Mining and Knowledge Discovery.* Chapman and Hall/CRC, Boca Raton, Florida, 15-56.

BOZDOGAN, H. (2006). *Information Complexity and Multivariate Learning Theory with Data Mining Applications.* Forthcoming book.

CAWLEY, G. and TALBOT, N.L.C. (2003): Efficient leave-one-out cross-validation of kernel Fisher discriminant classifiers. Paper submitted to Elsevier Science.

CHEN, M. C-F. (1976): Estimation of covariance matrices under a quadratic loss function. Research Report S-46, Department of Mathematics, SUNY at Albany, Albany, N.Y., 1-33.

CRISTIANINI, N. and SHAWE-TAYLOR, J. (2000): *An introduction to Support Vector Machines.* Cambridge University Press.

FRANC, V. and HLAVÁČ, V. (2004): Statistical Pattern Recognition Toolbox for Matlab. Research Report of CMP, Czech Technical University in Prague, No.8.

MERCER, J. (1909): Functions of positive and negative type and their connection with the theory of integral equations. *Philosophical Transactions Royal Society London, A209, 415-446.*

MIKA, S. (2002): Kernel Fisher discriminants. Ph.D. Thesis, Technical University of Berlin, Berlin, Germany.

MORGERA, S. D. (1985). Information theoretic covariance complexity and its relation to pattern recognition. *IEEE Transactions on Systems, Man, and Cybernetics, SMC-15, 608-619.*

PRESS, S.J. (1975): *Estimation of a normal covariance matrix.* University of British Columbia.

RIPLEY, B.D. (1994): Neural networks and related methods for classification (with discussion). *J. Royal Statistical Soc. Series B, 56: 409-456.*

SHURYGIN, A. M. (1983): The linear combination of the simplest discriminator and Fisher's one. In Nauka (Ed.), *Applied Statistics.* Moscow, 144-158 (in Russian).

VAN EMDEN, M.H. (1971): *An Analysis of Complexity.* Mathematical Centre Tracts, Amsterdam, 35.

VAPNIK, V. (1995): *The Nature of Statistical Learning Theory*, Springer-Verlag, New York.

Growing Clustering Algorithms in Market Segmentation: Defining Target Groups and Related Marketing Communication

Reinhold Decker, Sören W. Scholz, and Ralf Wagner

Business Administration and Marketing, Bielefeld University,
Universitätsstraße 25, D-33615, Germany

Abstract. This paper outlines innovative techniques for the segmentation of consumer markets. It compares a new self-controlled growing neural network with a recent growing k-means algorithm. A critical issue is the identification of the "right" number of clusters, which is externally validated by the *JUMP*-criterion. The empirical application counters several objections recently raised against the use of cluster analysis for market segmentation.

1 Introduction

Market segmentation by means of cluster analysis is rated to be one of the most fundamental and most useful techniques of last century's marketing management practice. But recently it has been seriously criticized. According to Fennel et al. (2003) neither demographic nor psychographic variables are highly correlated with actual consumption behavior. Sheth and Sisodia (1999) claimed, that the validity of market segmentation is restricted to markets consisting of only very few segments. According to their argumentation market segmentation becomes obsolete when subject to an increasing diversity of income, age, ethnicity, and lifestyle. Due to the heterogeneity of individual preferences in most consumer goods markets, the determination of an appropriate number of segments or clusters is a key question in almost any market segmentation study (Boone and Roehm (2002)). Moreover, Grapentine and Boomgaarden (2003) argue that difficulties in communicating with the target markets identified by clustering are the first of four foremost maladies of contemporary market segmentation. Based on this argumentation a substitution of segmentation by mass-customization would be desirable, but is impracticable in most categories of frequently purchased consumer goods. Punj and Steward (1983) systematize marketers' objections against the application of clustering algorithms into two streams of concerns: (1) methodical concerns (e.g. the theoretical understanding of different algorithms and properties of the available proximity measures) and (2) the weak guidelines regarding the choice of algorithms, particularly with respect to the danger of considering artifacts due to methodological options which may impair the validity of the results. In order to address the latter issue, they propose to compare the results of different clustering algorithms applied to the same data set. Facing these challenges the present paper aims

- to outline similarities and differences of modern segmentation techniques, in particular a growing k-means and a growing neural gas clustering algorithm,
- to demonstrate the usage of the $JUMP$-criterion (Sugar and James (2003)) and a test of multivariate normality (Mecklin and Mundfrom (2004)) for assessing the number of segments in a consumer market, and
- to exemplify the development of a communication concept related to the identified segments.

To achieve these aims we refer to a data set which describes attitudes and preferences concerning nutrition and consumption behavior. The respective market is highly fragmented because all four sources of diversity listed above contribute substantially to individual consumer behavior. To sketch the development of a cluster-related communication concept we employ the advertising principles of Armstrong (forthcoming).

The remainder of this paper is structured as follows: First, we outline three different methodologies that we are going to use for clustering consumer profiles, simultaneously determining an appropriate number of clusters in this market segmentation task. Subsequently, the results are compared and selected clusters are interpreted with respect to the immanent nutrition and consumption behavior. Finally we briefly address the development of communication strategies for these market segments. The paper concludes with a discussion of results and directions of further research.

2 Methodology

2.1 Determining the Optimal Number of Clusters by Means of the $JUMP$-Criterion

Let $\boldsymbol{x}_n = (x_{n1}, \ldots, x_{nl}, \ldots, x_{nL}) \in \mathbb{R}^L$, with $n \in \{1, \ldots, N\}$, denote the input data to be analyzed, i.e. the individual profiles or feature vectors of the consumers considered. Assuming the features used for clustering are uncorrelated (and, therefore, do not provide redundant information) the distortion $d(k)$ of a k cluster solution can be estimated by the sum of squared errors based on Euclidean distances. Since the true distortion in fragmented markets is unknown, it must be calculated using the data at hand. The estimated distortion $\hat{d}(k)$ decreases with an increasing number of clusters and, thus, needs to be corrected by a power transformation in order to assess the optimal number of clusters k^*. Figure 1 outlines the procedure applied to calculate the $JUMP$-criterion by using the estimated distortion (Sugar and James (2003)).

The procedure also includes the case $k^* = 1$ and, therefore, checks explicitly for the absence of a group structure in the data. The transformation

1: Run standard k-means for different numbers of clusters $k = 1, \ldots, K$.
2: Calculate $\hat{d}(k)$ for $k = 1, \ldots, K$ and define $\hat{d}(0) = 0$.
3: Compute the jumps $JUMP(k) = \hat{d}^{-y}(k) - \hat{d}^{-y}(k-1)$ with transformation power $y > 0$.
4: Determine the optimal number of clusters $k^* = \arg \max_{k \in \{1,\ldots,K\}} JUMP(k)$.

Fig. 1. Computation of the $JUMP$-criterion

power y used in step 3 is not fixed to a certain value. Thus, we define $y = \frac{L}{2}$ according to Sugar and James (2003). The $JUMP$-criterion will be used in the empirical part of the paper to externally validate the cluster number resulting from our modification of Hamerly and Elkan's (2003) G-means algorithm.

2.2 Modified Growing k-Means (GKM)

The basic principle of our growing k-means algorithm is to start with a small number of clusters and to expand this number until the consumer profiles assigned to each cluster fit a multivariate normal distribution that is centered at the cluster-centroids $\boldsymbol{\theta}_h$. The whole procedure is outlined in Figure 2.

1: Let k be the initial number of clusters (usually $k = 1$).
2: Apply the standard k-means algorithm to the given sets of consumer profiles.
3: Let $\{\boldsymbol{x}_1, \ldots, \boldsymbol{x}_{\tilde{n}}, \ldots, \boldsymbol{x}_{N_h}\}$ be the set of profiles assigned to centroid $\boldsymbol{\theta}_h$.
4: Test whether the profiles assigned to each cluster $h = 1, \ldots, k$ are multivariate normally distributed.
5: If the data meets the relevant criterion of multinormality, then stop with the respective optimal k^*, otherwise increase the number of clusters k by 1 and return to step 2.

Fig. 2. Modified growing k-means algorithm

To determine the optimal number of clusters, we consider the distribution of the observations (consumer profiles) assigned to the respective clusters. The basic idea is to assume that the output of a clustering procedure is acceptable if the centroids provide substantial information concerning–at least–the majority of the cluster members. According to Ding et al. (2002) the k-means algorithm can be regarded as a mixture model of independent identically distributed spherical Gaussian components. Thus, it is intuitive to presume that a cluster centroid should not only be the balance point but also the median as well as the modus of the multivariate distribution representing the cluster members. We refer to Mardia's measure of the standardized 4th central moment (kurtosis) to check whether the profiles of the members assigned to a cluster h adhere to a multivariate normal distribution or not.

The corresponding null hypothesis reads: The consumer profiles assigned to centroid $\boldsymbol{\theta}_h$ are multivariate normally distributed.

The measure is based on squared Mahalanobis distances, which are very helpful in detecting multivariate outliers (Mecklin and Mundfrom (2004)). The statistic for testing the above hypothesis is given by

$$\tilde{d}_h = \frac{\hat{d}_h - L(L+2)}{(8L(L+2)/N_h)^{1/2}} \qquad (h = 1, \ldots, k) \tag{1}$$

and

$$\hat{d}_h = \frac{1}{N_h} \sum_{\tilde{n}=1}^{N_h} \left((\boldsymbol{x}_{\tilde{n}} - \boldsymbol{\theta}_h)\boldsymbol{S}_h^{-1}(\boldsymbol{x}_{\tilde{n}} - \boldsymbol{\theta}_h)'\right)^2 \qquad (h = 1, \ldots, k), \tag{2}$$

with $\boldsymbol{S}_h$ denoting the cluster-specific sample covariance matrix and N_h equaling the number of observations assigned to cluster h.

The test statistic $\tilde{d}_h$ is asymptotically standard normally distributed with p-value $p_h = 2(1 - P_N(|\tilde{d}_h|))$, whereby P_N equals the cumulative probability of a standard normal distribution. The null hypothesis of multivariate normality is rejected for cluster h if p_h is sufficiently small. Therefore, we favor that k^* which meets the following condition:

$$k^* = \arg \max_{k \in \{1,\ldots,K\}} \left(\min_{h \in \{1,\ldots,k\}} p_h \right) \tag{3}$$

Thus, the algorithm stops growing when the distribution of the feature vectors of each cluster is, at least approximately, multivariate normal and tends towards small numbers of clusters.

2.3 Self-Controlled Growing Neural Network (SGNN)

Let $\mathcal{B}$ be a set of units u_h with $h \in \{1, \ldots, k = |\mathcal{B}|\}$, and $\mathcal{D}$ the set of connections between these units, capturing the current topological structure of an artificial neural network. Each unit is represented by a weight vector $\tilde{\boldsymbol{\theta}}_h = (\tilde{\theta}_{h1}, \ldots, \tilde{\theta}_{hl}, \ldots, \tilde{\theta}_{hL}) \in \mathbb{R}^L$. In fact, these weight vectors directly correspond to the centroids considered above. Figure 2.3 outlines the essential elements of the SGNN algorithm (Decker (2005)).

As a result of the adaptation process outlined in steps 2-5, we get the connection matrix $\boldsymbol{D}$, which describes the final topological structure of the neural network. Each unit u_h or rather the corresponding weight vector $\tilde{\boldsymbol{\theta}}_h$, represents a frequent (nutrition and consumption) pattern in the available survey data and is referred to as a prototype in the following.

1: Create an initial network comprising two non-connected units u_1 and u_2 with weight vectors $\tilde{\boldsymbol{\theta}}_1$ and $\tilde{\boldsymbol{\theta}}_2$ and initialize some internal control variables.
2: Select randomly one input vector $\boldsymbol{x}_n$ and calculate the Euclidean distances $\|\boldsymbol{x}_n - \tilde{\boldsymbol{\theta}}_h\|$ between $\boldsymbol{x}_n$ and each unit $u_h \in \mathcal{B}$.
3: Connect the best and the second best matching unit $u_{h_{Best}}$ and $u_{h_{Second}}$, if this is not yet the case, and set the age of this connection $a_{h_{Best}h_{Second}} = 0$.
4: If either the activity $v_{h_{Best}} = \exp(-\|\boldsymbol{x}_n - \tilde{\boldsymbol{\theta}}_{h_{Best}}\|)$ or the training requirement of the best matching unit exceeds an appropriate threshold then update the weight vectors of unit $u_{h_{Best}}$ and the neighbors to which it is connected, else create a new unit $u_{h_{New}}$ in between $\tilde{\boldsymbol{\theta}}_{h_{Best}}$ and $\boldsymbol{x}_n$.
5: Increase the age of all connections emanating from $u_{h_{Best}}$ by one, update the internal control variables, remove all the connections that are "older" than a predefined upper bound a_{Max}, and remove all disconnected units.
6: If the stopping criterion is not already reached, increase the adaptation step counter and return to step 2, otherwise create connection matrix $\boldsymbol{D}$.

Fig. 3. Outline of the SGNN algorithm

3 Market Segmentation by Means of Growing Clustering Algorithms

3.1 The Data

The data set to be analyzed was provided by the ZUMA Institute and is part of a sub-sample of the 1995 GfK ConsumerScan Household Panel Data. It comprises socio-demographic characteristics as well as attitudes and opinions with regard to the nutrition and consumption behavior of German households. We exemplarily consider the attitudes and opinions of $N = 4,266$ consumers measured by means of $L = 81$, mostly Likert-scaled items (with $1 \equiv$ 'I definitely disagree.', ..., $5 \equiv$ 'I definitely agree.').

3.2 Selected Results

According to the *JUMP*-criterion the optimal number of clusters for the data at hand is 37. The GKM algorithm also results in a 37-cluster solution, where sets of 174 to 394 individual consumer profiles are assigned to 15 meaningful clusters with centroids $\boldsymbol{\theta}_1, \ldots, \boldsymbol{\theta}_{15}$. Moreover, 22 outliers are assigned to single-object clusters. Applying the SGNN algorithm with $a_{Max} = 81 (= L)$ results in a neural network that comprises 14 prototypes $\tilde{\boldsymbol{\theta}}_1, \ldots, \tilde{\boldsymbol{\theta}}_{14}$. In fact, SGNN leads to very similar results.

Table 1 gives an exemplary description of those items that cause the highest distances between the GKM-cluster centroids $\boldsymbol{\theta}_3$ and $\boldsymbol{\theta}_9$ as well as between $\boldsymbol{\theta}_{10}$ and $\boldsymbol{\theta}_{14}$. The table also displays the corresponding weights resulting from applying the SGNN algorithm. Obviously, the cluster profiles represented by the SGNN prototypes match the clusters obtained from the GKM algorithm to a substantial extent (the correlations are significant at level $\alpha = 0.01$).

l	Brief item description	$\boldsymbol{\theta}_3$	$\tilde{\boldsymbol{\theta}}_7$	$\boldsymbol{\theta}_9$	$\tilde{\boldsymbol{\theta}}_4$	$\boldsymbol{\theta}_{10}$	$\tilde{\boldsymbol{\theta}}_5$	$\boldsymbol{\theta}_{14}$	$\tilde{\boldsymbol{\theta}}_2$
3	Likes to have company	4.28	4.31	3.65	3.64	4.65	4.54	3.90	3.96
15	Enjoys life to the full	3.15	3.62	2.68	3.26	3.72	4.26	3.59	3.22
21	Prefers traditional lifestyle	2.60	3.51	3.62	3.93	4.46	3.96	3.17	3.70
27	Quality-oriented food purchase	3.13	2.03	2.64	3.55	4.37	3.45	2.40	2.26
33	Prefers fancy drinks and food	3.59	3.15	1.70	1.83	4.18	3.75	1.90	2.22
34	Prefers hearty plain fare	2.89	2.91	4.40	4.14	4.55	4.19	3.23	3.50
35	Prefers whole food	2.31	2.28	1.63	1.89	3.65	3.37	1.53	1.52
42	Pays attention to antiallergic food	2.91	2.85	2.12	2.62	4.45	4.11	1.96	1.94
43	Rates oneself slimness-oriented	2.53	2.82	1.68	2.12	3.68	3.49	1.93	1.76
48	Eats vegetarian food only	1.32	1.37	1.10	1.19	1.82	1.65	1.22	1.10
59	Counts calories when eating	1.86	2.18	1.41	1.72	3.50	3.21	1.40	1.38
60	Prefers tried and tested recipes	2.13	2.52	4.31	4.14	4.50	4.06	3.08	3.06
	Pearson's correlation r (all items)	0.86		0.90		0.91		0.92	

Table 1. Parts of selected nutrition and consumption style prototypes

When looking at the largest distances of singular items, we find that the members of GKM-cluster 10 are on average highly involved in food consumption. The quality of food products (item 27) is of accentuated importance in comparison to the other GKM-clusters. This implies a strong preference for exclusive and fancy foods (item 33) as well as a strong preference for branded products (item 20). These consumers are characterized by health-oriented nutrition styles such as fat free foods (items 29 and 43), whole foods, or even vegetarian diets (item 48). The motivation for this behavior seems to stem from hedonistic (item 3) rather than well-founded health concerns. Their primary focus is on maintaining physical attractiveness and fitness (items 43, 59, and 69), which is a premise for their active and sociable lifestyle. As a result of these characterizations we can refer to the respondents represented by cluster centroid $\boldsymbol{\theta}_{10}$ as the *hedonistic consumers*.

Contrastingly, the consumers of cluster 14 (represented by $\boldsymbol{\theta}_{14}$) pay less attention to nutrition. They are less concerned about food products with additives or preservatives (item 44). Health-oriented food consumption behavior is less important in their lives. Moreover, they put emphasis on low prices (item 38). The origin and quality of the foods are secondary (item 27, 35, and 42). Consequently, we can label this cluster as the *uninvolved consumers*.

Cluster 3 (represented by $\boldsymbol{\theta}_3$) comprises consumers who have a preference for natural, unprocessed foods (items 35, 42, and 44). This behavior is accompanied by a distinctive fondness for exclusive and, partly, even fancy foods (item 33). The members of cluster 3 are less motivated by "external" reasons such as staying fit and attractive (items 43, 59, and 69), but primarily aim to avoid the consumption of any kind of non-natural food ingredients (items 35, 48, and 65). Nutrition and cooking play an important role in their life, so they take nutrition tips very seriously. We can refer to this cluster as the *health-oriented consumers.*

Cluster centroid $\boldsymbol{\theta}_9$ represents consumers who are critical of social or technical progress in their personal environment. This cluster stands out due to its traditional lifestyle (item 21 and 48). It shows a distinctive preference for domestic foods and likes to stick to home made cooking, especially hearty plain fare (item 34) on the basis of tried and tested recipes (item 60). Vegetarian food is strictly rejected (item 48). Being rather uninvolved, these consumers are price conscious and shy away from purchasing new products (item 5 and 9). Due to their distrust of advertising (item 11), these *traditional consumers* are also hardly interested in the branding of products.

3.3 Defining Related Marketing Communication

The advertising principles by Armstrong (forthcoming) advise us to cope with the strategic issues of deciding (1) which *information* should be communicated, (2) on the type of *influence* to focus on, (3) which *emotions* are associated with the offered products, and (4) how to increase the *mere exposure*[1] by providing the products with a suitable brand and image.

Considering the *hedonistic consumers* the information should be related to the quality of the products due to the high involvement of the consumers concerned. The influence should be based on rational arguments which lead the recipients to favor the offers. The emotions to be created are trust and prestige and the brand name should have semantic associations with self-confidence and wealth. The brand should be associated with luxury.

For the *uninvolved consumers* marketing communication should transmit the message "easy to prepare". The influences to focus on are "liking" and "attribution". Consequently, the communication should be persuasive but not pushing to avoid reactance. Suited emotions are pleasure and perhaps even provocation for the introduction of new offers to this segment. The brand's name should have a semantic association with a free and easy lifestyle and links should be made to modern convenience products.

4 Discussion and Implications

Several general objections have been raised against the concept of clustering-based market segmentation. This paper addresses these objections empirically by considering segments of the highly heterogeneous food market. Grasping the proposal of Punj and Steward (1983) we applied two modern growing clustering algorithms to a large empirical data set and obtained very similar results in terms of the identified clusters. Therefore, we argue that the objection to possible artifacts and the weaknesses in guidance as to which algorithms should be chosen, is less serious if data processing is done properly.

[1] The mere exposure effect is a psychological artifact utilized by advertisers: people express undue liking for things merely because they are familiar with them.

Although the validity of this implication is restricted to the case of sophisticated algorithms like those considered in this paper, it provides a good starting point for further research including different marketing data as well as other clustering algorithms.

Furthermore, this paper also addresses the problem of identifying the number of clusters. The *JUMP*-criterion and the test of multivariate normality performed well in our study and led to the same results.

Finally, we sketched the development of a related marketing communication for targeting the identified market segments properly. Of course, the development of marketing communication strategies is always a creative process, but nonetheless market segmentation has been found to provide useful information about the motives necessary to appeal on.

Summarizing these results we conclude that the objections against clustering for the purpose of data-based market segmentation are at least not of such general validity as claimed in recent literature.

References

ARMSTRONG, J. S. (forthcoming): *Persuasive Advertising.* Palgrave, New York.

BOONE, D.S. and ROEHM, M. (2002): Evaluating the Appropriateness of Market Segmentation Solutions Using Artificial Neural Networks and the Membership Clustering Criterion. *Marketing Letters*, 13 (4), 317–333.

DECKER, R. (2005): Market Basket Analysis by Means of a Growing Neural Network. *The International Review of Retail, Distribution and Consumer Research*, 15 (2), 151–169.

DING, C., HE, X., ZHA, H., and SIMON, H. (2002): Adaptive Dimension Reduction for Clustering High Dimensional Data. In: *Proceedings of the 2nd IEEE International Conference on Data Mining*, 147–154.

FENNEL, G., ALLENBY, G.M., YANG, S., and EDWARDS, Y.(2003): The Effectiveness of Demographic and Psychographic Variables for Explaining Brand and Product Category Use. *Quantitative Marketing and Economics*, 1 (2), 223–244.

GRAPENTINE, T. and BOOMGAARDEN, R. (2003): Maladies of Market Segmentation. *Marketing Research*, 15 (1), 27–30.

HAMERLY, G. and ELKAN, C. (2003): Learning the k in k-Means. In: *Advances in Neural Information Processing Systems*, 17.

MECKLIN, C. J. and MUNDFROM, D. J. (2004): An Appraisal and Bibliography of Tests for Multivariate Normality. *International Statistical Review*, 72 (1), 123–138.

PUNJ, G. and STEWARD, D.W. (1983): Cluster Analysis in Marketing Research: Review and Suggestions for Application. *Journal of Marketing Research*, 20 (May), 134–148.

SUGAR, C. A. and JAMES, G. M.(2003): Finding the Number of Clusters in a Dataset: An Information-theoretic Approach. *Journal of the American Statistical Society*, 98 (463), 750–762.

SHETH, J.N. and SISODIA, R.S.(1999): Revisiting Marketing's Lawlike Generalizations. *Journal of the Academy of Marketing Science*, 27 (1), 71–87.

Graphical Representation of Functional Clusters and MDS Configurations

Masahiro Mizuta

Information Initiative Center,
Hokkaido University, Japan
mizuta@iic.hokudai.ac.jp

Abstract. We deal with graphical representations of results of functional clustering and functional multidimensional scaling (MDS). Ramsay and Silverman(1997, 2005) proposed functional data analysis. Functional data analysis enlarges the range of statistical data analysis. But, it is not easy to represent results of functional data analysis techniques. We focus on two methods of functional data analysis: functional clustering and functional MDS. We show graphical representations for functional hierarchical clustering and functional k-means method in the first part of this paper. Then, in the second part, graphical representation of results of functional MDS, functional configuration is presented.

1 Introduction

Graphical representations of multidimensional data are key techniques for analyzing data and have been studied by a large number of statisticians all over the world since 1960's. We can use not only static graphics but dynamic graphics using computers. Most statistical packages have excellent graphical functions.

In most conventional data analysis methods, we assume that data are regarded as a set of numbers with some structures, *i.e.* a set of vectors or a set of matrices *etc.* Nowadays, we must often analyze more complex data. One type of complex data is the functional data structure; data themselves are represented as functions. Ramsay and Silverman have studied functional data analysis (FDA) as the analysis method to functional data since the 1990's. They have published excellent books on FDA (Ramsay and Silverman, 1997, 2002, 2005). They deal with various methods for functional data in these books.

We deal with graphical representations of results of functional data analysis. There are so many methods for functional data analysis. Among them, we pick up functional clustering and functional multidimensional scaling (MDS). Clustering is a very popular method for data mining and we must apply various types of data including functional data. MDS is also an effective method for data analysis, because the structure of dissimilarity data can be difficult to understand. It is not easy to interpret the results even if we adopt conventional MDS. We usually use scatter plot of configuration of objects and scatter plot of dissimilarities versus Euclid distances separately. But, the relations between the configuration and the residuals can not be investigated with these plots. Mizuta and Minami (1991) proposed another graphical representation for MDS. We extend this method to functional MDS in this paper.

2 Graphical Representations of Functional Clustering

The purpose of cluster analysis is to find relatively homogeneous clusters of objects based on measured characteristics. Sometimes, we divide methods of cluster analysis into two groups: hierarchical clustering methods and nonhierarchical clustering methods.

Hierarchical clustering refers to the formation of a recursive clustering of the objects data: a partition into two clusters, each of which is itself hierarchically clustered. We usually use datasets of dissimilarities or distances; $S = \{s_{ij}; i, j = 1, 2, \ldots, n\}$, where s_{ij} is the dissimilarity between objects i and j, and n is the size of the objects. Single Linkage is a typical hierarchical clustering method. We start with each object as a single cluster and in each step merge two of them together. In each step, the two clusters that have minimum distance are merged. At the first step, the distances between clusters are defined by the distance between the objects. However, after this, we must determine the distances between new clusters. In Single Linkage, the distance between two clusters is determined by the distance of the two closest objects in the different clusters. The results of hierarchical clustering are usually represented by dendrogram. The height of the dendrogram represents the distances between two clusters. It is well known that the results of Single Linkage and the minimal spanning tree (MST) are equivalent from the computational point of view.

Mizuta (2003a) proposed the algorithm of Single Linkage for functional dissimilarity data $S(t) = \{s_{ij}(t); i, j = 1, 2, \ldots, n\}$. The basic idea of the proposed method is that we apply conventional Single Linkage to $S(t)$ and get functional MST, say MST(t). Then we calculate functional configuration and adjust labels of objects. The results of functional single linkage are represented as motions of MST using dynamic or interactive graphics (Figure 1 (Left)). Nonhierarchical clustering partitions the data using a specific criterion. Most nonhierarchical clustering methods do not deal with a set of dissimilarities directly. They use a set of p-tuples: $Z = \{z_i, i = 1, 2, \ldots, n\}$. k-means method is a kind of nonhierarchical clustering procedure which starts with initial seeds points (centroids), then it assigns each data point to the nearest centroid, updates the cluster centroids, and repeats the process until the centroids do not change.

From the view points of functional data analysis, we assume that we have p-dimensional functions corresponding to n objects. We denote the functions for n objects depending on a variable t as $Z(t) = \{z_i(t)\}(i = 1, 2, \ldots, n)$. It is realistic that values $Z(t_j)(j = 1, 2, \ldots, m)$ are given. We restrict ourselves to two dimensional functional data and to the one dimensional domain. The number of clusters k is prespecified as a user parameter. The idea of functional k-means method is to repeat the procedure of the conventional k-means method. At first, we apply conventional k-means method to $Z(t_j)$ for each t_j. Then, we adjust the labels of clusters. Even if we fix clustering method, there is freedom of labeling. We denote $C_i(t)$ as the label of the i-th object at t for fixed K. We discuss about adjusting the labeling of $C_i(t_2)$ with fixed $C_i(t_1)$. $C_i^*(t_2)$ are the new labels of the objects that $\sum \sharp\{C_i(t_1) = C_i^*(t_2)\}$ takes the minimum value, where $\sharp$ indicates the size of the set. A simple method for adjusting the labels is to use the cluster centers of the previous clustering for initial guesses for the cluster centers. We must pay attention to the fact that even if two objects belong to the same cluster at t_j, it is possible that the two objects belong to different clusters at t_{j_2}. The results of the proposed functional k-means method can be represented graphically. We use the three dimensional space for the representation. Two dimensions are used for the given functional data, one is for t, and the clusters are shown with colors. We apply an artificial data set to the proposed functional k-means method and show the graphical representation. The data are two dimensional functional data of size 150. At first stage ($t = 0$), the data have relatively clear clusters and the clusters are destroyed along t. We set the number of the clusters at five. The results are shown in Figure 1 (Right). The figure is a snapshot of computer display. It is true that it is difficult to analyze the data from only this figure when the size of objects is too large. But, if we may use dynamic graphics, the results can be analyzed more effectively with interactive operations: rotations, slicing, zooming etc. We can find out the structures of the data.

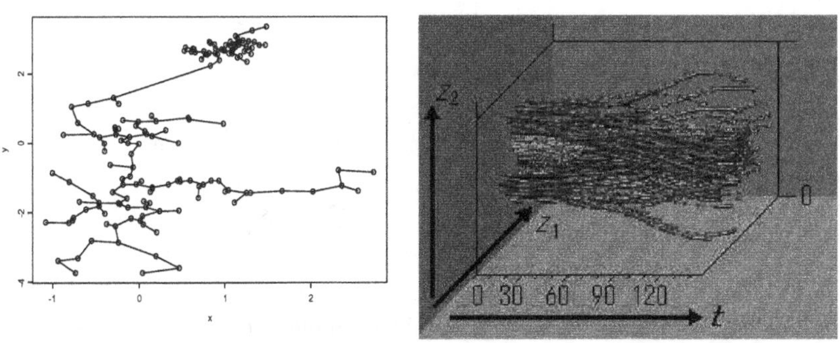

Fig. 1. Functional Single Linkage (Left), Functional k-means (Right)

3 Residual Plot for Conventional MDS

Multidimensional scaling (MDS) produces a configuration $\{\boldsymbol{x}_i\}$ for the results of the analysis of input dissimilarity data $\{s_{ij}\}$. We usually investigate the configuration with the scatterplot or scatterplot matrix of $\{\boldsymbol{x}_i\}$ and verify the relation of the input dissimilarity data and the configuration with plot of the input dissimilarities versus the distances among the configuration. We can evaluate the results globally with this plot, but can not find the objects whose positions are singular. So, Mizuta and Minami (1991) proposed a graphical method for MDS, named *residual plot* for MDS, which represents the configuration and residuals simultaneously. The graph is depicted in the following steps.

1. Calculate the distances $d_{ij}^2 = \| \boldsymbol{x}_i - \boldsymbol{x}_j \| \; (i, j = 1, 2, ..., n)$.
2. Transform s_{ij} to s_{ij}^* such that $s_{ij}^* = a + b s_{ij}$ where a, b are the coefficients of the simple regression function of $\{d_{ij}\}$ on $\{s_{ij}\}$ in order to justify the scale of $\{s_{ij}\}$.
3. Plot $\{\boldsymbol{x}_i\}$ on the scatterplot or scatterplot matrix and draw the segments from $\boldsymbol{x}_i$ in the direction of $\boldsymbol{x}_j$ (when $d_{ij} > s_{ij}^*$) or draw the dashed segments from $\boldsymbol{x}_i$ in the counter direction of $\boldsymbol{x}_j$ (when $d_{ij} \leq s_{ij}^*$) with length $\mid d_{ij} - s_{ij}^* \mid /2 \; (i, j = 1, 2, ..., n)$.

The lengths of the segments represent the residuals of the configuration (Figure 2). We can use this graph to represent asymmetric dissimilarity data.

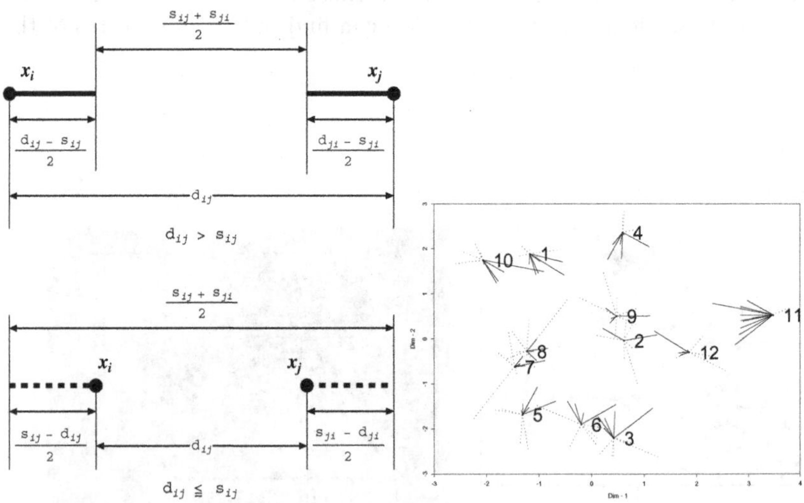

(a) Residual segments of the graphical method for MDS

(b) An example of the residual plot for MDS

Fig. 2. Residual plots of conventional MDS

4 Functional Multi Dimensional Scaling

Functional MDS is introduced here (Mizuta, 2000). We assume that the dissimilarity data among the n objects depend on two parameters, *i.e.* arguments u and v: $S(u, v) = \{s_{ij}(u, v)\}$. With a conventional MDS method for each argument, the (functional) configurations $X(u, v)$ are derived. Because the purpose of MDS is to investigate the relations among objects, the configurations of objects may be rotated (transformed with orthogonal matrix)[2]. The goal of the method is to find out a functional orthogonal matrix $Q(u, v)$ that adjusts $X(u, v)$ to $Q(u, v)X(u, v)$ in order to get almost continuous functions. For the sake of explanation, we assume that the dissimilarity data are given in lattice points: $\{S(u_k, v_l)\}(k, l = 1, \dots, m)$.

The criterion of the method is to minimize

$$\sum_{i=1}^{n}\sum_{l=1}^{m}\sum_{k=1}^{m} (|Q(u_{k-1}, v_l)x_i(u_{k-1}, v_l) - Q(u_k, v_l)x_i(u_k, v_l)|^2 \\ +|Q(u_k, v_{l-1})x_i(u_k, v_{l-1}) - Q(u_k, v_{l-1})x_i(u_k, v_l)|^2) \quad (1)$$

with respect to $\{Q(u_k, v_l)\}(k, l = 1, \dots, m)$ and where $Q(u_1, v_1) = \mathrm{I}$, *etc.* We can solve this optimization problem using elementary linear algebra.

We show an actual example of the proposed method and graphical representations of the results. The example data are facial data[3]. There are 6 persons, and we have 25 pictures under different viewing directions for each person. Similarities between two faces depend on viewing conditions (Figure 3(a)). The similarities between two persons for each condition are calculated with some degree of image matching method. We get $\{s_{ij}(u, v)\}; i, j = 1, \dots, 6; u, v = 1, \dots, 5$. Where u, v are related to orientations (rotation of the camera around the vertical axis and the horizontal axis respectively). Figure 3(b) shows the result of the proposed method. We can see the relations among 6 faces. Figure 3(c) reveals the changes of the configuration of the 5th person with u, v.

The proposed method can represent *functional configurations* of the objects. But, it is difficult to evaluate the functional configurations from the plot. The residual plots of the previous section can be extended for functional MDS. The residual plot for functional MDS is a simultaneous plot of functional configuration and functional residuals that are represented by segments.

---

[2] Some researchers claim that the results of MDS may not rotated because the axes imply information. But I do not take this istance in this method.

[3] The facial data in this paper are used by permission of Softpia Japan, Research and Development Division, HOIP Laboratory. It is strictly prohibited to copy, use, or distribute the facial data without permission.

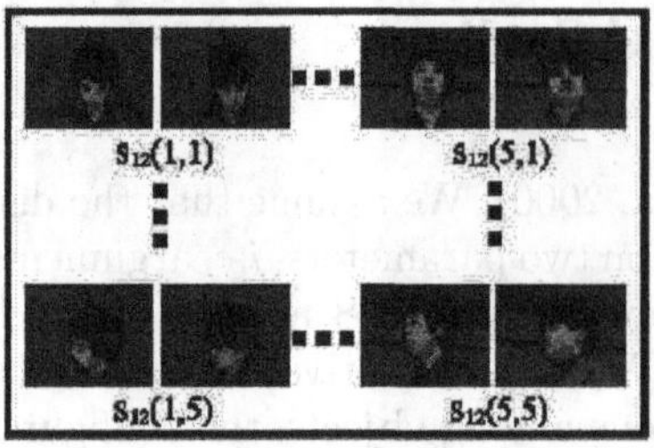

(a) Similarities between two objects at several directions

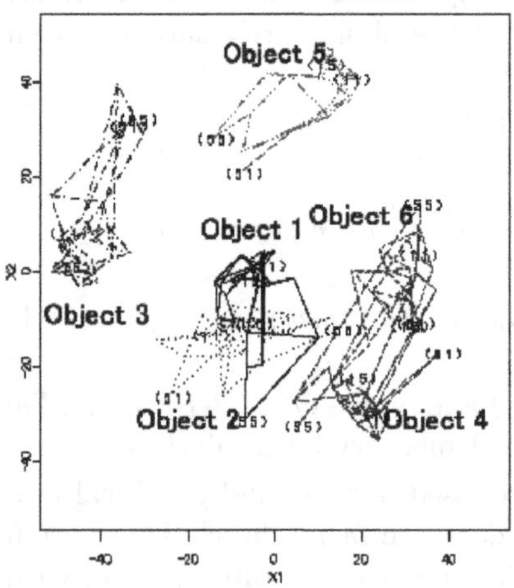

(b) (Functional) configuration

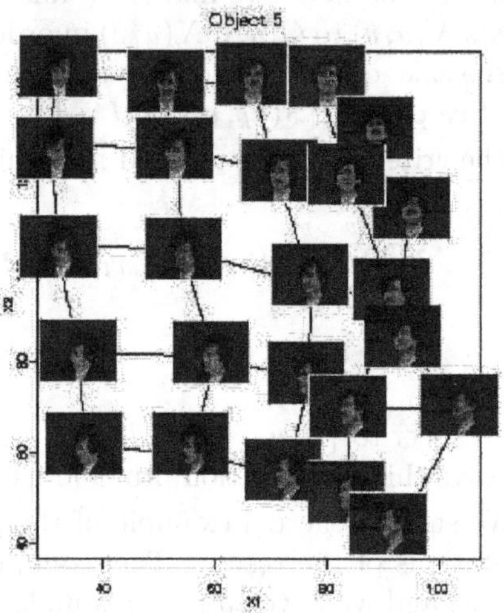

(c) Extraction of functional configuration of the object 5

Fig. 3. Functional MDS (Facial Data)

5 Concluding Remarks

We mention graphical methods mainly for functional data analysis. Needless to say, there are enormous subjects related to graphical representation of data that we do not mention up to here. For examples, the use of motion (*dynamic display*), business graph, statistical package, SVG, X3D, XML *etc.* *Data Analysis System with Visual Manipulation* is not a statistical graph but one of graphical methods in a wide sense. The concept of "Visual Language" can be applied to the field of data analysis.

References

MIZUTA, M. (2000): Functional multidimensional scaling. *Proceedings of the Tenth Japan and Korea Joint Conference of Statistics*, 77 – 82.

MIZUTA, M. (2002): Cluster analysis for functional data. *Proceedings of the 4th Conference of the Asian Regional Section of the International Association for Statistical Computing*, 219 – 221.

MIZUTA, M. (2003a): Hierarchical clustering for functional dissimilarity data. *Proceedings of the 7th World Multiconference on Systemics, Cybernetics and Informatics*, Volume V, 223 – 227.

MIZUTA, M.(2003b): K-means method for functional data. *Bulletin of the International Statistical Institute, 54th Session, Book* 2, 69 – 71.

MIZUTA, M. and MINAMI, H.(1991): A graphical representation for MDS. *Bulletin of the Computational Statistics of Japan*, 4, 2, 25 – 30.

RAMSAY, J.O. and SILVERMAN, B.W. (1997): *Functional Data Analysis.* New York: Springer-Verlag.

RAMSAY, J.O. and SILVERMAN, B.W. (2002): *Applied Functional Data Analysis – Methods and Case Studies –*. New York: Springer-Verlag.

RAMSAY, J.O. and SILVERMAN, B.W. (2005): *Functional Data Analysis.* 2nd edition, New York: Springer-Verlag.

Estimation of the Structural Mean of a Sample of Curves by Dynamic Time Warping

Isabella Morlini[1] and Sergio Zani[2]

[1] Dipartimento di Scienze Sociali, Cognitive e Quantitative, Università di Modena e Reggio Emilia, Italy, morlini.isabella@unimore.it
[2] Dipartimento di Economia, Università di Parma, Italy, sergio.zani@unipr.it

Abstract. Following our previous works where an improved dynamic time warping (DTW) algorithm has been proposed and motivated, especially in the multivariate case, for computing the dissimilarity between curves, in this paper we modify the classical DTW in order to obtain discrete warping functions and to estimate the structural mean of a sample of curves. With the suggested methodology we analyze series of daily measurements of some air pollutants in Emilia-Romagna (a region in Northern Italy). We compare results with those obtained with other flexible and non parametric approaches used in functional data analysis.

1 Introduction

This paper introduces a model based on Dynamic Time Warping (DTW) for computing the structural mean of discrete observations of continuous curves. Multivariate datasets consisting of samples of curves are increasingly common in statistics. One of the intrinsic problems of these functional data is the presence of *time* (or *phase*) *variability* beyond the well-known *amplitude variability*. Consider, for example, the problem of estimating the mean function of a sample of curves obtained at discrete time points, $\{x_n(t)\}$, $t = 1, \ldots, T$, $n = 1, \ldots, N$. The naive estimator $\bar{x}(t)$, the cross-sectional mean, does not always produce sensible results since it underestimates the amplitude of local maxima and overestimates the local minima. This is a severe problem because local extrema are often important features in physical, biological or ecological processes. This issue is illustrated in Fig. 1, where the carbon monoxide (CO) curves, as measured by three air quality monitory stations in Bologna (Italy), are plotted together. The time axis varies from January 1st 2004 to February 29th 2004. From an ecological point of view, it is important to estimate the average amplitude of peaks due to their effects on human health, but, as Fig. 1 shows, the cross sectional mean (the heavy line) underestimates some of these peaks. For example, the maximum mean value around February 4th is underestimated ignoring the time variation, since it occurs a few hours earlier for one of the three stations. The spurts may vary from site to site, not only in intensity, but also in timing, since weather-factors influencing the quantity of CO may be timed differently for the different spatial locations of the monitory stations. Various proposals deal with the

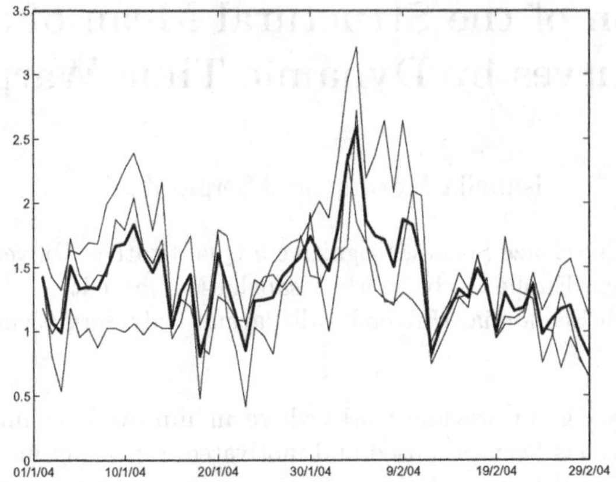

Fig. 1. CO curves and cross-sectional mean (heavy line).

time variability problem. Among others are landmark registration (Kneip and Gasser, 1992), continuous monotone registration (Ramsay and Li, 1998) and self modeling warping functions (Gervini and Gasser, 2004). The idea behind these methods is to estimate the warping functions $w_n(t)$ for aligning each series n, $(n = 1, \ldots, N)$ to a reference curve. Once the aligned values $x_n^*(t) = x_n(w_n(t))$ are computed, the structural mean is then estimated by $\mu(t) = \bar{x}^*(t)$, that is by the cross sectional mean of the warped series. In this paper we propose a nonparametric method for estimating *discrete* warping functions which does not involve landmark identification and is based on the DTW algorithm. The method relies on a minimization problem which can be solved efficiently by dynamic programming. However, the most attractive feature is that it can be directly applied to multivariate series: principal components and other multivariate analyses may be applied to the aligned series (Morlini, 2006). The paper is organized as follows. In Section 2 we briefly introduce the model and the algorithm implemented in Matlab. A discussion about differences with other methods drawn from DTW is also included. Section 3 presents an application regarding air pollutants in Emilia Romagna. In section 4 we give some final remarks.

2 Estimation of Discrete Warping Functions

To align two sequences $x_1(t)$ with $t = 1, \ldots, T_1$ and $x_2(t)$ with $t = 1, \ldots, T_2$ the classical DTW algorithm minimizes the following cost function:

$$\sum_{(i,j)\in w} (x_1(i) - x_2(j))^2 \tag{1}$$

Here $w = \{(i, j)\}$ is a warping path connecting $(1, 1)$ and (T_1, T_2) in a two dimensional square lattice which satisfies monotonicity and connectedness.

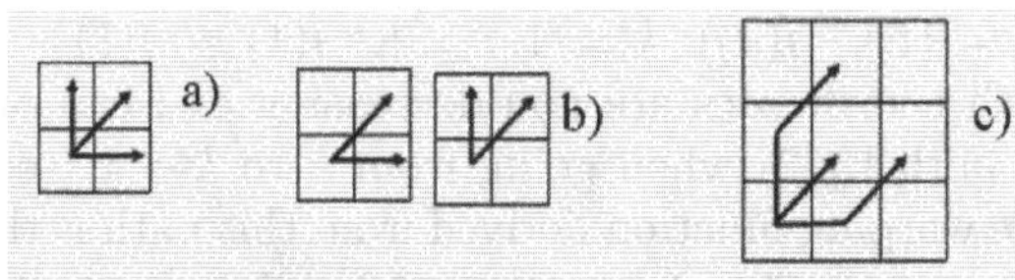

Fig. 2. Pictorial representations of step-paths a) The classical DTW step b) Steps with restrictions on the continuity constraints c) Step without connectedness.

This means that both coordinates of the parameterized path $w = \{i(k), j(k)$: $k = 1, \ldots, K$; $i(1) = j(1) = 1$; $i(K) = T_1$; $j(K) = T_2\}$ have to be non decreasing and they can only increase by 0 or 1 when going from k to $k+1$. Given N series $x_n(t)$, the DTW finds the time shift for each pair of series. In the structural mean framework, the aim is to align the family of curves $x_n(t)$ to an average time scale $t^* = 1, \ldots, T$ by minimizing a fitting criterion between the $x_n(t)$ and the value $y(t^*)$. For this aim, a reference curve y as long as N non decreasing warping functions should be defined. As shown in Morlini (2004), from the warping path we cannot draw two warping functions for aligning x_1 to x_2 and x_2 to x_1, since a single point on one series may map onto a large subsection of the other series. In order to find a non decreasing - discrete - warping function to align, for example, x_2 to x_1, we may relax the boundary condition, such that $i(K) = T_1$ and restrict the continuity constraint such that $i(k)$ can only increase by 1 when going from k to $k+1$ (the pictorial representation of this step in the warping path is reported in the left rectangle of Fig. 2). Note that this restriction may lead to unrealistic warping if the starting time for each series is arbitrary. The method is therefore convenient for series with a common starting time. If a windowing condition $|i(k) - j(k)| \leq u$ (with u being a given positive integer) is posted according to the concept that time-axis fluctuations should not lead to excessive differences in timing and in order to prevent over-warping, the connectedness must be lost in some points of the warping path. Once the bound $|i(k) - j(k)| = u$ is reached, the continuity constraint must be relaxed and $j(k)$ should increase by more than 1 when going from k to $k+1$, otherwise the path remains constantly on this bound. The index $j(k)$ can be set equal to $i(k+1)$ or to some integer in the interval $(j(k+1), i(k+1)]$.

Regarding, for example, the daily CO values (see the next Section) the positive integer u is the maximum number of days for which we assume the same weather-factors influencing air pollution may be timed differently for the different spatial locations of the monitoring gauges. In general, this value should be kept according to the series time scale and to the maximum temporal distance for which features in different subjects may be logically compared. Since the defined warping function is always below the line $w(t) = t$, $w(t) \leq t$ for every $t = 1, \ldots, T$ holds. If this condition is too restrictive, an alternative path for obtaining a non decreasing discrete warping function may be reached without connectedness. In this parameterized path, which is not new in literature, $w = \{i(k), j(k)): k = 1, \ldots, K; i(1) = j(1) = 1, i(K) = T_1, j(K) = T_2\}$, both coordinates are non decreasing and can only increase by 1 or 2 (but not both by 2) when going from k to $k+1$ (see Fig. 2c). The problem with the warping function derived from this path is that some values of the average time scale $t^* = 1, \ldots, T$ are missing since $w_i(t)$ is not defined for all $t = 1, \ldots, T$. If missing values are obtained by linear interpolation, the warping function is no longer discrete. In the application of next section we set $w_i(t) = w_i(t-1)$ for values of t not defined in the path. As remarked in Morlini (2004), DTW applied to raw data may produce incorrect alignments from a geometrical perspective when pairing two values x_1 and x_2 which are identical but the first is a part of a rising interval in series 1 and the second is a part of a falling interval in series 2. An additional problem is that the algorithm may fail to align a salient feature (i.e. a peak, a valley) simply because in one series this is slightly higher or lower than in the other one. We try to overcome these problems by smoothing the data before applying DTW and automatically selecting the smoothing parameter λ of the cubic interpolating splines by cross-validation (see Corbellini and Morlini, 2004). We estimate smooth interpolating functions in order to obtain less-noisy and shape dependent data but we maintain discrete cost and warping functions. This peculiarity characterizes this approach over other methods drawn from DTW for aligning curves. In Wang and Gasser (1997, 1999) some cost functions based on DTW are reviewed and a new functional of the interpolating splines and their derivatives is proposed. The main attractive feature of smoothing the data but maintaining a discrete cost function is perhaps the straightforward applicability of the results to multivariate series. This topic will be further illustrated in the next section. The problem concerning the reference vector may be solved with several possibilities (see Wang and Gasser, 1997). The reference vector should be close to the typical pattern of the sample curves and should have more or less the same features as most curves. In choosing this vector a trade off between accuracy and computational effort should be considered. Here we propose the following iterative method:

1. The longitudinal mean is considered as reference vector;
2. Every series is warped to this vector and the structural mean together with the total warping cost are computed;
3. The structural mean of step 2 is considered as the reference vector for the second cycle. The process is iterated until the warping cost has negligible changes, with respect to the previous ones.

Computation is not intensive since a few iterations are usually enough. However, if the relative shifts among curves are large, then the cross-sectional mean might be too atypical to start with, since the structure gets lost. A more convenient method, in this case, is to take each vector x_n as reference, to warp every other series to this vector, compute the total warping cost and choose the reference vector as the one corresponding to the maximum cost.

3 An Application to Air Pollutant Data

As an example to illustrate the technique presented in Section 2 for computing the structural mean, we consider measurements of the CO concentration in different sites in Emilia-Romagna. In the data set we have 1400 daily registrations (from January 1st 2001 to October 31st 2004) obtained by 67 air quality stations. A preprocessing step is made in order to replace missing and unreliable values (extreme points, as shown by boxplots, which are inconsistent with values registered in the two closest days) with moving averages of length 5. Fig. 3 and Fig. 4 report boxplots of pre-processed data for each site. Fig. 5 shows the structural means, for all the study period, obtained with DTW with restrictions on the continuity constraint and a 12 days maximum time shift, with self modeling registration (4 components) and with landmark registration. As landmarks we have chosen, graphically, two peaks in January and two peaks in December for each year. Individual identification of landmarks has been very difficult, time consuming and not so accurate since peaks in these and other months are much more than two. For these reasons, self modeling registration should be preferred as benchmark. Registered means for this method and modified DTW do not show significant differences. The linear correlation coefficient between the two series is equal to 0.95 and lines in Fig. 5 seem to overlap. Both methods are model-free and iterative. With a limited number of series, admittedly, the common starting reference vector clearly influences the results. Comparing the structural means computed by step b) and by step c) of Fig. 2 we note that rather than in amplitude, differences are in timing of the peaks. This gives an insight into the validity of both steps, since in many real data sets timing of the peaks in the mean curve does not have an objective location and is not exactly identifiable. Therefore the main goal is to estimate the real magnitude of the spurts rather than their timing.

Fig. 6 shows the 67 warping functions estimated by the different methods. Only the first 100 days are shown in order to have a clearer representation. Due to the windowing condition, functions estimated by DTW are not particularly irregular. They are, of course, step-wise functions. However, the steps are more visible in the case of restrictions on the continuity constraints rather than in the case without connectedness, since constant pieces are longer in the first case. Functions estimated by self modeling registration are always increasing and less wiggled. Landmark registration clearly produces illogical functions, too far from linearity, and leads to over-warping. As stated before, this is due to the difficulty of individualizing single landmarks in each series.

Even if the models based on DTW work well in this example and lead to results comparable with other well-established methods, the potential use of these models is better illustrated with multivariate series. To align the vector valued series $\mathbf{x}(t)$ to the reference curve $\mathbf{y}(t)$ the function minimized in the DTW algorithm becomes:

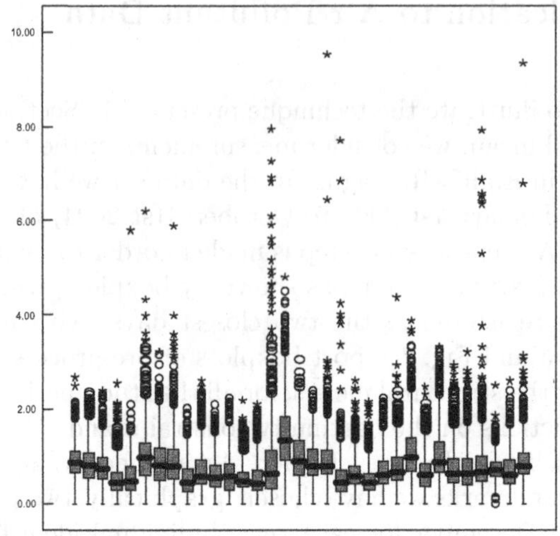

Fig. 3. Boxplots of pre-processed CO values for the first 33 monitory stations.

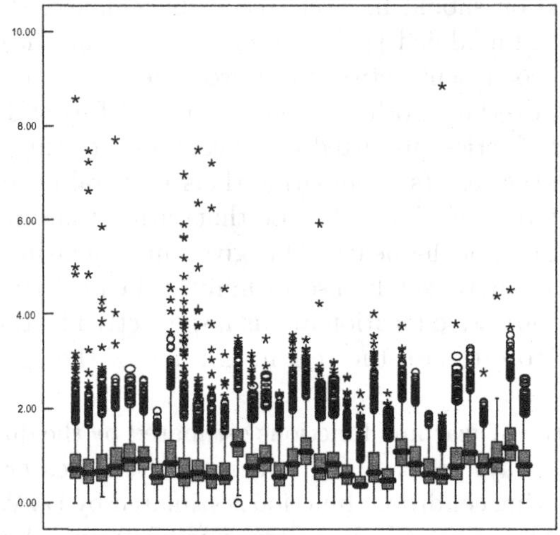

Fig. 4. Boxplots of pre-processed CO values for the last 34 monitory stations.

$$\sum_{(i,j)\in w} \|\mathbf{y}(i) - \mathbf{x}(j)\|^2 \tag{2}$$

where $\|.\|$ is the Euclidean norm. This objective function is the straightforward generalization of the univariate case. As an example, Fig. 7 shows the structural means of the daily values of sulfur dioxide (SO2), ozone (O3) and CO, computed considering vector-valued series in the DTW algorithm with restriction on the continuity constraint. The warping function is equal

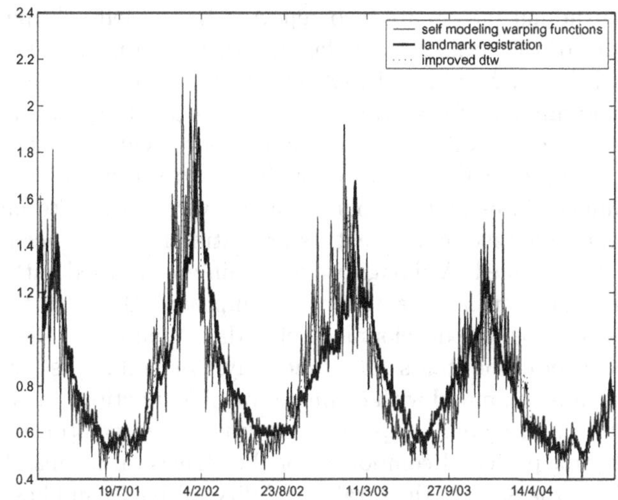

Fig. 5. Structural means of the CO curves obtained with different methods.

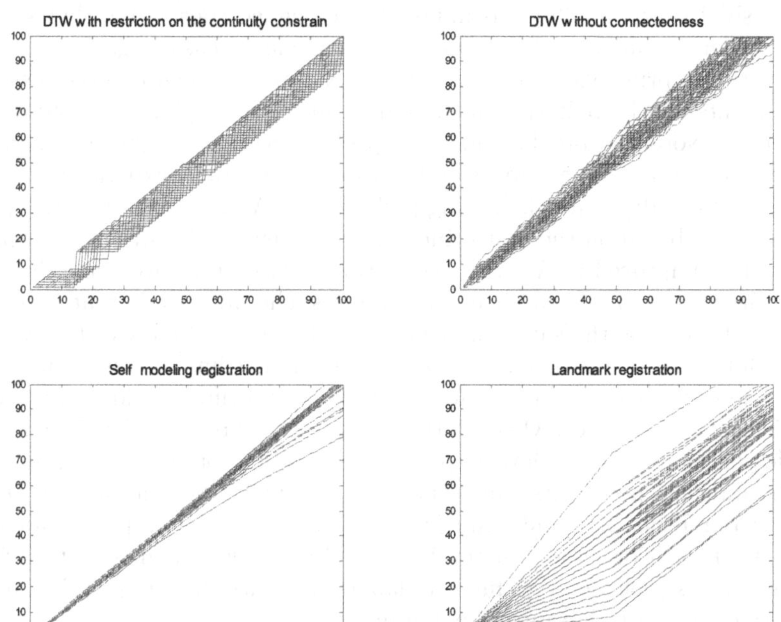

Fig. 6. Warping functions (from January 1st 2001 to April 10th 2001).

for each series and is computed considering simultaneously the values of the three pollutants, which are thought of different aspects of the same phenomenon. To reduce noise, raw series are interpolated with tensor product splines. Longitudinal means are also reported in the left-side panels. Here the amplitudes of the peaks are similar but structural means show less and better-defined peaks. Maxima of SO2 and CO (which are typically winter pollutants) and minima of O3 (which is a summer pollutant) are well aligned and the trend for each pollutant is also much more evident in the structural mean. Comparing, for CO, the graph of the structural mean obtained with the vector valued series and the one obtained with univariate analysis, we see that in the first case marker events are estimated more accurately and spurts are better defined. Although we are mainly concerned with the structural means estimation, a potential use of improved DTW with the vector valued series may be the development of a daily pollution index based on the values of several pollutants simultaneously. For pollutants for which the structural mean is computed with common warping functions, this index may be reached with a principal component analysis of the structural means. For example, the first principal component of the structural means of CO, SO2 and O3 explains the 85 % of the total variability (Table 2) and has significant correlations with all the three means (Table 3). This component may be then used as a quality index based on the three pollutants CO, SO2 and O3, simultaneously. Breakpoints for this index maybe defined considering that scores have zero mean and one of the three structural means has a negative correlation with the principal component. As an example, Table 4 reports possible breakpoints for the pollution index and frequency of days corresponding to each class. Note that breakpoints are considered with respect to the range of possible values and do not focus on health effects one may experience within a few hours or days after breathing polluted air. As a matter of fact, a pollution index based on the first principal component of the structural means reached by improved DTW with vector valued series may overcome the two main drawbacks of official air quality indexes The first one is that it is not realistic to assume the same air pollution index as valid all over the world, since different areas are characterized by different climatic conditions, and both the construction of the index and the breakpoints should be data dependent. The second drawback is that these idexes are usually referred to a single pollutant while the level of pollution should be considered with respect to the different pollutants simultaneously present in the atmosphere. Data reported in Tables 2, 3 and 4 illustrate the potential use of a principal components analysis over the structural means obtained with multivariate series. However, this topic deserves further elaboration since the analysis should be performed over a wide range of pollutants.

4 Conclusions

The algorithm based on DTW is shown to work well in the air pollutants example, and to lead to results at least comparable to those reached by other methods. Of course, with a limited number of series, the initial reference

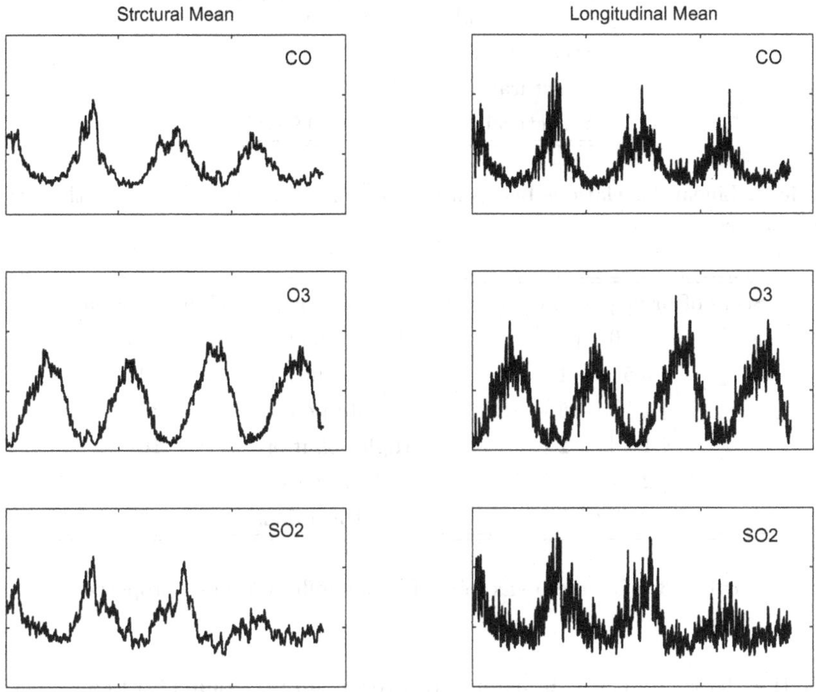

Fig. 7. Estimated means of some pollutants (period 01/01/2001 - 31/10/2004).

Component	Eigenvalues	Explained variability (%)
1	2.56	85.4
2	0.33	11.0
3	0.11	3.6

Table 1. Eigenvalues of the principal component analysis

vector (the cross sectional mean, both for the algorithms based on DTW and for self modeling registration) clearly influences the results. In literature, proposals for computing the structural mean of a sample of curves have been almost always applied to a greater number of series and with a limited length. Further elaboration is therefore needed in order to apply the method to examples with a greater number of series. Computational time substantially depends on the choice of the smoothing parameter of the interpolating splines. With the parameters estimated by cross validation, computational efforts are quite intensive. With parameters fixed a priori, equal for each se-

Variable	Component 1
Structural mean CO	0.96
Structural mean SO2	-0.91
Structural mean O3	0.90

Table 2. Linear correlations between the first principal component and the structural means

Score of principal component I	Air quality	Number of days
| 0.5 |	Good quality	360
| 0.5 | - | 1 |	Low pollution	526
| 1 | - | 1.5 |	Moderate pollution	376
| 1.5 | - | 2 |	High pollution	107
| 2 | - | 2.5 |	Hazardous	19
> | 2.5 |	Very Hazardous	12

Table 3. Possible breakpoints for the pollution index proposed

ries, the algorithm is much faster. However, computational efforts are always negligible with respect to the other methods used in this study.

References

CORBELLINI, A. and MORLINI, I (2004): Searching for the optimal smoothing before applying Dynamic Time Warping. *Proceedings of the XLII Meeting of the Italian Statistical Society,* pp. 47–50.

GERVINI, D. and GASSER, T. (2004): Self-modelling warping functions. *J.R. Statist. Society B, 66,* 959–971.

KNEIP, A. and GASSER, T (1992): Statistical tools to analyze data representing a sample of curves. *Annals of Statistics, 20, 1266–1305.*

MORLINI, I. (2004): On the dynamic time warping for computing the dissimilarities between curves. In: Vichi *et al.* (Eds): *New Developments in Classification and Data Analysis,* Springer, Berlin, pp. 63–70.

MORLINI, I. (2006): Searching for structure in air pollutants concentration measurements. *Environmetrics,* submitted.

RAMSAY, J.O. and LI, X. (1998): Curve registration. *J.R. Statist. Society B, 60, 351–363.*

WANG, K and GASSER, T. (1997): Alignment of curves by dynamic time warping. *The Annals of Statistics, 25, 3,* 1251–1276.

WANG, K and GASSER, T. (1999): Synchronizing sample curves nonparametrically. *The Annals of Statistics, 27, 2,* 439–460.

Sequential Decisional Discriminant Analysis

Rafik Abdesselam

CREM UMR CNRS 6211, Department of Economics and Management,
University of Caen, Esplanade de la paix, 14032 Caen, France
rafik.abdesselam@unicaen.fr

Abstract. We are describing here a sequential discriminant analysis method which aim is essentially to classify evolutionary data. This method of decision-making is based on the research of principal axes of a configuration of points in the individual-space with a relational inner product. We are in presence of a discriminant analysis problem, in which the decision must be taken as the partial knowledge evolutionary information of the observations of the statistical unit, which we want to classify. We show here how the knowledge from the observation of the global testimony sample carried out during the entire period, can be of particular benefit to the classifying decision on supplementary statistical units, of which we only have partial information about. An analysis using real data is here described using this method.

1 Introduction

The Sequential Decisional Discriminant Analysis (SDDA) proposed is based on the relational inner product notion in the individual-space. We distinguish two or more than two groups of individuals, defined *a priori* by a nominal variable on which the same continuous variables are measured at different times. This sequential method is especially conceived to study evolutionary data in a classifying aim of evolutionary new individuals. The main aim of this method is to obtain at time a better prediction performance of supplementary individuals during their evolution. The constraint being here to impose to the discriminant factors to belong to the subspace generated by all observations carried out during the period of information knowledge about the individual we want to classify. Contrary to the Partial Decisional Discriminant Analysis (PDDA) also proposed, the SDDA takes into account the global information brought by the testimony sample in the classification decision. It is possible to carry out such an analysis according to the evolution in time of the supplementary individual that has to be classified. Section 2 deals with a brief description of a Relational Euclidean Model and with a comparison between two proposed multiple discriminant analyses. The definition of a sequential method is given in section 3. Partial and sequential methods of discriminant analysis are illustrated and compared on the basis of real data.

2 Discriminant Analysis in a Relational Model

In this section, we propose in a Relational Euclidean Model two multiple discriminant analyses, then evaluate and compare their discrimination performance through an empirical study.

The Relational Euclidean Model is of some interest because the inner product in individual-space is relational, i.e. taking into account relationships observed between variables. We recall the notion of this inner product introduced by Schektman (1978).

$\{x^j(t); j = 1,p\}_{t=1,T}$, being T sets of p continuous same variables observed at different times and $\{y^k; k = 1,q\}$ the dummy variables of the modalities of a nominal variable y, observed at the end of the period that is at the last time T. Let us denote:
$E_t = E_{x(t)} = \mathbb{R}^p$ [resp. $E_y = \mathbb{R}^q$] the individual-subspace associated by duality to the p zero mean variables $\{x^j(t); j = 1,p\}$ [resp. zero mean dummy variables $\{y^k; k = 1,q\}$],
X_t the matrix of the explanatory data at time t associated to the set of variables $\{x^j(t); j = 1,p\}$, with n rows-individuals and p columns-variables,
$X^s = [X_1, \ldots, X_t, \ldots, X_s]$ the evolutionary data matrix: juxtaposition of tables X_t, it's named partial if s is less than T and global if s is equal T,
$Y_{(n,q)}$ the data matrix associated to the set of variables $\{y^k; k = 1,q\}$,
$E_x^T = \oplus\{E_t\}_{\{t=1,T\}} = \mathbb{R}^{Tp}$ [resp. $E_x^s = \oplus\{E_t\}_{\{t=1,s\}} = \mathbb{R}^{sp}$] the global [resp. partial] explanatory individual-subspace, associated by duality to the T [resp. s first] sets of explanatory variables $\{x^j(t); j = 1,p\}_{\{t=1,T\}}$ [resp. $\{x^j(t); j = 1,p\}_{\{t=1,s\}}$],
D_y the diagonal matrix of weights attached to the q modalities: $[D_y]_{kk} = \frac{n_k}{n}$ for all $k = 1,q$, where n_k be the number of individuals which have the kth modality of y,
$N_t = \{x_i(t) \in E_t\,;\, i = 1,n\}$ the configuration of the individual-points associated to X_t and N^T [resp. N^s] the configuration of the individual-points associated to the table X^T [resp. X^s] in E_x^T [resp. E_x^s],
M_y [resp. M_t] the matrix of inner product in the space E_y [resp. E_t].

Since, we are in the framework of a discriminant analysis, we choose Mahalanobis distance $M_t = V_t^+$ in all explanatory subspaces E_t and $M_y = \chi^2_y$ the Chi-square distance in E_y. So, M is a relational inner product, partitioned and well-balanced matrix, of order $Tp + q$, in the individual-space $E = E_x^T \oplus E_y$, according to all the couples of variables $\{x^j(t); j = 1,p\}_{t=1,T}$ and $\{y^k; k = 1,q\}$, if and only if:

$$\begin{cases} M_{tt} = M_t = V_t^+ & \textit{for all } t = 1,T \\ M_{tt'} = M_t[(V_tM_t)^{\frac{1}{2}}]^+ V_{tt'} M_{t'}[(V_{t'}M_{t'})^{\frac{1}{2}}]^+ = V_t^+ V_{tt'} V_{t'}^+ & \textit{for all } t \neq t' \\ M_{ty} = M_t[(V_tM_t)^{\frac{1}{2}}]^+ V_{ty} M_y[(V_yM_y)^{\frac{1}{2}}]^+ = V_t^+ V_{ty} \chi^2_y \end{cases} \quad (1)$$

where $V_t = {}^tX_tDX_t$, $V_y = {}^tYDY$ and $V_{ty} = {}^tX_tDY$ are the matrix of covariances, $D = \frac{1}{n}I_n$ is the diagonal weights matrix of the n individuals and I_n is a unit matrix with n order. $[(V_tM_t)^{\frac{1}{2}}]^+$ [resp. $[(V_yM_y)^{\frac{1}{2}}]^+$] is the Moore-Penrose generalized inverse of $(V_tM_t)^{\frac{1}{2}}$ [resp. $(V_yM_y)^{\frac{1}{2}}$] according to M_t [resp. M_y].

2.1 Relational Decisional Discriminant Analysis

Let $P^M_{E^s_x}$ the M-orthogonal projection operator in $E^s_x = \oplus\{E_t\}_{\{t=1,s\leq T\}}$ subspace and $N^s_g(x/y) = \{P^M_{E^s_x}(e_k(y)\,;\, k=1,q\} \subset E$ the configuration of the q centers of gravity points, where $\{e_k(y)\,;\, k=1,q\}$ is the canonical base in E_y.

Definition

The Relational Decisional Discriminant Analysis (RDDA) at time $s \in]1\,;T]$ of the configuration N^s consists in making, in a REM, the following PCA:

$$PCA\,[\,N^s_g(x/y) = \{P^M_{E^s_x}(e_k(y))\,;\, k=1,q\}\,;\, M\,;\, D_y\,]. \tag{2}$$

It's a multiple discriminant factorial analysis and named partial (PDDA) if $s \in]1\,;T[$, global (GDDA) if $s=T$, and it's equivalent to the classical and simple discriminant factorial analysis if $s=1$.

2.2 Sequential Decisional Discriminant Analysis

Decisiveness is the second aim of any discrimination procedure allowing to classify a new individual in one of the q groups, supposed here, *a priori* defined by all the values of the sets $\{x^j(t)\,;\, j=1,p\}_{t=1,T}$.

Using generators of inner products, we choose the "best" intra inner product $M_t(\alpha)$, $\alpha \in [0\,;1]$, in the explanatory subspaces E_t according to a maximum explained inertia criterion. So the nature of Euclidean inner product M_y in E_y is of no importance; however, we opt for the Chi-square distance for its use simplifies calculations. For M_t, in the subspace E_t, we use generators of within inner products $M_t(\alpha)$ to search the "best" one, denoted $M_t(\alpha^\star)$, which optimizes criterion (Abdesselam and Schektman (1996)).

In the context of the formal approach, we suggest these following simple formulas for choosing the inner product which maximizes the percentage of explained inertia:

$$\begin{cases} {}^1M_t(\alpha) = \alpha I_p + (1-\alpha)V_t^+ & for\,all\ t=1,T \\ {}^2M_t(\alpha) = \alpha I_p + (1-\alpha)V_t & with\quad \alpha \in [0,1]. \end{cases}$$

These generators will evolve from the symmetrical position ${}^1M_t(0) = V_t^+$ (Mahalanobis distance) towards the dissymmetrical position ${}^2M_t(0) = V_t$ in passing by the classical dissymmetrical position ${}^1M_t(1) = I_p = {}^2M_t(1)$, where I_p is the unit matrix in order p and V_t^+ the Moore-Penrose generalized inverse of V_t.

In this case, the partitioned and well-balanced matrix, of order $Tp+q$, associated to the relational inner product denoted $M^\star$ in the individual-space $E = E_x^T \oplus E_y$, is written like this:

$$\begin{cases} M^\star_{tt} = M_t(\alpha^\star) & for\,all\ t = 1, T \\ M^\star_{tt'} = M_t(\alpha^\star)[(V_t M_t(\alpha^\star))^{\frac{1}{2}}]^+ V_{tt'} M_{t'}(\alpha^\star)[(V_{t'} M_{t'}(\alpha^\star))^{\frac{1}{2}}]^+ & for\,all\ t \neq t' \\ M^\star_{ty} = M_t(\alpha^\star)[(V_t M_t(\alpha^\star))^{\frac{1}{2}}]^+ V_{ty} \chi^2_y & \end{cases}$$

Let $\hat{N}^s_g(x/y)$ the configuration of centre of gravity points associated to the configuration $\hat{N}^s = P^{M^\star}_{E^s_x}(N^T) \subset E$: the $M^\star$-orthogonal projection of global explanatory configuration N^T in the partial subspace $E^s_x \subset E$.

Property

The Sequential Decisional Discriminant Analysis (SDDA) at time $s \in [1; T-1]$ consists in making in E, the PDDA of the configuration $\hat{N}^s$, that is the following PCA:

$$PCA\ [\hat{N}^s_g(x/y)\,;\, M^\star\,;\, D_y]. \tag{3}$$

In (2) and (3), the configurations $N^s_g(x/y)$ and $\hat{N}^s_g(x/y)$ of the centre of gravity points are in the same subspace $E^s_x = \mathbb{R}^{sp}$ of the Euclidean individual space E.

3 Example - Apple Data

From the Agronomic Research National Institute - INRA, Angers, France, we kindly obtained an evolutionary data set that contains the measurements of $p = 2$ explanatory variables: the content of both sugar and acidity of $n = 120$ fruits of the same type of apple species at $T = 3$ different times - fifty days before and after the optimal maturity day - $\{t_1$: premature, t_2: maturity, t_3: postmaturity$\}$.

Note that for these repeated measurements of sugar and acidity, obviously it's not a matter of the same apple at different times but different apples of the same tree.

This sample of size $N = 120$ is subdivided into two samples: a basic-sample or "training set" of size $N_1 = 96$ (80%) for the discriminant rule and a test-sample or "evaluation set" of size $N_2 = 24$ (20%) for next evaluated the performance of this rule.

Global Decisional Discriminant Analysis

This first part concerns the performance of the discrimination rule of the GDDA, we use both explained inertia and misclassification rates as criteria for the two samples. At global time we obtain better discrimination results, indeed we show in Figure 1 that the $q = 3$ three groups-qualities of apples

$\{G_1, G_2, G_3\}$ are differentiated and well-separated. Note that these groups are *a priori* well-known at the end of the last period.

Table 1 and Figure 1 summarize the main results of the global discriminant analysis at time $\{t_1, t_2, t_3\}$, i.e. s = T = 3.

Inertia	GDDA
B-Between classes	1.7986
W-Within class	0.2014
T-Totale	2.0000
Explained inertia	89.93%
Samples	Misclassified
Basic-sample (80%)	0%
Test-sample (20%)	0%

Table 1. Explained inertia and misclassification criteria.

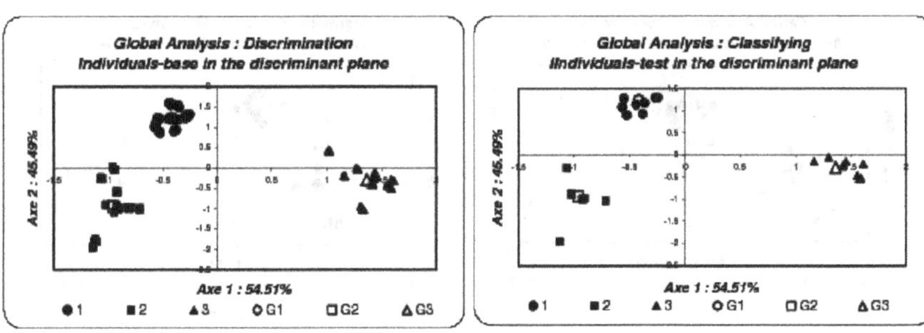

Fig. 1. Representations of the global analysis at time $s = T = 3$.

So, with this good percentages for the criteria: 89.93% of explained inertia and 0% of misclassified for both basic and test samples, we can considerate this basic-sample of evolutionary data as a good global testimony sample to make partial or sequential analyses at time $s = 1$ and/or $s = 2$, if we want to classify, into the three groups pre-defined at global time $s = T = 3$, new individuals which we only have partial information about, i.e. information at time $s = 1$ or $s = 2$.

Comparison of discrimination rules

This second part concerns the discrimination performances of the PDDA and SDDA methods. They are compared at time $\{t_1, t_2\}$, i.e. $s = 2$.

In Table 2, we show that the values of the discrimination and misclassification criteria of the SDDA are better than those of the PDDA according to those of the GDDA in Table 1.

PDDA	Criteria at time s = 2	SDDA
1.6347	B-Between classes	1.7858
0.3654	W-Within class	0.2143
2.0000	T-Total	2.0000
81.73%	Explained inertia	89.29%
Partial	Misclassified	Sequential
4.17%	Basic-sample (80%)	0%
0%	Test-sample (20%)	0%

Table 2. Comparison of the criteria

The graphical results in Figure 2 also show that the groups are more well-separated with the Sequential method than with the partial method.

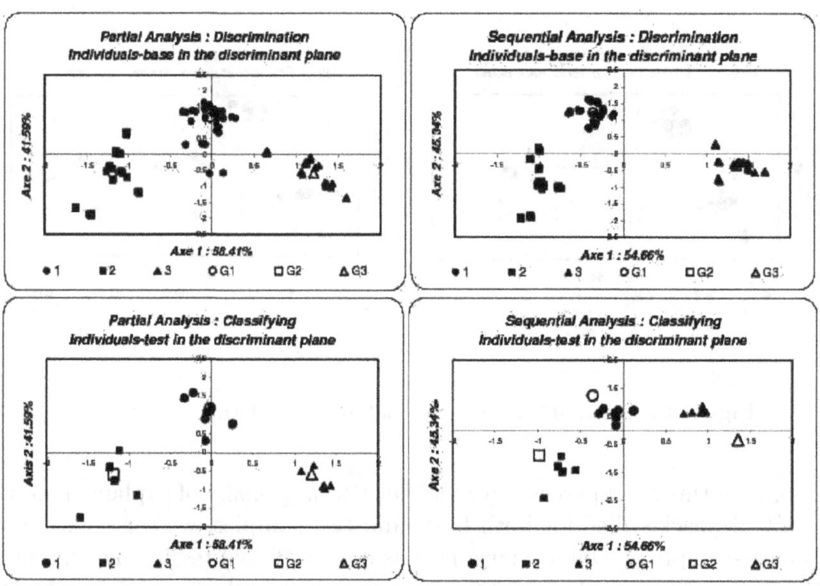

Fig. 2. Representations of the partial and sequential analyses at time $s = 2$.

Comparison of classification rules

This third part which is the main aim of this paper concerns the classification of new individuals at time $\{t_1\}$ and/or $\{t_1, t_2\}$. This objective is schematically illustrate in Figure 3.

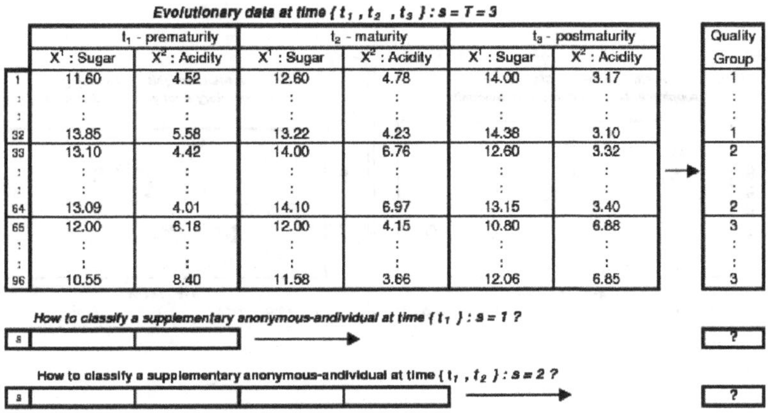

Fig. 3. Classifying new individuals

In Table 3 we summarize for each method the results of the affectation group and the distances between the centres of gravity and five anonymous supplementary individuals.

G_1	G_2	G_3	PDDA	New individual	SDDA	G_1	G_2	G_3
1.183	1.480	**1.183**	?	s_1	G_1	**1.163**	1.516	1.345
1.156	**1.156**	2.463	?	s_2	G_2	1.780	**0.495**	2.024
1.121	1.286	2.569	G_1	s_3	G_2	1.738	**0.515**	2.050
1.116	1.996	**1.007**	G_3	s_4	G_1	**0.915**	2.091	1.408
1.832	2.051	**0.371**	G_3	s_5	G_3	1.196	2.084	**1.123**

Table 3. Distances - Classification of anonymous individuals at time $s = 2$

Figure 4, shows the projection of these new individuals on the PDDA and SDDA discriminant plans at time $\{t_1, t_2\}$. An individual-point is affected to the group whose centre of gravity is the nearest.

On the discriminant plan of partial analysis, where we take into account only the values of the variables at times t_1 and t_2, the two supplementary individuals s_1, and s_2 are border points, they are unclassifiable. The individual-point s_3 is near the first group and the two other points, s_4 and s_5, are affected to the third group.

Whereas, on the discriminant plan of the sequential analysis - i.e. we take into account all information brought by the global testimony sample, as well as the values of the variables at time t_3, even if these values at time t_3 for these new individuals are unknown - we are led to classify s_1 and s_2 in one of the three groups more precisely, respectively in the first and the second group. As for s_3, it's allocated to the second group and finally, s_4 and s_5 are classified respectively in the first and third group.

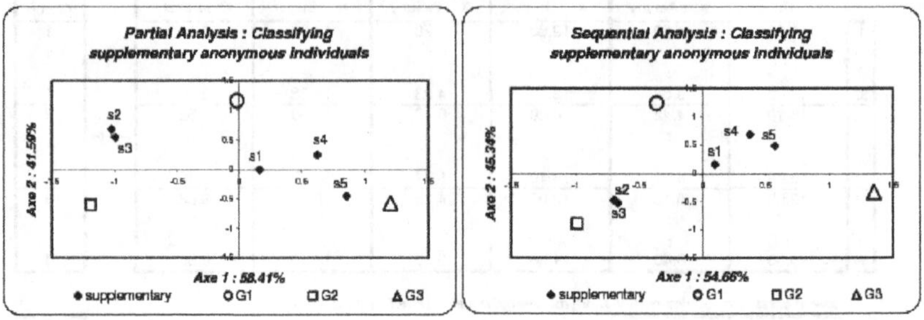

Fig. 4. Representations of the partial and sequential analyses at time $s = 2$.

4 Conclusion

In this paper, we have presented a Sequential Decisional Discriminant Analysis of evolutionary data in a relational Euclidian context in a discrimination and classification aim. This method is particularly adapted because it enables to take at time a classifying decision of a new evolutionary individual about which we have only a partial information. It's a simple and effective method which can be very useful in different fields especially in the medical field for diagnosis-making, where it is often important to anticipate before term in order to be able to intervene on time when necessary.

Acknowledgments: I am grateful to Professor Yves Schektman for his helpful suggestions and precious advice he provides me.

References

ABDESSELAM, R. and SCHEKTMAN, Y. (1996): Une Analyse Factorielle de l'Association Dissymétrique entre deux Variables Qualitatives. *Revue de Statistique Appliquée.* Paris, XLIV(2), 5–34.

ABDESSELAM, R. (2003): Analyse Discriminante Multiple Relationnelle. *Méthodes et perspectives en Classification.* Editeurs Y. Dodge et G. Melfi. Groupe de statistique. Presses Académiques Neuchâtel, Suisse, pp.39-42.

CASSIN, P. HUBERT, M. and VAN-DRIESSEN, K. (1995): L'analyse discriminante de tableaux évolutifs. *Revue de Statistique Appliquée.* 3, 73–91.

CHERKASSKY, V. and MULIER, F. (1998): Learning from Data: Concepts, Theory, and Methods. *Revue de Statistique Appliquée*, 3, 73–91.

DAZY, F. and LEBARZIC, J-F. (1996): *L'analyse des données évolutives. Méthodes et applications.* Editions technip.

ESCOFIER, B. and PAGÈS J. (1988): *Analyses factorielles simples et multiples : objectifs, méthodes et interprétation*, Dunod.
GEOFFREY, J. McLACHLAN (2004): *Discriminant Analysis and Statistical Pattern Recognition.* Wiley Series in Probability and Statistics, New Ed., 526p.
HAND, D.J. (1981): *Discrimination and Classification.* Wiley and Sons, New York.
HUBERT, M. and VAN-DRIESSEN, K. (2004): Fast and robust discriminant analysis. *Computational Statistics and Data Analysis*, 45, 301–320.
LACHENBRUCH, P. (1975): *Discriminant Analysis.* Hafner Press, New York.
SCHEKTMAN, Y. (1978): Contribution à la mesure en facteurs dans les sciences expérimentales et à la mise en œuvre des calculs statistiques. *Thèse de Doctorat d'Etat*, Université Paul Sabatier, Toulouse.

Regularized Sliced Inverse Regression with Applications in Classification

Luca Scrucca[1]

Dipartimento di Economia, Finanza e Statistica
Università degli Studi di Perugia, Italy
luca@stat.unipg.it

Abstract. Consider the problem of classifying a number of objects into one of several groups or classes based on a set of characteristics. This problem has been extensively studied under the general subject of discriminant analysis in the statistical literature, or supervised pattern recognition in the machine learning field. Recently, dimension reduction methods, such as SIR and SAVE, have been used for classification purposes. In this paper we propose a regularized version of the SIR method which is able to gain information from both the structure of class means and class variances. Furthermore, the introduction of a shrinkage parameter allows the method to be applied in under-resolution problems, such as those found in gene expression microarray data. The REGSIR method is illustrated on two different classification problems using real data sets.

1 Dimension Reduction Methods in Classification

Dimension reduction methods, developed in the context of graphical regression, have been recently used in classification problems. Cook and Yin (2001) discuss dimension reduction methods, such as sliced inverse regression (SIR, Li (1991)) and sliced average variance estimation (SAVE, Cook and Weisberg (1991)), as graphical methods for constructing summary plots in discriminant analysis. Both methods are able to recover at least part of the dimension reduction subspace $\mathcal{S}_{Y|\mathbf{X}}$, a subspace spanned by the columns of the $p \times d$ ($d \leq p$) matrix $\boldsymbol{\beta}$ such that $Y \perp\!\!\!\perp \mathbf{X} | \boldsymbol{\beta}^{\top}\mathbf{X}$. This implies that the $p \times 1$ predictor vector $\mathbf{X}$ can be replaced by the $d \times 1$ vector $\boldsymbol{\beta}^{\top}\mathbf{X}$ with no loss of regression information. Whatever $d < p$, we have effectively reduced the dimensionality of the problem. Furthermore, the estimated summary plot might provide useful information on how to construct a discriminant function.

The SIR method gains information on $\mathcal{S}_{Y|\mathbf{X}}$ from the variation on class means, so it might miss considerable information when classes differ not only on their location. SAVE, on the contrary, uses information from both class means and class variances. However, it requires the estimation of more parameters, and the summary plot it provides may not be as informative as that provided by SIR when the majority of statistical information useful for classification comes from the class means.

Cook and Yin (2001) discuss the connection between these two methods and a second-moment method called SIR-II (Li, 1991). In particular, the subspace identified by the SAVE method can be seen as the sum of the subspaces given by SIR and SIR-II methods. In this paper we propose a regularized version of the SIR algorithm, called REGSIR, which gains information from variation on both class means and variances. Regularization will depend on two parameters selected by minimizing an estimate of the misclassification error.

2 Regularized SIR

The SIR algorithm is based on the spectral decomposition of $\mathrm{Var}(\mathrm{E}(\mathbf{X}|Y))$, with the eigenvectors corresponding to the non-zero eigenvalues which span a subspace of the dimension reduction subspace, that is $\mathcal{S}_{\mathrm{E}(\mathbf{X}|Y)} \subseteq \mathcal{S}_{Y|\mathbf{X}}$. In the original proposal Y was replaced by a sliced version, but in the classification setting this is not required since Y is already in discrete form. The sample version of SIR is obtained through the eigen-decomposition of $\hat{\mathbf{\Sigma}}_X^{-1}\mathbf{M}$, where $\hat{\mathbf{\Sigma}}_X = \frac{1}{n}\sum_i (\boldsymbol{x}_i - \bar{\boldsymbol{x}})(\boldsymbol{x}_i - \bar{\boldsymbol{x}})^\top$ is the covariance matrix, $f_k = n_k/n$ is the proportion of cases in each class for $k = 1, \dots, K$ classes, each of size n_k ($n = \sum_k n_k$), and the kernel matrix

$$\mathbf{M} = \mathrm{Var}(\hat{\mathrm{E}}(\mathbf{X}|Y)) = \sum_k f_k(\bar{\boldsymbol{x}}_k - \bar{\boldsymbol{x}})(\bar{\boldsymbol{x}}_k - \bar{\boldsymbol{x}})^\top \tag{1}$$

The above matrix is also known as the between-group covariance matrix. Given the decomposition $\hat{\mathbf{\Sigma}}_X^{-1}\mathbf{M} = \mathbf{V}\mathbf{L}\mathbf{V}^\top$, SIR directions are given by $\hat{\boldsymbol{\beta}} = \hat{\mathbf{\Sigma}}_X^{-1/2}\mathbf{V}$ and SIR variates are computed as $\hat{\boldsymbol{\beta}}^\top\mathbf{X}$. It can be shown that canonical variables used in linear discriminant analysis (LDA) are equivalent to SIR variates but with different scaling (Li (2000)).

SIR-II gains information about $\mathcal{S}_{Y|\mathbf{X}}$ from the variation in class covariances. Thus, directions are estimated following the same procedure shown for SIR, but with a different kernel matrix given by

$$\begin{aligned} \mathbf{M} &= \mathrm{E}[\hat{\mathbf{\Sigma}}_{X|Y} - \mathrm{E}(\hat{\mathbf{\Sigma}}_{X|Y})]^2 \\ &= \sum_k f_k(\hat{\mathbf{\Sigma}}_{X|(Y=k)} - \hat{\mathbf{\Sigma}}_{\mathrm{E}(X|Y)})\hat{\mathbf{\Sigma}}_X^{-1}(\hat{\mathbf{\Sigma}}_{X|(Y=k)} - \hat{\mathbf{\Sigma}}_{\mathrm{E}(X|Y)})^\top \end{aligned} \tag{2}$$

where $\hat{\mathbf{\Sigma}}_{X|(Y=k)} = \frac{1}{n_k}\sum_{i:(y_i=k)} (\boldsymbol{x}_i - \bar{\boldsymbol{x}}_k)(\boldsymbol{x}_i - \bar{\boldsymbol{x}}_k)^\top$ is the within-class covariance matrix, and $\hat{\mathbf{\Sigma}}_{\mathrm{E}(X|Y)} = \sum_k f_k \hat{\mathbf{\Sigma}}_{X|(Y=k)}$ is the pooled within-class covariance matrix.

The proposed regularization approach aims at using information from differences both in class means and variances. Meanwhile, a shrunken version of $\hat{\mathbf{\Sigma}}_X$ is used to overcome ill- and poorly-posed inverse problems (Friedman, 1989). This latter goal is pursued defining the following convex combination:

$$\hat{\mathbf{\Sigma}}_X(\lambda) = (1-\lambda)\hat{\mathbf{\Sigma}}_X + \lambda \frac{\operatorname{tr}(\hat{\mathbf{\Sigma}}_X)}{p}\mathbf{I} \tag{3}$$

where the parameter λ ($0 \leq \lambda \leq 1$) controls the amounts of shrinkage applied, and $\mathbf{I}$ being the identity matrix. As $\lambda \to 1$, $\hat{\mathbf{\Sigma}}_X(\lambda)$ approaches a diagonal matrix with diagonal elements given by $\operatorname{tr}(\hat{\mathbf{\Sigma}}_X)/p$, i.e. the average eigenvalue of the covariance matrix.

The regularized kernel matrix is defined as

$$\mathbf{M}(\lambda,\gamma) = (1-\gamma)\mathbf{M}_I + \gamma\mathbf{M}_{II} \tag{4}$$

where $\mathbf{M}_I$ and $\mathbf{M}_{II}$ are, respectively, the kernel matrices in (1) and (2). The regularization parameter γ ($0 \leq \gamma \leq 1$) controls the convex combination of kernel matrices used by SIR and SIR-II. Finally, REGSIR directions are estimated through the spectral decomposition of $\hat{\mathbf{\Sigma}}_X(\lambda)^{-1}\mathbf{M}(\lambda,\gamma)$. The projection onto the estimated subspace is given by $\hat{\boldsymbol{\beta}}^\top \mathbf{X}$, where $\hat{\boldsymbol{\beta}} = \hat{\mathbf{\Sigma}}_X(\lambda)^{-1/2}\mathbf{V}$ and $\mathbf{V}$ is the matrix of eigenvectors with associated eigenvalues $(l_1, l_2, \ldots, l_d)$.

Depending on the value of the parameters (λ,γ), we may have several special cases. SIR is obtained setting $(\lambda,\gamma) = (0,0)$, while SIR-II for $(\lambda,\gamma) = (0,1)$. For any given value of γ, as $\lambda \to 1$ the covariance matrix tends to be diagonal, hence treating the predictors as uncorrelated. Holding λ fixed, the parameter γ allows to obtain intermediate models between SIR and SIR-II, i.e. models which use information coming from both class means and variances. In practice, both parameters are often unknown, so they are determined by minimizing an estimate of the misclassification error.

Once REGSIR directions have been estimated, cases can be projected onto $\mathcal{S}(\hat{\boldsymbol{\beta}})$ with Y as marking variable. Such graphical representation may have a major descriptive and diagnostic value in analyzing data. However, a formal classification rule can be stated as $\mathcal{C}(\boldsymbol{x}) = \arg_k \min \delta_k(\boldsymbol{x})$, with the discriminant score defined as

$$\delta_k(\boldsymbol{x}) = (\hat{\boldsymbol{\beta}}^\top \boldsymbol{x} - \hat{\boldsymbol{\beta}}^\top \bar{\boldsymbol{x}}_k)^\top \mathbf{W}_k^{-1} (\hat{\boldsymbol{\beta}}^\top \boldsymbol{x} - \hat{\boldsymbol{\beta}}^\top \bar{\boldsymbol{x}}_k) + \log|\mathbf{W}_k| - 2\log(\pi_k) \tag{5}$$

where $\mathbf{W}_k = \gamma\hat{\mathbf{\Sigma}}_{X|Y=k} + (1-\gamma)\hat{\mathbf{\Sigma}}_{\mathrm{E}(X|Y)}$ is a regularized pooled within-group covariance matrix. The first term in (5) is the Mahalanobis distance of $\boldsymbol{x}$ with respect to the centroid on $\mathcal{S}(\hat{\boldsymbol{\beta}})$, using $\mathbf{W}_k$ as scaling matrix. This matrix depends on a convex combination controlled by the parameter γ: for $\gamma = 0$, $\mathbf{W}_k$ equals the usual pooled within-group covariance matrix, while for $\gamma = 1$, each class has its own within-group covariance matrix. In this last situation, we are able to account for different degrees of dispersion within each class. The last terms in (5) are corrections, in analogy to Gaussian discriminant analysis, for different within-class covariances and class prior probabilities ($\sum_{k=1}^{K} \pi_k = 1$).

The classification rule based on the discriminant score in (5) can be applied on a test set, if available, or on a cross-validated set to provide an estimate of the misclassification error, as well as for selecting the optimal pair of (λ, γ) values.

3 Applications to Real Data Sets

In this section the proposed methodology is applied to remote sensing data on crops, which presents a multiclass problem with a complex structure, and to gene expression data, whose main feature is the very large number of variables (genes) relative to the number of observations (cells or samples).

3.1 Remote Sensing Data on Crops

Identifying and mapping crops is important for national and multinational agricultural agencies to prepare an inventory of what was grown in certain areas. This serves the purpose of collecting crop production statistics, yield forecasting, facilitating crop rotation records, mapping soil productivity, etc.

A sample of 36 observations were classified into five crops: clover, corn, cotton, soybeans, and sugar beets (SAS Institute (1999)). Table 3.1 reports the number of observations, the means and the log-determinant of the covariance matrices for the four predictors within each crop type. The dataset has a complex structure with quite different means and variances within group, and only 36 cases. For this reason it has been used in literature as an instance where quadratic discriminant analysis performs better than linear discriminant analysis (see Table 2).

	Clover	Corn	Cotton	Soybeans	Sugar beets
n_k	11	7	6	6	6
$\overline{x}_{1k}$	46.36	15.29	34.50	21.00	31.00
$\overline{x}_{2k}$	32.64	22.71	32.67	27.00	32.17
$\overline{x}_{3k}$	34.18	27.43	35.00	23.50	20.00
$\overline{x}_{4k}$	36.64	33.14	39.17	29.67	40.50
$\log(\lvert\hat{\mathbf{\Sigma}}_k\rvert)$	23.65	11.13	13.24	12.45	17.76

Table 1. Summary statistics for remote sensing data on crops

Figure 3.1 shows the misclassification error for REGSIR obtained by cross-validation (CV) on a regular grid of regularization parameters values. The minimum CV error is attained at $\gamma = 0.9$, whereas the parameter λ does not seem to play a crucial role here. Thus, most of statistical information for classification purposes is obtained from the kernel matrix of SIR-II, whereas the shrunken parameter is set to $\lambda = 0$, which means that the kernel matrix $\mathbf{M}(\lambda, \gamma)$ in equation (4) is scaled by $\hat{\mathbf{\Sigma}}_X$, the common estimate of the full covariance matrix. For $(\lambda, \gamma) = (0, 0.9)$ the REGSIR algorithm yields a train error equal to 0.44, which is smaller than the error attained by LDA and the other dimension reduction methods, albeit it is larger than the error provided by QDA. However, the leave-one-out CV error for REGSIR is equal to 0.50, which is smaller than the CV errors for QDA and the other methods used in the comparison (see Table 3.1).

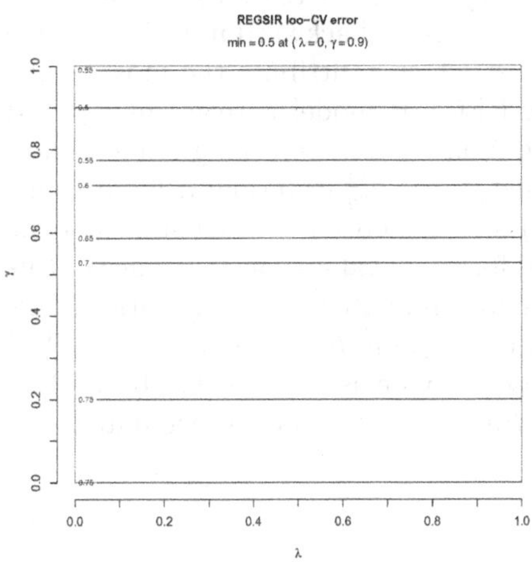

Fig. 1. Contour plot of the loo-CV error for re-gularization parameters on the grid $(0, 1) \times (0, 1)$

	train error	CV error
LDA	0.50	0.67
QDA	0.11	0.56
SIR	0.53	0.75
SIR-II	0.53	0.75
SAVE	0.53	0.75
REGSIR	0.44	0.50

Table 2. Misclassification errors for some classification methods applied to the crops dataset

3.2 Microarrays Gene Expression Data

The monitoring of the expression profiles of thousands of genes have proved to be particularly promising for biological classification, particularly for cancer diagnosis. However, DNA microarrays data present major challenges due to the complex, multiclass nature and the overwhelming number of variables characterizing gene expression profiles. Here we apply REGSIR to simultaneously develop a classification rule and to select those genes that are most important in terms of classification accuracy.

The REGSIR approach allows to overcome the under–resolution problem by setting the shrinkage parameter $\lambda > 0$. Moreover, since gene expression profiles for different groups mainly differ with respect to their average profiles, the parameter γ may be set to 0, so the regularization simplifies to a shrunken version of the ordinary SIR. Directions are estimated through the eigen-decomposition of $\hat{\mathbf{\Sigma}}_X(\lambda)^{-1}\mathbf{M}_I(\lambda)$, and the classification rule based on the score in equation (5) is applied to allocate the samples.

The inclusion of irrelevant genes often degrades the overall performances of estimated classification rules. Furthermore, identifying a small set of genes that is able to accurately discriminate the samples allows to employ a less expensive diagnostic assay in practical applications. Therefore, gene selection is a crucial step and, in the context of dimension reduction methods, it aims at identifying a subset of genes which is able to linearly explain the patterns variation in the estimated subspace.

The criterion adopted to select the, say, g relevant genes is based on R^2_g, the squared correlation coefficient between a set of g genes and the REGSIR variates $\hat{\boldsymbol{\beta}}^\top \mathbf{X}$, using the proportions $l_j / \sum_{i=1}^{d} l_j$ $(j = 1, \ldots, d)$ to reflect the importance of each estimated REGSIR variate. A backward iterative scheme is adopted: at each step only those genes which contribute the most to the overall patterns are retained, then the REGSIR model is re–estimated and the accuracy of the resulting classifier evaluated. The screening of redundant genes depends on the cut–off value used in the R^2_g criterion: a large value, say 0.999, implies that one or few genes are removed at each step. On the contrary, the process can be accelerated for small values, say 0.9, since in this case a large number of redundant genes are eliminated at each iteration. This process is repeated until the final subset contains $K - 1$ active genes. The classification accuracy of each gene subset may be assessed on the basis of its misclassification error on a test set, if available, or on a cross–validated set.

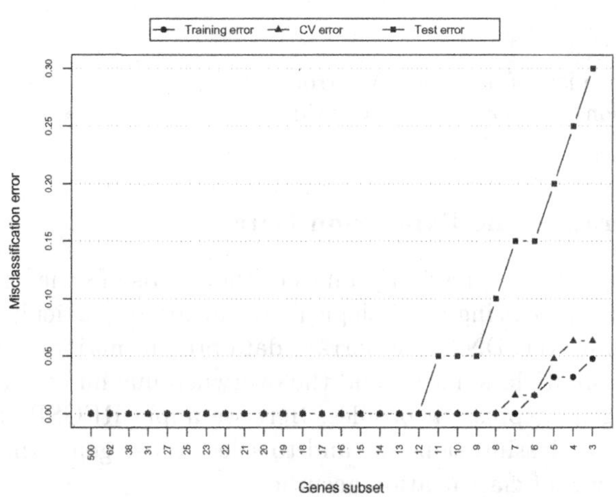

Fig. 2. Misclassification errors for genes subsets applied to the SRBCT data

This approach has been applied to data on small round blue cell tumors (SRBCT) of childhood (Khan et al. (2001)). Expression measurements for 2308 genes were obtained from glass-slide cDNA microarrays, and tumors were classified as: Burkitt lymphoma (BL), Ewing sarcoma (EWS), neuroblastoma (NB), and rhabdomyosarcoma (RMS). 63 cases were used as training samples and 25 as test samples, although five of the latter were not SRBCTs.

Since a large number of genes show a near constant expression levels across samples, we perform a preliminary screening of genes on the basis of the ratio of their between-groups to within-groups sum of squares (Dudoit et al. (2002)). Figure 2 shows the misclassification error rates obtained by REGSIR with $\lambda = 0.1$ for subsets of decreasing size obtained removing the redundant genes based on the $R_g^2 = 0.99$ criterion. Notably, the error rate is constantly equal to zero for many subsets, indicating that it is possible to remove several redundant genes without affecting the overall accuracy. Based on this plot, we might select the subset with, say, $g = 12$ genes as the "best" subset, because it is the smallest subset to achieve a zero error rate on both the test set and on 10–fold cross–validation. It is interesting to note that Khan et al. (2001) achieved a test error of 0% using a neural network approach and selected 96 genes for classification. Tibshirani et al. (2002) using shrunken centroids selected 43 genes, still retaining a 0% error on the test set. The REGSIR approach also achieves a 0% test error, but it uses far less genes.

Figure 3 shows the training samples projected onto the subspace spanned by the REGSIR directions estimated using the "best" 12 genes, along with decision boundaries. The tumor classes appear clearly separated, in particular along the first two directions. However, the 10-fold CV error considering only the first two directions rises from 0% to 5.5%.

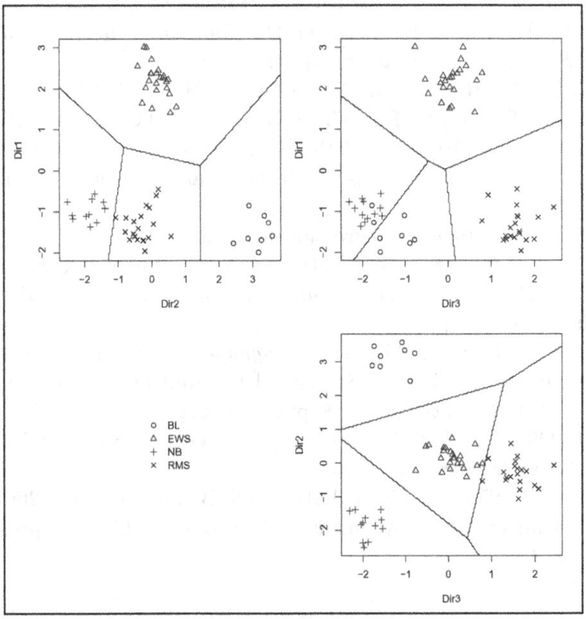

Fig. 3. Scatterplots of estimated SIR variates using the best subset of 12 genes for the SRBCT data with decision boundaries

4 Conclusions and Open Issues

The proposed REGSIR method appears to provide accurate classification rules on both real and artificial datasets we have studied so far. By a suitable choice of regularization parameters, it proves to be sufficiently flexible to adapt to several situations. However, further research is required to improve efficiency of the algorithm, particularly for the selection of regularization parameters, and a systematic comparison with other classification methods should be investigated on several simulated scenarios.

Finally, all the analyses and computations discussed in this paper have been conducted in R, a language and environment for statistical computing (R Development Core Team, 2005). Functions which implement the proposed methodology are freely available upon request from the author.

References

COOK, R.D. and YIN, X. (2001): Dimension Reduction and Visualization in Discriminant Analysis (with discussion). *Australian and New Zealand Journal of Statistics, 43, 147–199.*

COOK, R.D. and WEISBERG, S. (1991): Discussion of Li (1991). *Journal of the American Statistical Association, 86, 328–332.*

DUDOIT, S., FRIDLYAND, J. and SPEED, T.P. (2002): Comparison of Discrimination Methods for the Classification of Tumors Using Gene Expression Data. *Journal of the American Statistical Association, 97, 77–87.*

FRIEDMAN, J. H. (1989): Regularized Discriminant Analysis. *Journal of the American Statistical Association, 84, 165–175.*

KHAN, J., WEI, J.S., RINGNER, M., SAAL, L.H., LADANYI, M., WESTERMANN, F., BERTHOLD, F., SCHWAB, M., ANTONESCU, C.R., PETERSON, C. and MELTZER, P.S. (2001): Classification and Diagnostic Prediction of Cancers Using Gene Expression Profiling and Artificial Neural Networks. *Nature Medicine, 7, 673–679.*

LI, K. C. (1991): Sliced Inverse Regression for Dimension Reduction (with discussion). *Journal of the American Statistical Association*, 86, 316–342.

LI, K. C. (2000): *High Dimensional Data Analysis Via the SIR/PHD Approach.* Unpublished manuscript.

R Development Core Team (2005): *R: A Language and Environment for Statistical Computing.* R Foundation for Statistical Computing, Vienna, Austria. ISBN 3-900051-07-0, URL `http://www.R-project.org`.

SAS Institute (1999): *SAS/STAT User's manual. Version 8.0*, SAS Institute, Cary, NC, URL `http://v8doc.sas.com`.

TIBSHIRANI, R., HASTIE, T., NARASHIMAN, B. and CHU, G. (2002): Diagnosis of Multiple Cancer Types by Shrunken Centroids of Gene Expression. *PNAS, 99, 6567–6572.*

Part II

Multidimensional Data Analysis and Multivariate Statistics

Approaches to Asymmetric Multidimensional Scaling with External Information

Giuseppe Bove

Dipartimento di Scienze dell'Educazione,
Università degli Studi *Roma Tre*, Italy
bove@uniroma3.it

Abstract. In this paper some possible approaches to asymmetric multidimensional scaling with external information are presented to analyze graphically asymmetric proximity matrices. In particular, a proposal to incorporate external information in biplot method is provided. The methods considered allow joint or separate analyzes of symmetry and skew-symmetry. A final application to Morse code data is performed to emphasize advantages and shortcomings of the different methods proposed.

1 Introduction

Square data matrices whose rows and columns correspond to the same set of "objects", like proximities (e.g. similarity ratings), preferences (e.g. sociomatrices), flow data (e.g. import-export, brand switching), can be represented in low-dimensional spaces by scalar product or Euclidean distance models (MDS models). These models have to be suitably modified by increasing the number of parameters when not random asymmetry is present in the data (see e.g. Zielman and Heiser, 1996). In many applications additional information (external information) on the objects is available that could be conveniently incorporated in the data analysis. For instance, this allows to analyze the contribution of variables suggested from theoretical knowledge to the explanation of the relationships in the data. To this aim many methods were proposed in the context of symmetric MDS (see e.g. Borg and Groenen 1997, chapter 10), while a lack of proposals seems to characterize asymmetric MDS. In this paper some possible approaches to take into account external information in asymmetric MDS are considered, including a method based on the unique decomposition of the data matrix in its symmetric and skew-symmetric components recently proposed by Bove and Rocci (2004).

2 Direct Representation of Data Matrices by Biplot with External Information

Biplot (Gabriel, 1971) is a useful technique of data analysis to display graphically relationships between rows and columns of two-way matrices. This method was also considered to represent asymmetric proximities by *2n* points in a low-dimensional space (usually bidimensional). Approximate *r-dimensional* biplot of a square data matrix $\mathbf{X} = [x_{ij}]$, whose rows and columns correspond to the same set of n objects, is based on the approximate factorization $\mathbf{X} = \mathbf{AB}' + \boldsymbol{\xi}$, with r, the number of columns of $\mathbf{A}$ and $\mathbf{B}$, less than the rank of $\mathbf{X}$. The rows of matrices $\mathbf{A}$ and $\mathbf{B}$ provides coordinate vectors respectively for the n rows and columns of the data matrix and their scalar products approximate the entries of $\mathbf{X}$. This direct representation of the data matrix can also be used to analyze symmetry and skew-symmetry by sum and difference of the two scalar products corresponding to the entries x_{ij} and x_{ji}.

The *r-dimensional* biplot is obtained by minimizing the sum of squared residuals $\|\boldsymbol{\xi}\|^2$, i.e., by the singular value decomposition of the data matrix $\mathbf{X}$.

When a full column rank matrix of *external variables* $\mathbf{E} = [\mathbf{e}_1, \mathbf{e}_2, ..., \mathbf{e}_p]$ containing additional information on the n objects is available, we can try to incorporate in the analysis the external information in order to improve data interpretation (e.g. data theory compatible MDS). In this case we usually seek for a constrained solution whose dimensions are linear combination of the columns of $\mathbf{E}$, that is, we want the columns of matrices $\mathbf{A}$ and $\mathbf{B}$ to be in the subspace spanned by the columns of $\mathbf{E}$. In matrix notation $\mathbf{A} = \mathbf{EC}$ and $\mathbf{B} = \mathbf{ED}$ where $\mathbf{C}$ and $\mathbf{D}$ are matrices of unknown weights, so that

$$\mathbf{X} = \mathbf{AB}' + \boldsymbol{\xi} = \mathbf{ECD}'\mathbf{E}' + \boldsymbol{\xi} \tag{1}$$

The best least squares estimate for $\mathbf{C}$ and $\mathbf{D}$ is obtained by minimizing

$$h(\mathbf{C}, \mathbf{D}) = \|\mathbf{X} - \mathbf{ECD}'\mathbf{E}'\|^2 \tag{2}$$

This problem can be solved by noting that if we rewrite $\mathbf{E} = \mathbf{PG}$, where $\mathbf{P}'\mathbf{P} = \mathbf{I}$ and $\mathbf{G}$ is a square full rank matrix, then

$$h(\mathbf{C}, \mathbf{D}) = \|\mathbf{X} - \mathbf{ECD}'\mathbf{E}'\|^2 = \|\mathbf{X}\|^2 + \|\mathbf{P}'\mathbf{XP} - \mathbf{GCD}'\mathbf{G}'\|^2 + \\ - \|\mathbf{P}'\mathbf{XP}\|^2 \tag{3}$$

It is now clear that $h(\mathbf{C}, \mathbf{D})$ reach the minimum when $\mathbf{C} = \mathbf{G}^{-1}\mathbf{U}$ and $\mathbf{D} = \mathbf{G}^{-1}\mathbf{V}$ where $\mathbf{UV}'$ is the *r-dimensional* biplot of $\mathbf{P}'\mathbf{XP}$. We refer to Takane and Shibayama (1991) and Takane, Kiers and De Leeuw (1995) for other examples of component analysis with external information.

3 Generalized Escoufier and Grorud Model (GEG) with External Information

A known result of linear algebra is that each square matrix can be uniquely decomposed into the symmetric and the skew-symmetric components. For our data matrix, we can write: $\mathbf{X} = \mathbf{S} + \mathbf{K}$, where $\mathbf{S}$ is the symmetric matrix of averages $s_{ij} = \frac{1}{2}(x_{ij} + x_{ji})$ and $\mathbf{K}$ is the skew-symmetric matrix of differences $k_{ij} = \frac{1}{2}(x_{ij} - x_{ji})$. To analyse $\mathbf{S}$ and $\mathbf{K}$ jointly Escoufier and Grorud (1980) proposed the following EG model:

$$x_{ij} = \sum_{m=1}^{M} (a_{im}a_{jm} + b_{im}b_{jm} + a_{im}b_{jm} - b_{im}a_{jm}) + \varepsilon_{ij} \tag{4}$$

where ε_{ij} is a residual term. When $M = 1$ the model allows us to represent objects in a plane (*bimension*) having coordinates (a_{i1}, b_{i1}); for a pair of points in this plane, the scalar product describes the symmetric component, while twice the area of the triangle having the two points and the origin as vertices describes the absolute value of the skew-symmetric component, whose algebraic sign is associated with the orientation of the plane (positive counter-clockwise, negative clockwise). The previous EG model can be considered a constrained version of the GEG model proposed by Rocci and Bove (2002)

$$x_{ij} = \sum_{m=1}^{M} [\gamma(a_{im}a_{jm} + b_{im}b_{jm}) + \delta(a_{im}b_{jm} - b_{im}a_{jm})] + \varepsilon_{ij} \tag{5}$$

where $\gamma = 0$ if the data matrix is skew (*Gower decomposition*) and $\gamma = 1$ otherwise, while $\delta \geq 0$. When $\gamma = 1$ and $M = 1$ the GEG model is equivalent to Chino's GIPSCAL, while giving different weights to the skew-symmetric component we obtain Generalized GIPSCAL (Kiers and Takane, (1994)). GEG can be rotated simplifying the interpretation of bimensions. Bove and Rocci (2004) showed that the external information represented by the matrix $\mathbf{E}$ can be incorporated in the analysis by rewriting GEG in vector notation as

$$\mathbf{X} = \sum_{m=1}^{\mathbf{M}} [\gamma(\mathbf{a}_m\mathbf{a}'_m + \mathbf{b}_m\mathbf{b}'_m) + \delta(\mathbf{a}_m\mathbf{b}'_m - \mathbf{b}_m\mathbf{a}'_m)] + \boldsymbol{\xi} \tag{6}$$

By imposing $\mathbf{a}_m = \mathbf{E}\mathbf{c}_m$ and $\mathbf{b}_m = \mathbf{E}\mathbf{d}_m$, they obtain

$$\mathbf{X} = \gamma\mathbf{E}\mathbf{H}^{(s)}\mathbf{E}' + \delta\mathbf{E}\mathbf{H}^{(k)}\mathbf{E}' + \boldsymbol{\xi} \tag{7}$$

where: $\mathbf{H}^{(s)} = \sum_m(\mathbf{c}_m\mathbf{c}'_m + \mathbf{d}_m\mathbf{d}'_m)$ is symmetric, and $\mathbf{H}^{(k)} = \sum_m(\mathbf{c}_m\mathbf{d}'_m - \mathbf{d}_m\mathbf{c}'_m)$ is skew. In vector notation

$$\mathbf{X} = \gamma\sum_{ij} h_{ij}^{(s)}\mathbf{e}_i\mathbf{e}'_j + \delta\sum_{ij} h_{ij}^{(k)}(\mathbf{e}_i\mathbf{e}'_j - \mathbf{e}_j\mathbf{e}'_i) + \boldsymbol{\xi}, \quad i,j = 1,2,..,p \tag{8}$$

It follows from the last equation that the constrained representation of symmetry and skew-symmetry is a linear combination of the symmetry and skew-symmetry of the different planes determined by each pair of external variables.

4 Symmetric MDS of Skew-Symmetry with External Information

A very simple idea to analyze the size of skew-symmetry, disregarding its signs, is to perform symmetric MDS of the matrix **M** obtained with the absolute values of the entries of the skew-symmetric component matrix **K**. Even in this case the coordinate matrix **Y** by which to represent the matrix **M** can incorporate external information by imposing $\mathbf{Y} = \mathbf{EC}$, where **C** is a matrix of unknown weights. Solutions for this problem can be found, for instance, in De Leeuw and Heiser (1980) and can be easily applied even with standard statistical software (e.g. PROXSCAL, SPSS-Categories). This approach seems useful especially when methods for the analysis of symmetry and skew-symmetry like biplot or GEG fail to provide explicative constrained solutions.

5 Application to Morse Code Confusion Data

A Morse code signal is a sequence of up to five tones represented by dots · (a short beep of 0.05 sec.) and dashes – (a long beep of 0.15 sec.) separated by a silence of 0.05 sec. (e.g. the signal for A is short-silence-long, or, symbolically,·–). Rothkopf (1957) studied confusion rates on 36 Morse code signals (26 for the alphabet, 10 for the digits 0,..., 9). 598 subjects, who did not know Morse code, were required to state whether two signals listened one after another (separated by a quiet period of 1.4 sec.) were the same or different. The entries in the proximity matrix (36×36) are the percentages (rates) of roughly 150 subjects who responded "same" for the two signals they heard. The pairs of signals were presented in both orders, e.g. X following H (confusion rate 33%) and also H following X (confusion rate 6%). Symmetric MDS was performed on the symmetric part of the data by several authors. The obtained configuration could be partitioned by different criteria (e.g. number of elements in the signal, duration, type of composition, etc.) that could be used to explain the signal similarities. This suggested to Borg and Groenen (1997) to perform symmetric MDS externally constrained by two physical properties of the signals: signal type (the ratio of long versus short beeps) and signal length (varying from .05 to .95 seconds). They found that the difference in Stress of the constrained and the unconstrained solution was rather small (0.21 versus 0.18), so that the theory-consistent solution based on the two external variables seemed acceptable.

Even if the symmetric part of the Morse code data is dominant (96% of the total SSQ without the diagonal) many authors argued that asymmetry may still reveal interesting not random relations. For instance, Gower (1977, p. 111) argued that "... a two-component signal followed by a three component signal gives more confusion than the reverse. This tendency is more prevalent for pairs of similar signals". Moreover, Borg and Groenen (1997, p. 406-407), integrating skew-symmetry into the representation of the symmetrized data, noted that shorter Morse code signals were more often confused with longer ones than vice versa.

We applied the three approaches presented in the previous sections in order to detect external variables explaining asymmetry in the Morse code data. We focused on bidimensional representations in order to make easier the comparison with previous applications on these data.

Biplot and GEG class of models were applied constraining solutions by using different sets of external variables. The strong reduction in the fit of the skew-symmetry in the constrained solutions showed that these methods can not perform very well when the symmetric component is so relevant in the data. In fact, in the joint analysis of the two components the external variables tend to reflect much more symmetry than skew-symmetry.

Symmetric ordinal MDS of the absolute values of the skew-symmetric component of the data provided the configuration depicted in Figure 1, with Stress=0.33. Even if the fit value does not seem so good, much of the size

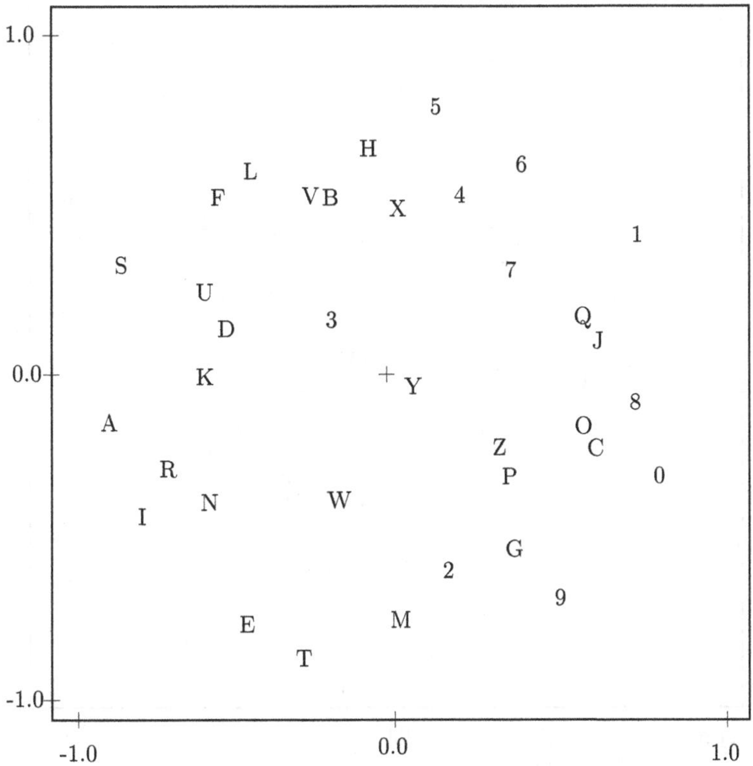

Fig. 1. Symmetric ordinal MDS of skew-symmetry

of relevant asymmetries can be easily detected in the configuration by point distances. For instance, the small distance between X (–··–) and H (····) or G (––·) and P (·––·) represents their large asymmetry.

The symmetric ordinal MDS solution was first constrained by signal type and signal length, used by Borg and Groenen (1997) for the symmetrized data. This caused quite an high increase in Stress, resulting equal to 0.42. The set of external variables was enlarged adding other four properties of the signal: number of dots, number of short/long beep inversion, presence of dot at the beginning, presence of dot at the end. The obtained configuration has Stress=0.37 and it is depicted in Figure 2. A comparison with the unconstrained solution reveals that many distances are adequately approximated, even if there are some differences (e.g the pairs 1-9, E-T, 2-M). Type, length and presence of dot at the beginning are the variables more correlated with first dimension, number of dots is the variable more correlated with second dimension.

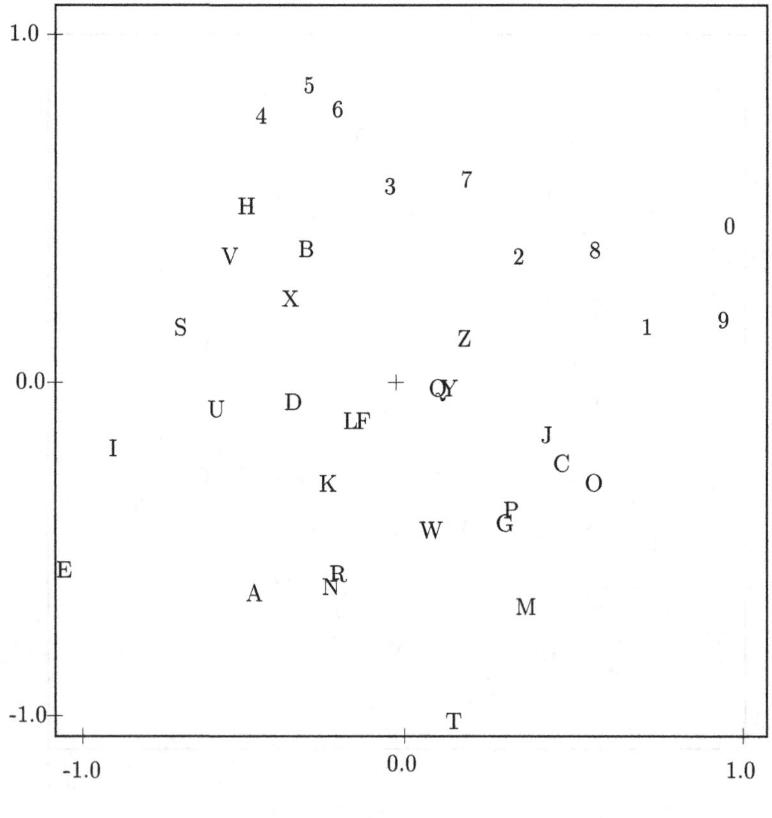

Fig. 2. Symmetric ordinal MDS of skew-symmetry with external variables

Quite peculiar are the aligned positions of the ten digits from left-top to right-bottom corner, given to the correlations between variables and dimensions.

In summary, these results seem to suggest that theory-consistent solutions for symmetry and asymmetry in these Morse code data need two different not disjoint sets of external variables. In particular, length and type cannot explain completely asymmetry, that occurs in several ways according to the peculiar position of dots and dashes inside the signal.

6 Conclusions

We have shown different possible approaches to incorporate external information in asymmetric MDS. The choice between the different proposals also depends on the relevance of symmetric and skew-symmetric components. In some applications a direct representation of the data can make approaches like biplot preferable. In these situations even unfolding models incorporating external information could be considered. On the other hand, a direct analysis of symmetry and skew-symmetry performed by models like GEG could be most appealing when the two components have an easy interpretation and improve the analysis of the relationships between objects.

When methods for joint analysis of symmetry and asymmetry fail to reveal theory-consistent explanations of the proximities, separate analysis of the two components should be preferred. Future developments of this study could regard a more detailed analysis of biplot with external information for skew-symmetric matrices.

Acknowledgments

Special thanks go to Roberto Rocci for his contribution to the results of section 2.

References

BORG, I. and GROENEN, P. (1997): *Modern multidimensional scaling. Theory and applications.* Springer, Berlin.

BOVE, G. and ROCCI, R. (2004): A method of asymmetric multidimensional scaling with external information. In: *Proceedings of the XLII Scientific Meeting of the Italian Statistical Society.* CLEUP, Padova, 631-634.

DE LEEUW, J. and HEISER, W.J. (1980): Multidimensional scaling with restrictions on the configuration. In: P.R. Krishnaiah (ed.): *Multivariate analysis V.* North Holland, Amsterdam, 501-522.

ESCOUFIER, Y. and GRORUD, A. (1980): Analyse factorielle des matrices carrées non-symétriques. In: E. Diday et al. (eds.): *Data Analysis and Informatics.* North Holland, Amsterdam, 263-276.

GABRIEL, K.R. (1971): The biplot graphic display of matrices with application to principal component analysis. *Biometrika, 58, 453-467.*

GOWER, J.C. (1977): The analysis of asymmetry and orthogonality. In: J.R. Barra et al. (eds.):*Recent Developments in Statistics.* North Holland, Amsterdam, 109-123.

KIERS, H.A.L. and TAKANE, Y. (1994): A generalization of GIPSCAL for the analysis of nonsymmetric data. *Journal of Classification, 11, 79-99.*

ROCCI, R. and BOVE, G. (2002): Rotation techniques in asymmetric multidimensional scaling. *Journal of Computational and Graphical Statistics, 11, 405-419.*

ROTHKOPF, E.Z. (1957): A measure of stimulus similarity and errors in some paired-associate learning tasks. *Journal of Experimental Psychology, 53, 94-101.*

TAKANE, Y. and SHIBAYAMA, T. (1991): Principal component analysis with external information on both subjects and variables. *Psychometrika, 56, 97-120.*

TAKANE, Y., KIERS, H.A.L., DE LEEUW, J. (1995): Component analysis with different sets of constraints on different dimensions. *Psychometrika, 60, 259-280.*

ZIELMAN, B. and HEISER, W. J. (1996): Models for asymmetric proximities. *British Journal of Mathematical and Statistical Psychology, 49, 127-146.*

Variable Architecture Bayesian Neural Networks: Model Selection Based on EMC

Silvia Bozza and Pietro Mantovan

Dipartimento di Statistica
Università Ca' Foscari di Venezia, Italy
silvia.bozza@unive.it
pietro.mantovan@unive.it

Abstract. This work addresses the problem of selecting appropriate architectures for Bayesian Neural Networks (BNN). Specifically, it proposes a variable architecture model where the number of hidden units are selected by using a variant of the real-coded Evolutionary Monte Carlo algorithm developed by Liang and Wong (2001) for inference and prediction in fixed architecture Bayesian Neural Networks.

1 Introduction

A crucial problem which arises when dealing with Bayesian Neural Networks is that of determining their most appropriate size expressed in terms of number of computational units and/or connections. In fact, too small a network may not be able to learn the sample data, whereas one that is too large may give rise to overfitting phenomena and cause poor "generalization" performance, *i.e.* the performance of the model on out-of-sample data is unsatisfactory (Neal (1996)). A few solutions have been proposed in the literature to solve this problem. Rios Insua and Müller (1998) proposed a reversible jump algorithm to move between architectures having a different number of hidden units. Müller and Rios Insua (1998) proposed a Markov chain Monte Carlo scheme for inference and prediction within fixed-architecture feedforward neural networks. The scheme is also extended to the variable architecture case, through the introduction of a geometric prior on the number of hidden units, providing a procedure to identify sensible architectures. Liang and Wong (2001) proposed a real-coded Evolutionary Monte Carlo (EMC) algorithm for inference and prediction within fixed architecture models. This is an extension of the EMC algorithm proposed by Liang and Wong (2000) for binary coded chromosomes. The algorithm works by simulating a population of Markov chains in parallel where a different temperature is attached to each chain. The population is evolved by applying three genetic operators: mutation, crossover and exchange. The most attractive features of genetic algorithms and simulated annealing are thus incorporated into the framework of Markov chain Monte Carlo. Bozza *et al.* (2003) proposed to incorporate a binary coded EMC step in the overall Markov chain Monte Carlo scheme that allows the Bayesian learning. In this work, we propose to extend the real-coded EMC algorithm to variable architectures models.

The paper is organized as follows. The Bayesian Neural Network (BNN) model is introduced in Section 2. Section 3 illustrates the real-coded Evolutionary Monte Carlo algorithm for fixed architecture models, while in Section 4 the evolutionary algorithm is extended to the variable architectur case. Section 5 presents the experimental results and, finally, Section 6 concludes the paper.

2 The Bayesian Neural Network (BNN) Model

Let us consider the following feed-forward neural network consisting of L input units with an extra "bias" unit w_0 permanently clamped at $+1$, $\mathbf{w} = (w_0, w_1, ..., w_l, ..., w_{L-1})$, one intermediate layer of M hidden units, and a final layer of K output units, $\mathbf{y} = (y_1, ..., y_k, ..., y_K)$:

$$y_k^{(n)} = \sum_{j=1}^{M} \beta_{kj} \psi\left(\gamma_j' \mathbf{w}^{(n)} + \eta_j\right) \tag{1}$$

where $n = 1, ..., N$ is the number of observations, $\beta_{kj} \in \mathbb{R}$ and $\gamma_j \in \mathbb{R}^L$ denote respectively the connection weight from the hidden unit j to the output unit k and the connection weights from the input units to the hidden unit j, and η_j is the bias term of the hidden unit j. For the sake of simplicity, we set the η_j's equal to 0. The function $\psi(\cdot)$ represents a tanh activation function. Other types of activation functions can be chosen, such as logistic activation function (see Bishop (1995)).

This neural network model can be analyzed from a Bayesian viewpoint. Specifically, it can be viewed as a nonlinear regression of a response y_k on covariates $\mathbf{w} = (w_0, w_1, ..., w_l, ..., w_{L-1})$:

$$y_k^{(n)} = \sum_{j=1}^{M} \beta_{kj} \psi\left(\gamma_j' \mathbf{w}^{(n)}\right) + \epsilon_k^{(n)}, \qquad \epsilon_k^{(n)} \sim N(0, \sigma^2). \tag{2}$$

From this, it follows that the conditional distribution of the response $y_k^{(n)}$ is Gaussian.

The prior distributions are assumed to be $\gamma_j \sim N(0, \sigma_\gamma^2 I)$, $\beta_{kj} \sim N(0, \sigma_\beta^2)$, $\forall j = 1, ..., M$, $\forall k = 1, ..., K$, $\sigma^{-2} \sim Ga(\nu, \delta)$.

Without loss of generality, we assume $K = 1$. The log-posterior (up to an additive constant) of the model is:

$$\log \pi(\boldsymbol{\beta}, \boldsymbol{\gamma}, \sigma^{-2} \mid D) \propto -\left(\frac{N}{2} + \nu - 1\right) \log(\sigma^2)$$

$$-\frac{1}{2\sigma^2} \left\{ 2\delta + \sum_{n=1}^{N} \left[y^{(n)} - \sum_{j=1}^{M} \beta_j \psi(\gamma_j' \mathbf{w}^{(n)}) \right]^2 \right\} - \sum_{j=1}^{M} \frac{\beta_j^2}{2\sigma_\beta^2} - \sum_{j=1}^{M} \sum_{i=0}^{L} \frac{\gamma_{ij}^2}{2\sigma_\gamma^2}.$$

where $\boldsymbol{\beta} = (\beta_1, ..., \beta_M)', \boldsymbol{\gamma} = (\gamma'_1, ..., \gamma'_M)', D = \{(y^{(1)}, \mathbf{w}^{(1)}), .., (y^{(N)}, \mathbf{w}^{(N)})\}$.

The posterior distribution is a complex function, thus it is not possible to obtain the marginal posteriors analitically. The complexity is due mainly by the nonlinearity and the multimodality. Multiple modes occur because prior and likelihood, and hence the posterior, are invariant with respect to arbitrary relabeling of the nodes. The posterior is also invariant with respect to the simultaneous change of the signs of β_{kj} and γ_j, since $\psi(-z) = -\psi(z)$. This problem can be solved by imposing an arbitrary ordering of the nodes, as in Müller and Insua (1998). Liang and Wong (2001) outline how the evolutionary algorithm performs well in presence of multimodality and nonlinearity, and imposes no constraint on the parameter space. A real coded Evolutionary Monte Carlo algorithm is then implemented to explore the posterior distribution.

3 Real-Coded Evolutionary Algorithm for Fixed Architecture BNN

Let us consider how the evolutionary algorithm works. Suppose we want to sample from: $f(\mathbf{x}) \propto \exp\{-H(\mathbf{x})/t\}$, where $\mathbf{x}$ is a real-coded vector of the unknown parameters of the network (chromosome), $H(\mathbf{x})$ is a fitness function (generally the log-posterior), and t is a scale parameter (temperature). So, in this case, $H(\mathbf{x}) = -\log \pi(\boldsymbol{\gamma}, \boldsymbol{\beta}, \sigma^{-2} \mid D)$.

A sequence of distributions $f_1(\mathbf{x}), ..., f_{N_p}(\mathbf{x})$ is constructed as follows:

$$f_i(\mathbf{x}) = \frac{1}{Z(t_i)} \exp\{-H(\mathbf{x})/t_i\} \qquad i = 1, ..., N_p$$

with $Z(t_i) = \sum_{\mathbf{x}_i} \exp\{-H(\mathbf{x}_i)/t_i\}$. The temperatures $t_1, .., t_{N_p}$ form a ladder with ordering $t_1 > ... > t_{N_p}$. Let $\mathbf{X} = \{\mathbf{x}_1, ..., \mathbf{x}_{N_p}\}$ denote a population of samples, where $\mathbf{x}_i$ is a sample from $f_i(\mathbf{x})$, and N_p is the population size. In the EMC algorithm, the Markov chain state is augmented as whole population instead of a single sample, and the Boltzmann distribution of the population is

$$f(\mathbf{X}) = \frac{1}{Z(\mathbf{t})} \exp\left\{-\sum_{i=1}^{N_p} H(\mathbf{x}_i)/t_i\right\}. \tag{3}$$

where $Z(\mathbf{t}) = \prod_{i=1}^{N_p} Z(t_i)$. The population is updated by mutation, crossover and exchange operators.

In the mutation operator one chromosome, say $\mathbf{x}_k$, is uniformly chosen from the current population. A new chromosome is generated by adding a random vector: $\mathbf{z}_k = \mathbf{x}_k + \mathbf{e}_k$. The random vector is usually chosen to have a moderate acceptance probability $(20\% - 50\%)$. The new population $\mathbf{Z} = \{\mathbf{x}_1, ..., \mathbf{z}_k, ..., \mathbf{x}_{N_p}\}$, is accepted with probability $\min(1, r_m)$ according to the

Metropolis-Hastings rule,

$$r_m = \frac{f(\mathbf{Z})}{f(\mathbf{X})} \frac{T(\mathbf{X}|\mathbf{Z})}{T(\mathbf{Z}|\mathbf{X})} = \exp\{-(H(\mathbf{z}_k) - H(\mathbf{x}_k))/t_k\} \frac{T(\mathbf{X}|\mathbf{Z})}{T(\mathbf{Z}|\mathbf{X})},$$

where $T(\cdot|\cdot)$ denotes the transition probability between populations. This operator is symmetric, *i.e* the transition probability from $\mathbf{x}$ to $\mathbf{z}$ is the same as that from $\mathbf{z}$ to $\mathbf{x}$.

In the crossover operator, one chromosome pair, say $\mathbf{x}_i$ and $\mathbf{x}_j$, is selected from the current population $\mathbf{X}$ according to some selection procedure, such as random selection or roulette wheel selection. In the random selection, a chromosome is uniformly selected from the current population. In the roulette wheel selection each individual is assigned a weight and then is selected with a probability proportional to its weight. In this case, each individual is assigned a weight proportional to its "Boltzmann probability"

$$p(\mathbf{x}_i) = \exp(-H(\mathbf{x}_i)/t_s)/\mathbf{Z} \tag{4}$$

where $\mathbf{Z} = \sum_{i=1}^{N_p} \exp(-H(\mathbf{x}_i)/t_s)$ and t_s is the selection temperature. Two offsprings $\mathbf{z}_i$ and $\mathbf{z}_j$ are generated by recombining the selected chromosomes $\mathbf{x}_i$ and $\mathbf{x}_j$ according to some crossover operators: real crossover (one-point, k-point and uniform) and the snooker crossover. In this work, only the one-point real crossover is considered. First, an integer crossover point is drawn uniformly; then $\mathbf{z}_i$ and $\mathbf{z}_j$ are constructed by swapping the gene to the right of the crossover point between parents.
A new population $\mathbf{Z} = \{\mathbf{x}_1, ..., \mathbf{z}_i, ..., \mathbf{z}_j, ..., \mathbf{x}_{N_p}\}$ is proposed and accepted with probability $\min(1, r_c)$ according to the Metropolis-Hastings rule,

$$r_c = \exp\{-(H(\mathbf{z}_i) - H(\mathbf{x}_i))/t_i - (H(\mathbf{z}_j) - H(\mathbf{x}_j))/t_j\} \frac{T(\mathbf{X}|\mathbf{Z})}{T(\mathbf{Z}|\mathbf{X})},$$

with $T(\mathbf{Z}|\mathbf{X}) = P((\mathbf{x}_i, \mathbf{x}_j) \mid \mathbf{X}) \, P((\mathbf{z}_i, \mathbf{z}_j) \mid (\mathbf{x}_i, \mathbf{x}_j))$.
Real crossover is symmetric, that is $P((\mathbf{z}_i, \mathbf{z}_j) \mid (\mathbf{x}_i, \mathbf{x}_j)) = P((\mathbf{x}_i, \mathbf{x}_j) \mid (\mathbf{z}_i, \mathbf{z}_j))$. So, $T(\mathbf{X}|\mathbf{Z})/T(\mathbf{Z}|\mathbf{X}) = P((\mathbf{z}_i, \mathbf{z}_j) \mid \mathbf{Z}) / P((\mathbf{x}_i, \mathbf{x}_j) \mid \mathbf{X})$. Throughout the paper, the parental chromosomes are chosen as follow. The first chromosome $\mathbf{x}_i(\mathbf{x}_j)$ is selected according to a roulette wheel procedure with Boltzmann weights; the second chromosome $\mathbf{x}_j(\mathbf{x}_i)$ is chosen randomly from the rest of the population. The selection probability is then

$$P((\mathbf{x}_i, \mathbf{x}_j) \mid \mathbf{X}) = \frac{1}{(N-1)Z(\mathbf{X})} \left[\exp\{-H(\mathbf{x}_i)/t_s\} + \exp\{-H(\mathbf{x}_j)/t_s\}\right].$$

In the exchange operator, given the current population $\mathbf{X}$ and the attached temperature ladder $\mathbf{t}$, $(\mathbf{X}, \mathbf{t})$, the new population $\mathbf{Z}$ is obtained by making an exchange between $\mathbf{x}_i$ and $\mathbf{x}_j$ without changing the temperatures, that is $(\mathbf{Z}, \mathbf{t}) = (\mathbf{x}_1, t_1, ..., \mathbf{x}_j, t_i, ..., \mathbf{x}_i, t_j, ..., \mathbf{x}_{N_p}, t_{N_p})$. The new population is accepted with probability $\min(1, r_e)$

$$r_e = \frac{f(\mathbf{Z})}{f(\mathbf{X})} \frac{T(\mathbf{X}|\mathbf{Z})}{T(\mathbf{Z}|\mathbf{X})} = \exp\left\{(H(\mathbf{x}_i) - H(\mathbf{x}_j)) \left(\frac{1}{t_i} - \frac{1}{t_j}\right)\right\}.$$

Typically the exchange is performed only on two individuals with neighboring temperatures, that is $| i - j | = 1$. It can be shown that $T(\mathbf{Z}|\mathbf{X}) = T(\mathbf{X}|\mathbf{Z})$.

The real coded evolutionary algorithm, with the genetic operators defined above, can be summarized as follows. Given an initial population $\mathbf{X} = \{\mathbf{x}_1, ..., \mathbf{x}_{N_p}\}$ initialized at random, and a temperature ladder $\mathbf{t} = \{t_1, ..., t_{N_p}\}$, one iteration comprises the following steps:

1. Apply mutation, one-point real crossover operators to the population with probability p_m, $1 - p_m$.
2. Try to exchange $\mathbf{x}_i$ and $\mathbf{x}_j$ for N_p pairs (i, j), with i sampled uniformly on $1, ..., N_p$ and $j = i \pm 1$.

In the mutation step, each chromosome of the population is mutated independently. In the crossover step, about 50% of chromosomes are chosen to mate. Note that the crossover operator works in an iterative way; that is, each time, two parental chromosomes are chosen from the updated population by the previous crossover operation.

4 EMC Algorithm for Variable Architecture BNN

The evolutionary algorithm presented in the earlier Section is now modified to allow the selection of the complexity. We consider the Bayesian Neural Network introduced in Section 2, (2), with an unknown number of hidden units.

The initial population of models $\mathbf{X}$ is composed by several subpopulations, $\mathbf{X}_1$,...,$\mathbf{X}_{M^*}$, each of them represents a Bayesian Neural Network with a different number of hidden units, where M^* denotes the maximum number of hidden nodes. The presence or the absence of a hidden unit is indicated by the correspondent connection weights: when the hidden unit is not present in the model they are set equal to zero. As an example, let us consider a BNN with two input units, one output unit and a maximum number of hidden units set equal to 4. The initial population in this case is composed by four subpopulations, containing the following models:

$$\begin{aligned}
\mathbf{X}_1 &: (\gamma_{11}, \gamma_{21}, \beta_1;\ 0,\ 0,\ 0;\ 0,\ 0,\ 0;\ 0,\ 0,\ 0;\ \sigma^2) \\
\mathbf{X}_2 &: (\gamma_{11}, \gamma_{21}, \beta_1;\ \gamma_{12}, \gamma_{22}, \beta_2;\ 0,\ 0,\ 0;\ 0,\ 0,\ 0;\ \sigma^2) \\
\mathbf{X}_3 &: (\gamma_{11}, \gamma_{21}, \beta_1;\ \gamma_{12}, \gamma_{22}, \beta_2;\ \gamma_{13}, \gamma_{23}, \beta_3;\ 0,\ 0,\ 0;\ \sigma^2) \\
\mathbf{X}_4 &: (\gamma_{11}, \gamma_{21}, \beta_1;\ \gamma_{12}, \gamma_{22}, \beta_2;\ \gamma_{13}, \gamma_{23}, \beta_3;\ \gamma_{14}, \gamma_{24}, \beta_4;\ \sigma^2)
\end{aligned}$$

The genetic algorithms presented in the earlier section are adapted to the variable architecture case.

The mutation operator works by adding a random quantity to the non-zero gene of the selected chromosomes. In this way the mutation operator do not activate hidden units not present in the selected chromosomes. A peculiar attention is dedicated to the gene representing the signal error: if a negative value is proposed, this is discarded and another value is proposed.

The crossover operator allows to propose models of intermediate complexity among the ones with the selected chromosomes. A one-point crossover is implemented: the integer crossover point is sampled among multiples of the total number of the input and output units. So, if a model presents two input units and one output unit, then the integer crossover point is a multiple of three. In this way, the offsprings represent only fully connected units. For example, let us consider a pair of selected chromosomes $\mathbf{x}_i$ and $\mathbf{x}_j$ from different subpopulations, say for example subpopulation $\mathbf{X}_1$ and $\mathbf{X}_4$. Then $\mathbf{z}_i$ and $\mathbf{z}_j$ are constructed by swapping the gene to the right of the crossover point between parents (excluding the gene corresponding to the signal error, which will be swapped last). The two offsprings $\mathbf{z}_i$ and $\mathbf{z}_j$ might represent models with one up to four hidden units, depending on the integer crossover point.

The problem with this solution is that the Markov chain tends to get stuck in subspaces correspondent to the intermediate models. This is because, at each step, there will be proposed and eventually accepted models with an intermediate architecture. To avoid this, an alternative crossover operator is introduced. The alternative crossover operator works as follows. Once a chromosome pair is selected from the current population, we consider the one which represents the model with the lower number of hidden units activated. The integer crossover point is sampled between multiples of the total number of input and output units up to the number of connections of the smaller model. The two offsprings are obtained by swapping the parental genes in correspondence of the integer crossover point; the gene after the integer crossover point are set equal to zero. In this way, the two offsprings represent the lower architecture.

The exchange operator works as in the earlier Section.

The algorithm proposed works as follows. The EMC algorithm for fixed architectures is run for each subpopulation, initialized at random, as in Section 3. The initial population is then given by several subpopulations, which are not initialized at random, but follow from a first run of the algorithm. The EMC for selecting the architecture works as follows:

1. Apply mutation, one-point real crossover, alternative crossover with probability p_m, p_c, p_{ac}.
2. Try to exchange $\mathbf{x}_i$ and $\mathbf{x}_j$ for N_p pairs (i,j), with i sampled uniformly on $1, ..., N_p$ and $j = i \pm 1$.

5 A Simulated Example

To test the effectiveness of the proposed procedure, we performed experiments on a simulated data set. We simulated $y_1, y_2, ..., y_{100}$ from (2) with $M = 2$, $\gamma_1 = (\gamma_{10}, \gamma_{11}) = (2, -1)$, $\gamma_2 = (\gamma_{20}, \gamma_{21}) = (1, 1.5)$, $\boldsymbol{\beta} = (\beta_1, \beta_2) = (20, 10)$, $\sigma = 1$, and the input pattern $\mathbf{w} = (1, z_t)$ where $z_t = t \cdot 0.1$ for $t = 1, 2, ..., 100$.

1	2	3	4	5	6	7	9
0.165	0.400	0.180	0.155	0.060	0.020	0.015	0.005

Table 1. Frequency of the different models in the final population.

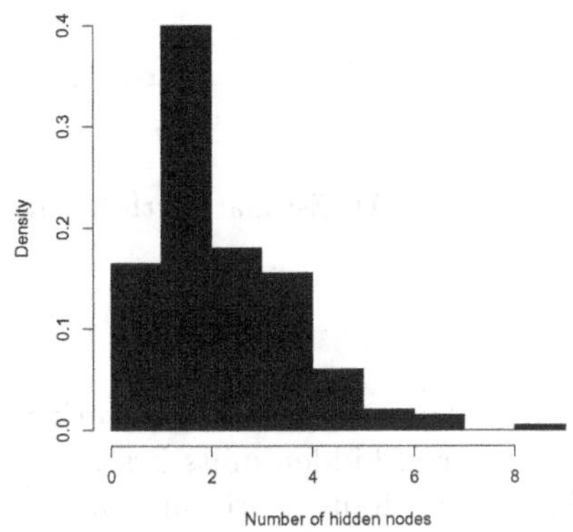

Fig. 1. Frequency of the different models in the final population.

This example is identical to the one in section 5.1 of Liang and Wong (2001) except for the signal error, which there was fixed at 0.1. The prior parameters are set as $\sigma_\beta = 20$, $\sigma_\gamma = 5$, $\nu = 0.01$, $\delta = 0.01$. The population size is $N_p = 200$. The mutation rate is 0.25, the crossover rate is 0.65, and the alternative crossover rate is 0.15.

The population was initialized with subpopulations of the same dimension representing architectures with an increasing number of hidden nodes (from a minimum of 1 to a maximum of 10).

We ran the algorithm five times independently. Each run comprised 10,000 iterations. The overall acceptance rates of the mutation, crossover, alternative crossover, and exchange operators were respectively 0.32, 0.32, 0.15, and 0.83.

The population evolved toward a population of models with 2 hidden units. Figure 1 shows the frequency of the different models in the final population. The proposed algorithm selected neural models having two hidden units with probability equal to 0.4. Table 1 shows the frequency of the different models in the final population.

Figure 2 shows the maximum a posteriori (MAP) estimate of the regression line obtained from the model in the final population which presented the higher fitness among those with two hidden units.

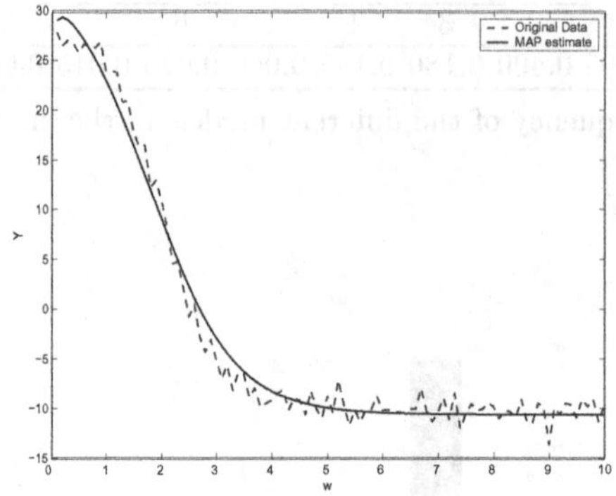

Fig. 2. The Original Data and the MAP Estimate of the Nonlinear Regression Line.

6 Conclusions

In this work we propose a Bayesian Neural Network model with variable architecture where the number of hidden units is selected by using a variant of the real-coded Evolutionary Monte Carlo algorithm. Opportune genetic operators are introduced to allow the selection of the optimal architecture. The dimension of the population and the definition of the genetic operators are critical points which need to be addressed with extreme attention. The results obtained on synthetic data show that the proposed algorithm allows to select models with a small number of hidden units.

References

BISHOP, C.M. (1995): *Neural Networks for Pattern Recognition*, Oxford University Press.

BOZZA, S., Mantovan, P., Schiavo, R.A. (2003): Evolutionary model selection in Bayesian Neural Networks. In M. Schader, W. Gaul, and M. Vichi, editors, *Between Data Science and Applied Data Analysis*, Studies in Classification, Data Analysis, and Knowledge Organization. Springer.

LIANG, F. and WONG, W.H. (2000): Evolutionary Monte Carlo: Applications to C_p Model Sampling and Change Point Problem. *Statistica Sinica, 10, 317–342.*

LIANG, F. and WONG, W.H. (2001): Real-Parameter Evolutionary Monte Carlo with Applications to Bayesian Mixture Models. *Journal of the American Statistical Association, 96, 653–666.*

MÜLLER, P. and RIOS INSUA, D. (1998): Issues in Bayesian Analysis of Neural Network Models. *Neural Computation, 10, 749–770.*

NEAL, R.M. (1996): *Bayesian Learning for Neural Networks.* Lecture Notes in Statistics. Springer.

RIOS INSUA, D. and MÜLLER, P. (1998): Feedforward Neural Networks for Nonparametric Regression. In Dey D.K., Müller, P. and Sinha D. (Eds.) *Practical Nonparametric and Semiparametric Bayesian Statistics.* Springer, *181–194.*

Missing Data in Optimal Scaling

Pier Alda Ferrari and Paola Annoni

Dep. of Economics, Business and Statistics; University of Milan, Italy
pieralda.ferrari@unimi.it, paola.annoni@unimi.it

Abstract. We propose a procedure to assess a measure for a latent phenomenon, starting from the observation of a wide set of ordinal variables affected by missing data. The proposal is based on Nonlinear PCA technique to be jointly used with an *ad hoc* imputation method for the treatment of missing data. The procedure is particularly suitable when dealing with ordinal, or mixed, variables, which are strongly interrelated and in the presence of specific patterns of missing observations.

1 Introduction

The paper draws on a practical problem which deals with the evaluation of vulnerability degree of a number of valuable historical-architectural buildings located in a Northern Italian Region. The starting point is a wide set of ordinal variables which describe the buildings status and are affected by the problem of missing data. The underlying hypothesis of the analysis is that *vulnerability* represents the latent factor that cannot be measured directly but only indirectly by many variables, whose categories represent different aspects and different levels of the latent dimension. To solve the problem we adopt a statistical approach to extract latent factors, where buildings are statistical units (objects) and risk factors are variables. In this context we are interested in the major latent factor that describes buildings vulnerability. Furthermore, since we are dealing with ordinal data, the final indicator should be consistent with ordinal properties of the original data-set. The final result is the assessment of individual risk for each building in the sample.

The proposal to the problem is Nonlinear Principal Component Analysis, Nonlinear PCA (Gifi (1990)). The procedure is particularly suitable in this context because it incorporates the measurement level of variables into the analysis. In particular the method computes optimal 'quantification' of variable categories preserving categories ordination as required. In addition, data-set is affected by the problem of missing data, as it is often the case when dealing with this type of data. When missing data are present, additional attention is required to the analyst. Many methods are available for missing data treatment (Little & Rubin (2002)). From the point of view of nonlinear data analysis, up to the present one can distinguish among three different options. In the paper we discuss these different standard options. However, since in our case standard options do not yield any satisfactory ordination of the buildings, we propose an alternative method particularly suitable to treat our specific pattern of missing data, but it could be adapted to handle more general cases.

2 Nonlinear PCA

The problem can be termed as measuring vulnerability of a building. Clearly vulnerability cannot be directly measured. The basic hypothesis of the analysis is that every observed variables can be mapped into a single real number, that expresses building vulnerability. Observed variables concern different parts of the building, here called structural component (foundations, vertical and horizontal structures, etc.). For each structural component, observed variables characterize different types (structural damages, biological degradation, etc.) and different levels (damage severity, urgency of intervention or damage extension) of various aspects of vulnerability. Every variable is intended to describe different aspects of the same one-dimensional phenomenon (vulnerability) that eventually we want to quantify.

Let X be the latent variable to be measured on n units, i.e. in our case the level of damage of each building. If m ordinal variables are observed on each unit with ordered categories $\mathbf{c}_j = \{c_{jl} : l = 1,\ 2, \ldots\ k_j\}$, let $\mathbf{G}_j$ be the indicator matrix nxk_j for variable j. The goal is here to get the value of x_i for $i = 1, \ldots,\ n$ as linear combination of monotonically transformed categories:

$$\begin{aligned} & x_i = \textstyle\sum_{j=1}^{m} \lambda_j \alpha(\mathbf{G}_j(i)\mathbf{c}_j) \\ & \text{with } \ \alpha(\mathbf{G}_j(i)\mathbf{c}_j) \in \Re \ \text{ and } \ \alpha(c_{jl}) \geq \alpha(c_{jq}) \ \text{ if } \ c_{jl} \geq c_{jq} \end{aligned} \tag{1}$$

where notation $\mathbf{G}_j(i)\mathbf{c}_j$ indicates the category of variable j observed on unit i. Coefficient λ represents the loading assigned to variables and non linear transformation α represents proper quantification of variable categories. To this purpose the choice consists of a particular exploratory analysis: Nonlinear Principal Component Analysis. It allows to synthesize variables in one or more dimensions and simultaneously preserves measurement levels of qualitative ordinal data. The technique optimally computes category quantifications and variable loadings. In particular Nonlinear PCA finds p orthogonal axes that optimally fit the data where each column of the data matrix (each variable) is monotonically transformed in such a way that the axes optimally fit transformed data (Gifi (1990)). As we are interested in the first and major dimension, in the following $p = 1$. The starting point for the derivation of one dimensional Nonlinear PCA is the following loss function σ:

$$\sigma(\mathbf{x}, \mathbf{q}_1 \ldots, \mathbf{q}_m) = \frac{1}{m} \sum_{j=1}^{m} tr(\mathbf{x} - \mathbf{G}_j \beta_j \mathbf{q}_j)'(\mathbf{x} - \mathbf{G}_j \beta_j \mathbf{q}_j) \tag{2}$$

Vector $\mathbf{x}$, of dimension $nx1$, contains the object scores; $\mathbf{q}_j$ is the vector of order k_j that contains optimal category quantifications for variable j

and β_j is the loading of variable j. Nonlinear PCA computes the minimum of loss function $\sigma(\mathbf{x}, \mathbf{q}_1 ....., \mathbf{q}_m)$ under various normalization conditions and also under a particular restriction termed 'rank-1 restriction', that allows the analysis to take into account the measurement level of the variables (Michailidis et al. (1998)). This way of computing variable transformations is called *optimal scaling* because the transformations are chosen so as to minimize the loss function. In our case the one dimension Nonlinear PCA solution for object scores represents building vulnerability. To evaluate the validity of the latent dimension hypothesis, a scree plot for a high dimensional solution is built. From the very steep trend of the curve it emerges that the first eigenvalue is effectively much larger than the others. Furthermore all loadings in the first dimension are of the same sign. These facts together indicate that the hypothesis of one-dimensional data is reasonable and object scores can be considered as vulnerability indexes. Each building can then be reordered and compared with each other according to the assigned score x_i, while β_j and $\mathbf{G}_j\mathbf{q}_j$ are respectively loadings λ_j and quantified categories $\alpha(\mathbf{G}_j\mathbf{c}_j)$ as indicated in (1). If data-set is non affected by missing data, loadings β_j are correlations between object scores and quantified variables, hence, they can be nicely interpreted as "ordinary" loadings in standard PCA.

3 Missing Data: Standard Nonlinear PCA Options Comparison

Data-set under examination is characterized by several missing data which are due to the physical absence of a particular structural component of the building (for example a vertical structure). For this reason buildings that lack one or more components do present missing data for all the variables that describe the decay status of that component. Since the final goal is quantitative comparison among buildings, it is here necessary to make buildings directly comparable. In other words all scores are to be computed on the basis of the same set of variables and therefore a proper imputation of missing values has to be set.

When missing data are present, loss function σ includes an incomplete indicator matrix $\mathbf{G}_j$ and a binary matrix $\mathbf{M}_j$:

$$\sigma(\mathbf{x}, \mathbf{q}_1 ....., \mathbf{q}_m) = \frac{1}{m} \sum_{j=1}^{m} tr(\mathbf{x} - \mathbf{G}_j \beta_j \mathbf{q}_j)' \mathbf{M}_j (\mathbf{x} - \mathbf{G}_j \beta_j \mathbf{q}_j) \qquad (3)$$

where $\mathbf{M}_j$, of order nxn, has the role of binary indicator of missing observations for variable j. Normalization restrictions on object scores are (Michailidis et al. (1998)):

$$\begin{array}{ll} \mathbf{x}'\mathbf{M}_*\mathbf{x} = m \cdot n & \mathbf{M}_* = \sum_{j=1}^{m} \mathbf{M}_j \\ \mathbf{u}_n'\mathbf{M}_*\mathbf{x} = 0 & \end{array} \qquad (4)$$

with $\mathbf{u}_n$ vector of ones with dimension n, and object scores are:

$$\hat{\mathbf{x}} = \mathbf{M}_*^{-1} \sum_{j=1}^{m} \beta_j \mathbf{G}_j \mathbf{q}_j \tag{5}$$

Many methods are available for the treating of missing data and they all depend on researchers' assumption about the process that underlies *missingness.* In nonlinear data analysis one can distinguish among the following three options (Gifi (1990)): (*i*) the indicator matrix is left incomplete or (*ii*) it is completed adding single or (*iii*) multiple columns for each variable with missing data. First option (*missing data passive*) implies that the row of $\mathbf{G}_j$ is a zero row if the corresponding object has a missing observation for variable j. Option (*ii*) is called *missing data single category* and option (*iii*) is called *missing data multiple category.* They both imply a complete indicator matrix, so that missing data are treated as if they are a category themselves. In the single mode one extra column is added to $\mathbf{G}_j$ with entry '1' for each object with missing data on the j-th variable. The multiple category mode adds to $\mathbf{G}_j$ as many extra columns as there are objects with missing data on the j-th variable. The first option discards missing observations from the computation, whilst the others require strong assumptions regarding the pattern of missing data. In addition, when ordinal or numerical data are under examinations, further complications are added to the algorithm used to solve Nonlinear PCA, which has to take into account the measurement levels of variables. If missing data are few and sparse, no substantial impact is expected to affect results. On the contrary, if missing data show a specific pattern, results can be sensibly influenced by the specific option which is adopted to treat missing data.

To show impacts of options (*i*) and (*ii*), implemented in the SPSS-CATPCA module, on Nonlinear PCA results we set up a simple artificial case. It includes eight ordinal variables and twenty objects. Variables are defined to have different number of categories, from two to four categories. A single one-dimensional Nonlinear PCA solution is firstly computed for the complete data-set without missing data (reference case). Afterward, four values on three units are eliminated to artificially simulate missing data. Two Nonlinear PCA solutions are computed for the 'missing case' where missing data are treated using option (*i*) and option (*ii*) respectively. The third option is non considered here because it is certainly the most impacting one. Loadings for the three cases (no missing data, missing data treatment (i) and missing data treatment (ii)) are computed and relative differences of loadings for option (i) and option (ii) with respect to the reference case are shown in Figure 1. It must be noted that the two different options act in different ways: even if loadings β_j are always positive for all the three models, relative differences are not always of the same sign. This means that option (i) could assign to certain variables more weight than the non-missing solution whilst option (ii) the other way down (see for example VAR2). This behavior could obvi-

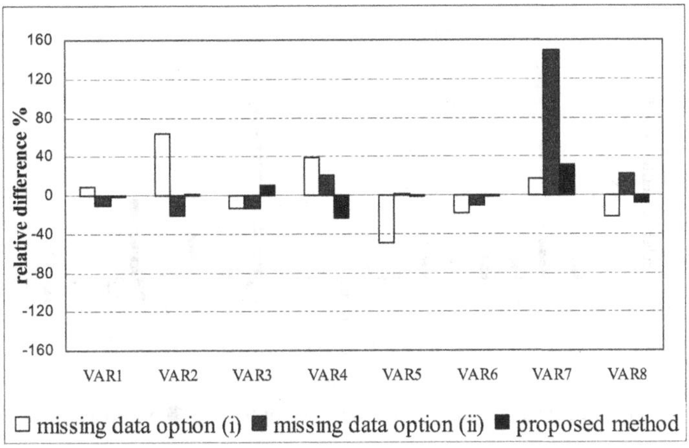

Fig. 1. Comparison among loadings β_j from Nonlinear PCA standard missing data treatment.

ously induce misleading results. Category quantifications are compared only for some representative variables and the two options act in a similar way, they both overestimate or underestimate quantifications. The overall effect of the two different options on object scores can be seen in Figure 2. Since the effect of different missing data treatments on loadings β_j is not of the same sign, it follows that where option (i) underestimates a score it often occurs that option (ii) overestimates the same score. Furthermore, the effect of treatment (ii) is reasonably stronger than the one of treatment (i), with most of the longest bars in the picture belonging to the second option. This example highlights that, even with a simple artificial case, standard missing data treatments in Nonlinear PCA could have serious impacts on results. In addition, when missing data are present and apart from specific missing data treatments, loadings β_j are no more interpretable as correlation coefficients, thus loosing their useful role.

4 Missing Data: Proposed Method

In our case standard missing data options are not suitable and we need to seek an alternative treatment of missing observations.

Data affected by missing data are generally classified as MCAR, MAR or NMAR according to the missing-data mechanism. If $\mathbf{Y}$ defines the complete

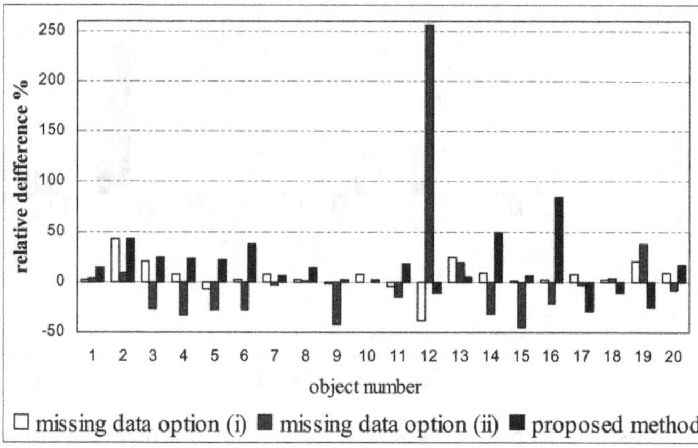

Fig. 2. Comparison among object scores by Nonlinear PCA standard missing data treatment.

data and **M** the missing-data indicator matrix, missing-data mechanism is characterized by the conditional distribution of **M** given **Y**. If *missingness* does not depend on the values of **Y**, neither missing nor observed, data are called missing completely at random (MCAR). If *missingness* depends only on the observed part of **Y** and not on the part that is missing, missing-data mechanism is called missing at random (MAR). Finally the mechanism is called not missing at random (NMAR) if the distribution of **M** depends on the missing values in **Y** (Little & Rubin (2002)). It is important to note that the assumption MCAR does not necessarily mean that missing-data pattern itself is random, but rather that *missingness* does not depend on the data values. In particular, also a monotone pattern can be missing at random as this is our case. In fact our missing data show a specific pattern, which is close to be monotone, but the probability that a variable is missing for a building does not depend on neither observed nor unobserved variables, but it depends only on the physical lack of a structural component for that building, not on its latent level of damage. We are then empowered to treat our data as MCAR and choose among various method available for monotone missing data treatment (Srivastava (2002)). Classical methods are nevertheless generally based on the assumption of multivariate normality of the data. For the case under examination, as for most problems that handle a great amount of data, it is very difficult to assume an underling distribution and, in any case, our framework is descriptive and not probabilistic. It is then necessary to find valid alternatives. The solution is here an imputation method which enables to assess a damage indicator which is consistent for all the buildings in the data-set, with correct handling of missing observations. To this aim two steps are performed in succession. Firstly, the original data-set is divided into two strata according to the presence of at least one missing structural component for the building. The first stratum contains only buildings for which all the variables are observed (*reference subset*) and it is reduced to 363 units; the second stratum contains only buildings that lack one ore more structural components (*imputation subset*), and it contains 458 units. In spite of the reduction, *reference subset* preserves representative properties of the original data-set. Nonlinear PCA is then applied to the homogeneous *reference subset*, obtaining category quantifications and loadings for all the variables involved. In the *imputation subset* observed variables are assigned quantified categories and loadings obtained by Nonlinear PCA applied to the *reference subset*. Missing quantifications for unobserved variables in the *imputation subset* are instead imputed starting from the most similar building in the *reference subset*. Similarity between building u_i in the *reference subset* and building u_k^* in the *imputation subset* is defined by a weighted Minkowski distance:

$$d(u_i, u_k^*) = (\sum_j \beta_j |\mathbf{G}_j(i)\mathbf{q}_j - \mathbf{G}_j(k)\mathbf{q}_j|^r)^{\frac{1}{r}} \quad r \geq 1 \tag{6}$$

where index j runs among all the variables actually observed on unit u_k^*. Our choice is $r = 1$, hence the adopted distance is Manhattan distance weighted by variable loadings computed by Nonlinear PCA on the *reference subset*. The distance choice is Manhattan since Euclidean distance overweights high differences (Little & Rubin (2002)) thus leading to a undesirable alteration of effective loadings of structural components. Unobserved variables are then assigned quantifications of the corresponding variables observed on the nearest building in the *reference subset*.

The overall damage indicator for unit u_k^* in the *imputation subset* can now be computed as:

$$x_k^* = m^{-1} \sum_{j=1}^{m} \beta_j \mathbf{G}_j(k)\mathbf{q}_j \tag{7}$$

with some imputed and some observed $\mathbf{G}_j(k)\mathbf{q}_j$. Ordination of units follows straightforward by using object scores just computed for the whole set of buildings.

Figures 1 and 2 show relative differences respectively for loadings and object scores of proposed method with respect to the reference case. Comparison with standard options (i) and (ii) highlights the overall advantages of the proposal. In the proposed procedure, if two buildings obtain similar scores on common observed structures they are expected to be, and they actually are, classified with similar total scores. This is not always guaranteed by Nonlinear PCA with standard options, in the presence of missing data patterns of the type under examination. Furthermore, since loadings β_j are computed on the basis of the *reference subset*, which is not affected by missing data, they recover their useful meaning of correlation coefficients between object scores and optimally quantified variables. It is then possible to assess variable importance in defining the latent dimension.

It is important to note that in this case the use of a reduced data-set (the *reference subset*) does not lead to any loss of information since buildings in the *reference subset* are highly representative of the overall data-set. It is nevertheless our intention to extend, in near future, the proposed approach in order to include more general situations, such for example those cases where the *reference subset* does not satisfy required representative conditions.

5 Conclusions

Proposed method provides a measurement instrument for latent factors which are described by categorical variables. To this aim it computes category quantifications that preserve original order and it solves the problem of missing data. The proposal is consistent with the geometric and descriptive approach typical of Nonlinear PCA. In particular it treats missing data according to the relevance of the variables with missing data and, at the same time, it preserves all the characteristics of Nonlinear PCA without missing data, such as the classical meaning of variable loadings (β_j). When applied to the vulnerability problem it provides scores generally lower than the ones from standard Nonlinear PCA, thus having a smoothing effect on vulnerability indicator (Ferrari et al. (2005)).

References

FERRARI, P.A., ANNONI, P. and URBISCI, S. (2005): A proposal for setting-up indicators in the presence of missing data: the case of vulnerability indicators. *UNIMI - Research Papers in Economics, Business, and Statistics. Statistics and Mathematics. Working Paper 1. http://services.bepress.com/unimi/statistics/art1.*.

GIFI, A. (1990):*Nonlinear Multivariate Analysis.* Wiley, New York.

LITTLE, R.J.A. and RUBIN, D.B. (2002):*Statistical Analysis with missing data.* IInd edition; John Wiley & Sons, Inc.

MICHAILIDIS, G. and de LEEUW, J. (1998): The Gifi System of Descriptive Multivariate Analysis. *Statistical Science, Vol. 13, 4, 307–336.*

SRIVASTAVA, M.S. (2002):*Methods of Multivariate Statistics.* John Wiley & Sons, Inc.

Simple Component Analysis Based on RV Coefficient

Michele Gallo[1], Pietro Amenta[2], and Luigi D'Ambra[3]

[1] Department of Social Science, University of Naples "L'Orientale", Italy mgallo@unior.it
[2] Department of Analysis of Economic and Social Systems, University of Sannio, Italy amenta@unisannio.it
[3] Department of Mathematics and Statistics, University of Naples "Federico II", Italy dambra@unina.it

Abstract. Among linear dimensional reduction techniques, Principal Component Analysis (PCA) presents many optimal properties. Unfortunately, in many applicative case PCA doesn't produce full interpretable results. For this reason, several authors proposed methods able to produce sub optimal components but easier to interpret like Simple Component Analysis (Rousson and Gasser, (2004)). Following Rousson and Gasser, in this paper we propose to modify the algorithm used for the Simple Component Analysis by introducing the RV coefficients (SCA-RV) in order to improve the interpretation of the results.

1 Introduction

When the number of observed variables is very large, it may be advantageous to find linear combination of the variables having the property to account for most of the total variance. Unfortunately, in many applicative cases, like in case of patient satisfaction data, variables are all positive (or negative) correlated and consequently the first PCA principal component may correspond to overall size. In this case, PCA doesn't produce full interpretable results. To resolve this kind of problem, several authors proposed methods which are able to produce sub optimal components, compared to the classical PCA ones, but easier interpretable.

In the literature three different algorithms have been proposed. The first one is based on replacement of some elements of the correlation matrix with others such that to have simpler and generally more interpretable results (Hausman (1982); Vines (2000); Rousson and Gasser (2004); Zous et al. (2004)). Other techniques are based on a rotation method of the loading matrix (Jolliffe (2003)). Another proposal is developed according to the centroid method (Choulakian et al. (2005)).

2 Simple Component Analysis - RV

Simplicity does not assure more interpretable components but it allows to analyze the case when several variables measure different aspects of a same theme with all correlation matrix elements are positive (negative). A correlation matrix with this structure gives more interpretability problems in PCA. In order to get more simple and generally more interpretable components sometimes a suboptimal solution is preferred to the optimal solution of PCA. Unfortunately, there is a trade-off between simplicity and optimality. When simplicity is searched by rotated principal components (called "block-components"), the optimality is worsen because the block-components could be correlated and less variability is extracted from the original variables.

Analogously to SCA, in our approach we shall distinguish between two kinds of components or constraints: block-components (components whose non zero loadings have all the same sign) and difference-components (components which have some strictly positive and some strictly negative loadings). Each difference-component involves only original variables of the same block which is a contrast of the original variables.

Moreover, SCA-RV gives the possibility to choose the number of block-components and difference-components. So the correlation between them is cut off whereas correlations between variables in the same block are systematically larger than those between variables of different blocks.

Let Y be a matrix with p standardized random variables $(Y_1, \ldots, Y_p)$, such that $C = Y'Y$ is the correlation matrix with rank q $(q \leq p)$. Moreover, let p_j (with $j = 1, \ldots, q$) be the column of a $p \times q$ projection matrix P. PCA points out a solution P with the following major properties: (a) the columns of P are orthogonal; (b) the projected data YP are uncorrelated; (c) the vector p_1 is chosen to maximize the variance of Yp_1 and p_j is chosen to maximize the variance of Yp_j with $p_j'p_{j'} = 0$ and $p_j'p_{j'} = 1$(for $j \neq j'$ and $j = 1, \ldots, q$ and $j' = 2, \ldots, q$). All these properties are desirable and in this sense PCA is a reduction technique with optimal features. When all variables measure several aspects of a theme and there is a direct link between them then all elements of C can be strictly positive. In this case all the elements of p_j are of same sign (Perron Frobenius theorem). This structure of the correlation matrix gives more problems of interpretability of PCA results. In order to get more simple and often more interpretable components sometimes a suboptimal solution could be preferred to the optimal solution of PCA.

Let $C_j = C - CP_j(P_j'CP_j)^{-1}P_j'C$ the jth residual matrix where P_j is the matrix containing the first j columns of P. Rousson and Gasser (2004) proposed a procedure that maximizes $p_1'Cp_1 + \sum_{j=2}^{q} p_j'C_{(j-1)}p_j$ divided by the sum of variances of original variables. This criterion called corrected sum of variance (CSV) assures equivalent results to PCA only in case of uncorrelated components, while it is a penalized version of PCA criterion for correlated components.

For systems which might be neither orthonormal nor uncorrelated, a better criterion called Best Linear Predictor (BLP) maximizes $p_1'C^2p_1 + \sum_{j=2}^{q} p_j'C_{(j-1)}^2 p_j$ always divided by the sum of variances of original variables. It does not penalize correlation between components because it measures the optimality of linear combinations of components. For this reason BLP criteria should be used only when the simple components are supposed to be combined together. Seeking a system of q simple components with b blocks maximizing the first criterion of optimality, the two stages SCA algorithm provides an approximation to the optimal system of simple components. First stage of SCA classifies p variables into b disjoint blocks. With fixed values of b and q, P has the b block-components into the first b columns and the difference-components in the last $(q - b)$ columns. The approximate block-structure in the correlation matrix leads to a maximal within block correlations and in the meantime to a minimal between blocks correlations. Rousson and Gasser (2004) solved this problem with an agglomerative hierarchical procedure based on a dissimilarity measure between clusters called *median* linkage alternative to the possible *single* or *complete* linkages. Our proposal modifies this first stage criterion. Instead of using an agglomerative hierarchical procedure based on simple correlation coefficient, which can lead to very different solution with a choice of a possible different link criterion, we propose to use the RV vectorial correlation coefficient (Robert and Escoufier (1976)). This coefficient gives a measure of similarity of the two configurations, taking into account the possibly distinct metrics to be used on them to measure the distances between points (Amenta, 1993).

Let $W_V D = VQ_1V'D$ and $W_Z D = ZQ_2Z'D$ be the scalar products matrices associated to matrices V and Z, respectively, with Q_1 and Q_2 metric matrices and D weight diagonal matrix. The measure is computed as $RV(V, Z) = tr(W_V D W_Z D)/\sqrt{tr(W_V D)^2 tr(W_Z D)^2}$. This measure respects all the four conditions for a vectorial correlation coefficient proposed by Renyi (1959). RV can be viewed as a special case of a more general correlation coefficient framework (Ramsay et al. (1984)) based on the singular value decompositions of two matrices. Several correlation coefficients belonging to Ramsay's framework, result to be independent to the spectrum effect as well as others are not constrained to the direction effects induced by the correlation matrices. The first stage of SCA-RV can be synthesized in three steps:

1. Start with p blocks $B_1, ..., B_p$ where each block contains one of the original variables;
2. Select two blocks B_I and B_J for which a measure of RV is maximum and aggregate them into a new block $B_{(I,J)}$;
3. If b blocks remain then stop the loop, otherwise go back to step 2.

Similarly to SCA, the agglomeration process could continue until the correlation between block components is larger than a prefixed value.

Second stage of SCA-RV algorithm is developed according SCA.

We remark that when the loss of extracted variability is small and the correlation between the components are low then it is advantageous to use SCA-RV for practical use.

3 The Patient Satisfaction Evaluation by SCA-RV

Often in patient satisfaction analysis the first component of PCA corresponds to overall size, so it can be considered as a sum or an average of the original variables. To show how the SCA-RV analysis produces easier interpretable results with respect to PCA and Varimax we use a dataset relative to the 1022 patient satisfaction evaluations at a Neapolitan Hospital. These data obtained by a Servqual questionnaire are based on 15 items measuring five latent dimensions (Babakus and Mangold (1992)). Items V1 – V3 give "tangible" dimension, items V4 – V6 form the "reliability" dimension while items V7 – V9 give "responsiveness" dimension. Items V10 – V13 form the "assurance" dimension and items V14 – V15 give "empathy" dimension, respectively. Before to analyze it, data was quantified by the "Rating Scale Models" procedures (Wright and Masters, 1982).

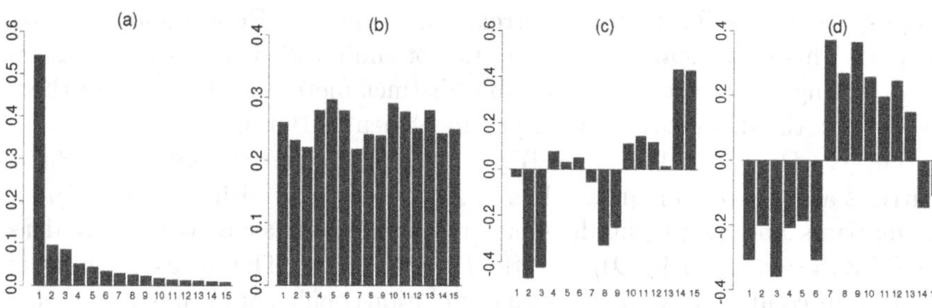

Fig. 1. Relative importance of principal components (a). Loadings of the first (b), second (c) and third (d) component of PCA, respectively.

First three PCA components explain the 74.7% of the original variance (Figure 1.a). This is a very good result but it is not full interpretable due the overall satisfaction meaning of the first principal component (54.5%)(Figure 1.b). Second component (12.9%) highlights a contrast between the set formed by the V2, V3, V8 and V9 variables versus the *empathy* dimension (Figure 1.c). This component is not easy to interpret because this contrast is not confirmed by Babakus' model. Finally, third component (7.3%) highlights a contrast between *tangibles*, *reliability* and *empathy* versus *responsiveness* and *assurance* (Figure 1.d).

More interpretable results are given by Varimax where the *assurance* dimension is given on the second component (Figure 2.b). *Tangibles* dimension is given on the third components (Figure 2.c). *Responsiveness* dimension is

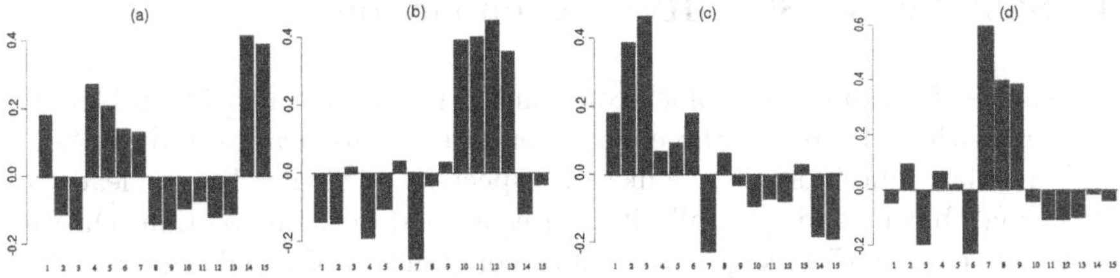

Fig. 2. Loadings relative to first (a), second (b), third (c) and fourth (d) component of Varimax, respectively.

given on the fourth component (Figure 2.d). More complicate is the interpretation of the first component (Figure 2.a) where *empathy* and *realibility* dimensions are not full determinate because are hidden from the items V1 and V2.

Differently full interpretable results are given from SCA-RV (Figure 3). Block 1 contains three items relative to the *tangibles*, three items relative to *reliability* and two items relative to the *empathy* and it represents the 32.6% of original variability (Figure 3.a). Block 2 contains all four items relative to the *assurance* and it represents the 20.6% of original variability (Figure 3.b). Block 3 contains all three items relative to the *responsiveness* and represents the 14.6% of original variability (Figure 3.c). Finally, by the first contrast is possible to observe that on first block the items relative to *tangibles* set again the item relative to *empathy* (Figure 3.d).

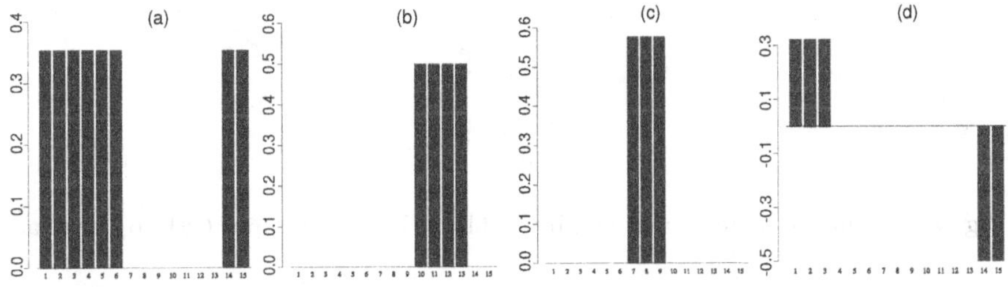

Fig. 3. Loadings of the first (a), second (b) and third (c) component of SCA-RV, respectively. Loadings of the first difference-component of SCA-RV (d).

Of course SCA-RV components are correlated and it does not extracts a maximum of the variability of the original variables. In these cases, it is possible give up to these two remarks since the more variability obtained by PCA and Varimax give irrelevant information and SCA-RV results are full coherent with the conceptual model used in the study of customer satisfaction.

4 SCA Versus SCA-RV: a Comparison

The choice of the number of block-components to retain is a tricky problem in SCA algorithm. There are three agglomerative hierarchical procedures "single", "median" and "complete" linkage, respectively. "Single" linkage leads to form small blocks and generally it assures a good optimal solution. On the contrary, "complete" linkage forms block-components of equal size but with a worst optimal solution. How many component we have to retain is a common problem for PCA and SCA. The SCA and SCA-RV goal of searching for several interpretable components is an advantage for the choice of the number of components. SCA and SCA-RV use all the interpretable components checking the optimality results.

Figure 4 shows the optimality observed for the three different linkage procedures of SCA and SCA-RV according to *CSV* and *BLP* criteria. By using the *BLP* criteria, only for three block SCA-RV gives a result for *single* and *median* linkages while for all other cases it assures always best results. With *CSV* criteria, *single* linkage generally gives better results while for two and three blocks better results are given by *median* linkage.

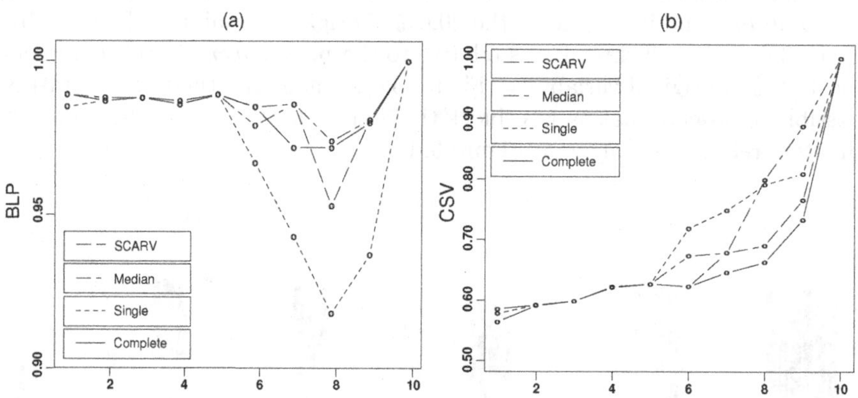

Fig. 4. Optimality values versus number of blocks by *BLP* (a) and *CSV* (b) criteria.

Another thorny problem is that the agglomerative hierarchical procedures provide very different matrices of loadings. A comparision of loadings matrices with 5 to 2 blocks is given in Table 1. For the *single* linkage it is possible to see how the original variables tend to join in single block. On the contrary, SCA-RV and *complete* linkage of SCA are inclined to obtain blocks with the same number. SCA-RV has always better or equal optimal results respect to SCA with *complete* linkage for both *CSV* and *BLP* criteria. Moreover SCA-RV provides always better coherent results than SCA with with respect of patient satisfaction studies.

		B1	B2	B3	B4	B5	B1	B2	B3	B4	B1	B2	B3	B1	B2
SCA-RV	V1	0	1	0	0	0	1	0	0	0	1	0	0	1	0
	V2	0	0	0	0	1	0	0	0	1	1	0	0	1	0
	V3	0	0	0	0	1	0	0	0	1	1	0	0	1	0
	V4	0	1	0	0	0	1	0	0	0	1	0	0	1	0
	V5	0	1	0	0	0	1	0	0	0	1	0	0	1	0
	V6	0	1	0	0	0	1	0	0	0	1	0	0	1	0
	V7	0	0	1	0	0	0	0	1	0	0	0	1	0	1
	V8	0	0	1	0	0	0	0	1	0	0	0	1	0	1
	V9	0	0	1	0	0	0	0	1	0	0	0	1	0	1
	V10	1	0	0	0	0	0	1	0	0	0	1	0	0	1
	V11	1	0	0	0	0	0	1	0	0	0	1	0	0	1
	V12	1	0	0	0	0	0	1	0	0	0	1	0	0	1
	V13	1	0	0	0	0	0	1	0	0	0	1	0	0	1
	V14	0	0	0	1	0	1	0	0	0	1	0	0	1	0
	V15	0	0	0	1	0	1	0	0	0	1	0	0	1	0
SCA-Median	V1	1	0	0	0	0	1	0	0	0	1	0	0	1	0
	V2	0	0	1	0	0	0	0	0	1	0	0	1	0	1
	V3	0	0	1	0	0	0	0	0	1	0	0	1	0	1
	V4	1	0	0	0	0	1	0	0	0	1	0	0	1	0
	V5	1	0	0	0	0	1	0	0	0	1	0	0	1	0
	V6	1	0	0	0	0	1	0	0	0	1	0	0	1	0
	V7	0	0	0	0	1	0	0	1	0	0	1	0	1	0
	V8	0	0	0	1	0	0	0	1	0	0	1	0	1	0
	V9	0	0	0	1	0	0	0	1	0	0	1	0	1	0
	V10	0	1	0	0	0	0	1	0	0	1	0	0	1	0
	V11	0	1	0	0	0	0	1	0	0	1	0	0	1	0
	V12	0	1	0	0	0	0	1	0	0	1	0	0	1	0
	V13	0	1	0	0	0	0	1	0	0	1	0	0	1	0
	V14	1	0	0	0	0	1	0	0	0	1	0	0	1	0
	V15	1	0	0	0	0	1	0	0	0	1	0	0	1	0
SCA-Single	V1	1	0	0	0	0	1	0	0	0	1	0	0	1	0
	V2	0	0	1	0	0	1	0	0	0	1	0	0	1	0
	V3	0	0	1	0	0	1	0	0	0	1	0	0	1	0
	V4	1	0	0	0	0	1	0	0	0	1	0	0	1	0
	V5	1	0	0	0	0	1	0	0	0	1	0	0	1	0
	V6	1	0	0	0	0	1	0	0	0	1	0	0	1	0
	V7	0	0	0	0	1	0	0	0	1	0	0	1	0	1
	V8	0	0	0	1	0	0	0	1	0	0	1	0	1	0
	V9	0	0	0	1	0	0	0	1	0	0	1	0	1	0
	V10	1	0	0	0	0	1	0	0	0	1	0	0	1	0
	V11	1	0	0	0	0	1	0	0	0	1	0	0	1	0
	V12	1	0	0	0	0	1	0	0	0	1	0	0	1	0
	V13	1	0	0	0	0	1	0	0	0	1	0	0	1	0
	V14	0	1	0	0	0	0	1	0	0	1	0	0	1	0
	V15	0	1	0	0	0	0	1	0	0	1	0	0	1	0
SCA-Complete	V1	0	1	0	0	0	0	1	0	0	1	0	0	0	1
	V2	0	0	0	0	1	0	0	0	1	1	0	0	0	1
	V3	0	0	0	0	1	0	0	0	1	1	0	0	0	1
	V4	0	1	0	0	0	0	1	0	0	1	0	0	0	1
	V5	0	1	0	0	0	0	1	0	0	1	0	0	0	1
	V6	0	1	0	0	0	0	1	0	0	1	0	0	0	1
	V7	0	0	1	0	0	0	0	1	0	0	0	1	1	0
	V8	0	0	1	0	0	0	0	1	0	0	0	1	1	0
	V9	0	0	1	0	0	0	0	1	0	0	0	1	1	0
	V10	1	0	0	0	0	1	0	0	0	0	1	0	1	0
	V11	1	0	0	0	0	1	0	0	0	0	1	0	1	0
	V12	1	0	0	0	0	1	0	0	0	0	1	0	1	0
	V13	1	0	0	0	0	1	0	0	0	0	1	0	1	0
	V14	0	0	0	1	0	1	0	0	0	0	1	0	1	0
	V15	0	0	0	1	0	1	0	0	0	0	1	0	1	0

Table 1. Matrices of loadings of 5, 4, 3 and 2 blocks obtained by SCA-RV with *single*, *median* and *complete* linkage.

5 Conclusions and Perspectives

PCA is a useful tool to use when the number of observed variables is large. If there is a direct link between all these variables then all elements of correlation matrix can have the same sign and PCA doesn't produce full interpretable results. In order to overcome this problem, SCA focuses on simplicity and seeks optimal simple components. We propose to modify the SCA algorithm by introducing the RV coefficients. The example as well as several simulation studies lead to highlight that SCA-RV has always better or equal optimal results with respect to SCA with *complete* linkage for *CSV* and *BLP* criteria. In the meantime, it provides better coherent results with respect to patient satisfaction studies.

RV can be viewed as a special case of a more general correlation coefficient framework (Ramsay et al. (1984)) based on the singular value decompositions of two matrices. A comparative study of the performance effects of several correlation matrices within SCA is under investigation.

Another interesting direction is the application of this technique to methods other than PCA (e.g.: PLS, canonical correlation analysis, etc.).

Acknowledgement

Authors thank the anonymous referees for their comments.

References

AMENTA, P. (1993): *Il coefficiente di correlazione lineare tra matrici di dati nel contesto multivariato.* XVII Convegno A.M.A.S.E.S., Ischia.

BABAKUS, E. and MANGOLD, G. (1992): Adapting the servqual scale to hospital services: an empirical investigation. *Health Services Research*, 26, 6, 767-786.

CHOULAKIAN, H.A., D'AMBRA, L. and SIMONETTI, B. (2005): Hausman principal component Analysis. In: 29th Annual International Conference of the German Classification Society (GfKl 2005), Magdeburg (Germany).

HAUSMAN, R.E. (1982): Constrained multivariate analysis, in: S.H Zanckis. and J.S. Rustagi, (Eds.): *Optimisation in Statistics.* North Holland. Amsterdam.

JOLLIFFE, I.T. (2002): *Principal Component Analysis*, New York, Springer

JOLLIFFE, I.T. and UDDIN, M. (2003): A modified principal component technique based on the Lasso. *Journal of Computation and Graphical Statistics*, 12, 531 - 547.

RAMSAY, J., TEN BERGE, J. and STYAN, G. (1984): Matrix correlation. *Psychometrika*, 49, 403-423.

RENYI, A. (1959): *On measures of dependence.* Technical Report 10, Acta Mathematica of the Academy of Science of Hungary.

ROBERT, P. and ESCOUFIER, Y. (1976): A unifying tool for linear multivariate statistical methods: the rv-coefficient. *Applied Statistics*, 25, 257-265.

ROUSSON, V. and GASSER, T. (2004), Simple Component Analysis. *Applied Statistics*, 53, pp. 539-555

VINES, S.K. (2000): Simple principal components. *Applied Statistics*, 49, 441 - 451.
WRIGHT, B.D. and MASTERS, G.N. (1982): Rating scale analysis. Rasch Measurement, MESA.
ZOUS, H., HASTIE, T. and TIBSHIRANI, R. (2004): *Sparse principal component analysis*. Manuscript on http://www-stat.stanford.edu/~ hastie/pub.htm.

Baum-Eagon Inequality in Probabilistic Labeling Problems

Crescenzio Gallo[1] and Giancarlo de Stasio[2]

[1] Dipartimento di Scienze Economiche, Matematiche e Statistiche,
Università di Foggia, Italy
c.gallo@unifg.it
[2] Ufficio Statistico,
Università di Foggia, Italy
g.destasio@unifg.it

Abstract. This work illustrates an approach to the study of labeling, aka "object classification". This kind of parallel computing problem well suites to AI applications (pattern recognition, edge detection, etc.) Our target consists in simplifying an overly computationally costly algorithm proposed by Faugeras and Berthod; using Baum-Eagon theorem, we obtained a reduced algorithm which produces results comparable with other more complex approaches.

1 Introduction

Our work aims to study the possible applications of Baum-Eagon inequality (Baum and Eagon (1967)) to the "labeling" problems, which consist in assigning classes (labels) to objects. For example, let us consider an image whose included objects' contours we want to outline (*edge detection*). In this case, the objects are pixels of which the image is made of, and the labels (classes) assignable to every pixel can be "contour pixel", "not-contour pixel".

Many authors have faced this problem; in particular, an approach from Faugeras and Berthod (1981) require every object to be related with one or more neighbor ones. This situation can be represented by a graph, in which nodes are objects and edges represent existing relations between objects (Hummel and Zucker (1983)). Such concept can be exemplified considering a phrase containing an ambiguous word: to get its meaning, it may suffice to understand the meaning of neighbor words (*context*). The fact that a word in the phrase allows to go back to the ambiguous word's meaning shows a certain relation between them. Generally, in a phrase the words nearest to the ambiguous one are those useful for its meaning's discovery.

More recently, some authors have faced this problem adopting different approaches. Divino, Frigessi and Green (2000) propose a non parametric method to estimate the potential interactions in a Markov Random Field; such method can be applied to some parts of a "biased" image taken from sensors subject to noise (e.g. from satellite). Kleinberg and Tardos (2002) deal with the so-called "metric labeling problem"; it consists in the control of a cost function $c(p, a)$, which represents the cost deriving from the assignment of a label a to an object p. The evaluation of such a cost comes from a likelihood estimate (of having the label a assigned to the object p) starting from an initial observation. To $c(p, a)$ is added the cost bound to a weight measuring, in some way, the "influence" brought to the object p by other objects related to it (in the case of an image, the influence of neighbor pixels).

The assignment of a label to an object depends on the labels currently assigned to the related objects: in other words, the context of the object under examination is taken into account. To formalize all this, let us consider, at the beginning, N objects $a_1, a_2, \ldots, a_N$ and L labels $\lambda_1, \lambda_2, \ldots, \lambda_L$. It is necessary to suppose to be able to define a set of initial probabilities, which represent the probability of assigning each label to an object. Elements of such a set are indicated by $p_i(\lambda_k)$, for $i = 1, \ldots, N$ and $k = 1, \ldots, L$, and represent the probability to assign the label λ_k to the i-th object.

Contextual Faugeras and Berthod's information is represented by a conditional probability set $p_{i,j}(\lambda_k|\lambda_l)$, where $i, j = 1, \ldots, N$ and $k, l = 1, \ldots, L$, representing the probability of assigning label λ_k to the object a_i, currently having neighbor object a_j assigned label λ_l. The object a_j must belong to the set $V_i(\lambda_k)$, which is the set of objects related to a_i, the object currently having label λ_k assigned to it. In many applications, objects related to a specific one do not depend on the label currently assigned to it; in such a case, the set $V_i(\lambda_k)$ will be simply denoted by V_i (aka "homogeneous case"). In practical problems, the initial probabilities suffer from two lacks, i.e.:

1. **Inconsistency**. In practice, they do not verify the relationship

$$p_i(\lambda_k) = \sum_{j \in V_i(\lambda_k)} \sum_{l=1}^{L} p_{ij}(\lambda_k|\lambda_l) p_j(\lambda_l) \tag{1}$$

   In other words, initial probabilities are not compatible with conditional probabilities.
2. **Ambiguity**. The initial probabilities are ambiguous if, for at least one $i = 1, \ldots, N$, there exists at least one $l = 1, \ldots, L$ such as vector $\overline{p}_i = [p_i(\lambda_1), \ldots, p_i(\lambda_l), \ldots, p_i(\lambda_L)] \neq [0, \ldots, 1, \ldots, 0]$ (i.e., there is an ambiguity for an object when it tends to fall in more than one class).

2 Consistency and Ambiguity Functions

Faugeras and Berthod define two functions C_1 and C_2 measuring, respectively, consistency and ambiguity. Consistency is measured through the formula: $C_1 = \frac{1}{2N} \sum_{i=1}^{N} \|\overline{p}_i - \overline{q}_i\|^2$, where $\overline{q}_i$ is a vector having, for each i, the form $[q_i(\lambda_1), q_i(\lambda_2), \ldots, q_i(\lambda_L)]$. Fixed i and k, the values $q_i(\lambda_k)$ are given by the following formula: $q_i(\lambda_k) = Q_i(\lambda_k) / \sum_{l=1}^{L} Q_i(\lambda_l)$, where $Q_i(\lambda_k)$ is given by:

$Q_i(\lambda_k) = \frac{1}{|V_i(\lambda_k)|} \sum_{j \in V_i(\lambda_k)} \sum_{l=1}^{L} p_{ij}(\lambda_k|\lambda_l) p_j(\lambda_l)$. The values $q_i(\lambda_k)$ represent an estimate of the probability $p_i(\lambda_k)$ on the basis of the set of conditional probabilities $p_{ij}(\lambda_k|\lambda_l)$ (Faugeras and Berthod (1981)); Faugeras and Berthod's consistency is guaranteed by the equivalence $p_i(\lambda_k) = q_i(\lambda_k)$[1]. From this, there is the need to minimize the function C_1 (which just represents the Euclidean distance between $\overline{p}_i$ and $\overline{q}_i$). The factor $\frac{1}{2N}$ is for bounding C_1 between 0 and 1.

Ambiguity is measured through the following function:

$$C_2 = \frac{L}{L-1}\left[1 - \frac{1}{N}\sum_{i=1}^{N} \|\overline{p}_i\|^2\right]$$

where $\overline{p}_i$ is the probability vector $[p_i(\lambda_1), p_i(\lambda_2), \ldots, p_i(\lambda_L)]$. Let us observe that in C_2 the factor in square brackets represents the entropy function; the factor $\frac{L}{L-1}$ also here serves to bound C_2 between 0 and 1. Entropy function has its minimum when vector $\overline{p}_i = [p_i(\lambda_1), p_i(\lambda_2), \ldots, p_i(\lambda_L)] = [0, \ldots, 1, \ldots, 0]$, i.e. it is totally unambiguous. In this case, too, the aim is to find C_2's minimum because it guarantees a non-ambiguous labeling. From C_1 and C_2 derives the function named *Global Criterion* $C = \alpha C_1 + (1-\alpha)C_2$, where $0 \leq \alpha \leq 1$. The value α is a constant which represents the relative weight we want to assign to C_1 and C_2; an higher value of α favours C_1 (i.e. consistency), vice versa C_2 (ambiguity).

The search for C's minimum represents the "weak point" of Faugeras and Berthod's algorithm, because this is implemented with the gradient projection method and requires quite complex operations (Rosen (1960)), as well as a relatively high computational cost. More precisely, the algorithm passes from a labeling x_n to the next x_{n+1} according to the formula:

$$x_{n+1} = x_n + \rho_n u_n$$

where u_n is the negative of C's gradient in x_n, and ρ_n is a positive number calculated in such a way to minimize $C(x_{n+1})$. In this case the problem consists in the fact that the searched minimum is linearly bounded by

$$\begin{cases} \sum_{k=1}^{L} p_i(\lambda_k) = 1 \text{ (i=1,...,N)} \\ p_i(\lambda_k) \geq 0 \end{cases} \tag{2}$$

This involves the computation, at every iteration, of a projection operator P_n. The computation of P_n becomes necessary because the negative of u_n gradient may point out of (2) hyperplane.

[1] Tipically, in image processing problems holds the "homogeneous" case, and then $V_i(\lambda_k) = V_i$

The complexity of Faugeras and Berthod's algorithm leads to a difficult implementation, even with the use of parallel computing architectures; in fact, the work done by every single processor remains heavy. Its complexity becomes high especially in the non-homogeneous case (though this last is rarely applied) in which the same authors do not define the number of computations necessary to obtain ρ_n. So, we aim to simplify the algorithm's complexity, exploiting Baum-Eagon's theorem. It applies to homogeneous polynomials of degree d; another theorem (Baum and Sell (1968)) removes this limitation.

Baum-Eagon's Theorem (Baum and Eagon (1967)): Let $P(x) = P(\{x_{ij}\})$ an homogeneous polynomial with nonnegative coefficients in variables $\{x_{ij}\}$ verifying

$$x_{ij} \geq 0, \qquad \sum_{j=1}^{L} x_{ij} = 1, \qquad i = 1, \ldots, N.$$

Let

$$F(x_{ij}) = \frac{x_{ij} \frac{\partial P}{\partial x_{ij}}(x)}{\sum\limits_{j=1}^{L} x_{ij} \frac{\partial P}{\partial x_{ij}}(x)}$$

Then $P(F(x)) > P(x)$ until $F(x) = x$.

Faugeras and Berthod's C function is a quasi-homogeneous polynomial[2] of degree two, if we consider the *homogeneous* case (see section 1). In practice, because C polynomial does not generally have nonnegative coefficients (as required by the previous theorem), it is necessary to transform C in such a manner that the theorem be applied to another polynomial C' with nonnegative coefficients.

Formally, polynomial C has the form: $C = \sum\limits_{i,j} \sum\limits_{k,l} k_{ijkl} x_{ik} x_{jl}$, where k_{ijkl} are the (not all nonnegative) coefficients of the polynomial, $x_{ik} x_{jl}$ $(i, j = 1, \ldots, N; k, l = 1, \ldots, L)$ are the unknown factors of the polynomial. Baum-Eagon's theorem leads to an increasing transformation, so it searches relative maximum points. The case of C is different, because we must minimize instead of maximize. So, instead of minimizing C, we equivalently maximize $-C$. Then, let:

$$C^{(-)} = -C = -\sum_{i,j} \sum_{k,l} k_{ijkl} x_{ik} x_{jl}.$$

$C^{(-)}$ is still a polynomial with not all nonnegative coefficients. It is possible to make $C^{(-)}$'s coefficients nonnegative increasing each coefficient k_{ijkl} by the quantity

$$m = \min \left\{ \min_{i,j,k,l} \{k_{ijkl}\}, 0 \right\} \tag{3}$$

[2] Except for a constant, which disappears after the application of the partial derivatives $\frac{\partial P}{\partial x_{ij}}(x)$

so $C^{(-)}$ becomes:

$$\begin{aligned}C^{(T)} &= -\sum_{i,j}\sum_{k,l}(k_{ijkl}+m)x_{ik}x_{jl}\\ &= -\sum_{i,j}\sum_{k,l}(k_{ijkl}x_{ik}x_{jl}+mx_{ik}x_{jl})\\ &= -\sum_{i,j}\sum_{k,l}k_{ijkl}x_{ik}x_{jl} - m\sum_{i,j}\sum_{k,l}x_{ik}x_{jl}\\ &= C^{(-)} - mN^2\end{aligned}$$

where it is to be considered that holds the relation: $\sum_{k=1}^{L} x_{ik} = 1$, for each $i = 1, \ldots, N$.

Applying Baum-Eagon's theorem to $C^{(T)}$, we have:

$$C^{(T)}(x) < C^{(T)}(F(x))$$

from which:

$$\left[C^{(-)}(x) - mN^2\right] < \left[C^{(-)}(F(x)) - mN^2\right],$$

$$C^{(-)}(x) < C^{(-)}(F(x)),$$

$$C(x) > C(F(x)).$$

The above steps then show that, shifting polynomial coefficients by a constant quantity and applying the theorem, we obtain the growing of both $C^{(T)}$ and $C^{(-)}$, which coincides with the decreasing of the Faugeras and Berthod's C function.

3 Experimental Results

Baum-Eagon's theorem application requires, from a practical point of view, the writing of the function C in polynomial form. The development of the function

$$C = \alpha C_1 + (1-\alpha)C_2$$

leads to the following (quasi-homogeneous) polynomial form:

$$\begin{aligned}C = &\frac{3\alpha L - \alpha - 2L}{2N(L-1)}\sum_{i=1}^{N}\sum_{k=1}^{L}x_{ik}^2\\ &-\frac{\alpha}{N}\sum_{i=1}^{N}\sum_{k=1}^{L}x_{ik}\sum_{j\in V_i}\sum_{l=1}^{L}p_{ij}(\lambda_k|\lambda_l)x_{jl}\\ &+\frac{\alpha}{2N}\sum_{i=1}^{N}\sum_{k=1}^{L}x_{ik}\sum_{j\in V_i}\sum_{l=1}^{L}p_{ij}(\lambda_k|\lambda_l)x_{jl}\sum_{\omega\in V_i}\sum_{v=1}^{L}p_{\omega i}(\lambda_v|\lambda_k)\\ &+M\end{aligned}$$

where:

- L is the number of labels,
- N is the number of objects,
- V_i is the set of objects related to the object i[3],
- $M = \frac{(1-\alpha)L}{L-1}$,
- $x_{ik} = p_i(\lambda_k)$.

From the above we note the quasi-homogeneity of C (in the variables x_{ik}) apart of constant M, that does not influence the computation of the partial derivatives, used for the implementation of the algorithm.

The algorithm which implements the above mentioned method must initially find a constant value such as it makes all polynomial coefficients non-negative, so obtaining a new polynomial to which is possible to apply Baum-Eagon theorem. Such a constant value is obtained by increasing every coefficient by a quantity equal to the minimum m as in (3).

Labeling through Baum-Eagon theorem has been tested on the *thresholding* problem. This one consists in the transformation of an image, made of a matrix of different grey level tones pixels, into another image made only of black and white pixels, as shown in Figure 1. Formalizing the problem, the

Fig. 1. Sample image - initial and final labeling.

image is made of $n \times m$ objects and $L = 2$ labels, corresponding to "*light pixel*" and "*dark pixel*". The initial probability set is computed according to a method suggested by Rosenfeld and Smith (1981): let d and l be, respectively, the toward "*dark*" and toward "*light*" grey levels; let z_i be the grey level of the i-th pixel. Then, for that pixel we'll have the following initial probabilities: $p_{i,dark} = (l - z_i)/(l - d)$ and $p_{i,light} = (z_i - d)/(l - d)$.

As to the conditional probabilities set, Peleg and Rosenfeld (1978) suggest a method based on statistical computation. Initially, there is an estimate of the probability that every pixel has a certain label λ; this is realized through the formula:

$$\overline{P}(\lambda) = \frac{1}{N} \sum_{(x,y)} P_{(x,y)}(\lambda)$$

[3] We assume the "homogeneous" case, so $V_i(\lambda_k) = V_i$

where N is the number of pixels, and pairs (x, y) are the coordinates of every pixel. Then, it is computed the joint probability of every pair (x, y) and $(x+i, y+j)$ of neighbor points have assigned, respectively, labels (λ, λ') according to the formula:

$$P_{ij}(\lambda, \lambda') = \frac{1}{N} \sum_{(x,y)} P_{(x,y)}(\lambda) P_{(x+i,y+j)}(\lambda')$$

From the two above derives the conditional probabilities formula:

$$P_{ij}(\lambda|\lambda') = \frac{P_{ij}(\lambda, \lambda')}{\overline{P}(\lambda')}$$

which has been used in our experimental tests to set the initial probabilities.

4 Conclusions

The goodness of Baum-Eagon approach has been hypothesized in various environments, especially in probabilistic labeling problems, well suited to AI with parallel computing architecture. Starting with Faugeras and Berthods overly computationally complex algorithm, we developed a simplified version using Baum-Eagon inequality, and reached positive experimental results, which encourage us to refine and carry on trying more complex test sets. In Appendix A is reported the detailed development of the Baum-Eagon C function in a quasi-homogeneous polynomial form of degree two.

A Development of the Baum-Eagon C Function

In order to obtain the final form of $C = \alpha C_1 + (1-\alpha) C_2$ we first consider $C_1 = \frac{1}{2N} \sum_{i=1}^{N} \sum_{k=1}^{L} \left[p_i(\lambda_k) - \sum_{j \in V_i} \sum_{l=1}^{L} p_{ij}(\lambda_k|\lambda_l) p_j(\lambda_l) \right]^2$. In order to obtain a form of C_1 more suited to the partial derivatives needed for the development of our algorithm, let us consider the case of image processing (which we are treating in our work). In this situation, from the Euclidean distance it is obvious to derive a sort of "reciprocity" in the sequence of indexes involved in the formula, in the sense that if a pixel i is at distance d from a pixel j, obviously the same holds for j respect to i.

So we may write C_1 in its final form: $C_1 = \frac{1}{2N} \sum_{i=1}^{N} \sum_{k=1}^{L} p_i(\lambda_k)^2 - \frac{1}{N} \sum_{i=1}^{N} \sum_{k=1}^{L} p_i(\lambda_k) \sum_{j \in V_i} \sum_{l=1}^{L} pij(\lambda_k|\lambda_l) p_j(\lambda_l) + \frac{1}{2N} \sum_{i=1}^{N} \sum_{k=1}^{L} \sum_{j \in V_i} \sum_{l=1}^{L} p_{ij}(\lambda_k|\lambda_l) p_j(\lambda_l) \sum_{\omega \in V_i} \sum_{v=1}^{L} p_{\omega i}(\lambda_v|\lambda_k) p_i(\lambda_k)$.

Let us consider the development of C_2. We obtain: $C_2 = \frac{L}{L-1} \left[1 - \frac{1}{N} \sum_{i=1}^{N} \|\overline{p}_i\|^2 \right] = \frac{L}{L-1} \left[1 - \frac{1}{N} \sum_{i=1}^{N} \sum_{k=1}^{L} p_i(\lambda_k)^2 \right]$. Multiplying C_1

and C_2 respectively by α and $(1-\alpha)$ we obtain the polynomial form of "Global Criterion" C function. Developing in detail we have:
$C = \alpha C_1 + (1-\alpha)C_2 = \frac{\alpha}{2N}\sum_{i=1}^{N}\|\overline{p}_i - \overline{q}_i\|^2 +$
$(1-\alpha)\frac{L}{L-1}\left[1 - \frac{1}{N}\sum_{i=1}^{N}\|\overline{p}_i\|^2\right] = \frac{3\alpha L-\alpha-2L}{2N(L-1)}\sum_{i=1}^{N}\sum_{k=1}^{L}p_i(\lambda_k)^2 -$
$\frac{\alpha}{N}\sum_{i=1}^{N}\sum_{k=1}^{L}p_i(\lambda_k)\sum_{j\in V_i}\sum_{l=1}^{L}p_{ij}(\lambda_k|\lambda_l)p_j(\lambda_l) +$
$\frac{\alpha}{2N}\sum_{i=1}^{N}\sum_{k=1}^{L}\sum_{j\in V_i}\sum_{l=1}^{L}p_{ij}(\lambda_k|\lambda_l)p_j(\lambda_l)\sum_{\omega\in V_i}\sum_{v=1}^{L}p_{\omega i}(\lambda_v|\lambda_k)p_i(\lambda_k) +$
$\frac{(1-\alpha)L}{L-1}$.

From the above formula it is simple to get the partial derivatives:

$$\frac{\partial C}{\partial p_i(\lambda_k)} = \frac{3\alpha L-\alpha-2L}{N(L-1)}p_i(\lambda_k) - \frac{\alpha}{N}\sum_{j\in V_i}\sum_{l=1}^{L}p_{ij}(\lambda_k|\lambda_l)p_j(\lambda_l) +$$

$$\frac{\alpha}{2N}\sum_{j\in V_i}\sum_{l=1}^{L}p_{ij}(\lambda_k|\lambda_l)p_j(\lambda_l)\sum_{\omega\in V_i}\sum_{v=1}^{L}p_{\omega i}(\lambda_v|\lambda_k).$$

References

BAUM, L.E. and EAGON, J.A. (1967): An inequality with application to statistical estimation for probabilistic function of markov processes and to a model for ecology, *Bull. Amer. Math. Soc., 73, 361–363.*

BAUM, L.E. and SELL, G.R. (1968): Growth transformation for functions on mainfolds, *Pacific Journal of Mathematics, 27, 2, 211–227.*

FAUGERAS, O.D. and BERTHOD, M. (1981): Improving consistency and reducing ambiguity in stochastic labeling: an optimization approach, *IEEE Trans. Patt. Anal. and Mach. Intell., PAMI-3, 4, 412–424.*

DIVINO, F., FRIGESSI, A. and GREEN, P.J. (2000): Penalized pseudolikelihood inference in spatial interaction models with covariates, *Scandinavian Journal of Statistics, 27, 445–458.*

HUMMEL, R.A. and ZUCKER, S.W. (1983): On the foundations of relaxation labeling processes, *IEEE Trans. Patt. Anal. and Mach. Intell., PAMI-5, 3, 267–286.*

KLEINBERG, J. and TARDOS, E. (2002): Approximation algorithms for classification problems with pairwise relationships: metric labeling and markov random fields, *Journal of the ACM, 49, 616–639.*

PELEG, S. and ROSENFELD, A. (1978): Determining compatibility coefficients for curve enhancement relaxation processes, *IEEE Trans. Syst., Man. and Cybernetics, SMC-8, 7, 548–555.*

ROSEN, J.B. (1960): The gradient projection method for nonlinear programming-part i: Linear constraints, *J. Soc. Appl. Math., 8, 181–217.*

ROSENFELD, A. and SMITH, R.C. (1981): Thresholding using relaxation, *IEEE Trans. Patt. Anal. and Mach. Intell., PAMI-3, 5, 598–606.*

Monotone Constrained EM Algorithms for Multinormal Mixture Models

Salvatore Ingrassia[1] and Roberto Rocci[2]

[1] Dipartimento di Economia e Metodi Quantitativi,
Università di Catania, Italy
s.ingrassia@unict.it

[2] Dipartimento SEFEMEQ,
Università di Tor Vergata, Italy
roberto.rocci@uniroma2.it

Abstract. We investigate the spectral decomposition of the covariance matrices of a multivariate normal mixture distribution in order to construct constrained EM algorithms which guarantee the monotonicity property. Furthermore we propose different set of constraints which can be simply implemented. These procedures have been tested on the ground of many numerical experiments.

1 Introduction

Let $f(\mathbf{x};\psi)$ be the density of a mixture of k multinormal distribution $f(\mathbf{x};\psi) = \alpha_1 p(\mathbf{x};\boldsymbol{\mu}_1,\boldsymbol{\Sigma}_1) + \cdots + \alpha_k p(\mathbf{x};\boldsymbol{\mu}_k,\boldsymbol{\Sigma}_k)$ where the α_j are the mixing weights and $p(\mathbf{x};\boldsymbol{\mu}_j,\boldsymbol{\Sigma}_j)$ is the density function of a $q-$multivariate normal distribution with mean vector $\boldsymbol{\mu}_j$ and covariance matrix $\boldsymbol{\Sigma}_j$. Finally, we set $\psi = \{(\alpha_j,\boldsymbol{\mu}_j,\boldsymbol{\Sigma}_j),\ j=1,\ldots,k\} \in \boldsymbol{\Psi}$, where $\boldsymbol{\Psi}$ is the parameter space. It is well known that the log-likelihood function $\mathcal{L}(\psi)$ coming from a sample of N i.i.d. observations with law $f(\mathbf{x};\psi)$ is unbounded and presents many local spurious maxima, see McLachlan and Peel (2000); however under suitable hypotheses in Hathaway (1985) a constrained (global) maximum-likelihood formulation has been proposed which presents no singularities and a smaller number of spurious maxima by imposing the following constraint satisfied by the true set of parameters

$$\min_{1\le h\neq j\le k} \lambda_{\min}(\boldsymbol{\Sigma}_h\boldsymbol{\Sigma}_j^{-1}) \ge c, \qquad c\in(0,1] \tag{1}$$

where $\lambda_{\min}(\mathbf{A})$ is the smallest eigenvalue of $\mathbf{A}$. Such constraints are difficult to apply in algorithms like the EM, where the estimates of the covariance matrices are iteratively updated; for this aim they have been reformulated as

$$a \le \lambda_i(\boldsymbol{\Sigma}_j) \le b \qquad i=1,\ldots,q, \quad j=1,\ldots,k \tag{2}$$

where $\lambda_i(\mathbf{A})$ is the ith eigenvalue of $\mathbf{A}$, in non increasing order, and a and b are positive numbers such that $a/b = c$, see Ingrassia (2004). In this way a set of stronger constraints are obtained; in fact, the inequalities

$$\lambda_{\min}(\boldsymbol{\Sigma}_h \boldsymbol{\Sigma}_j^{-1}) \geq \frac{\lambda_{\min}(\boldsymbol{\Sigma}_h)}{\lambda_{\max}(\boldsymbol{\Sigma}_j)} \geq \frac{a}{b} = c \,, \qquad 1 \leq h \neq j \leq k \tag{3}$$

show that (2) implies (1).

In this paper we analyze the eigenvalue and eigenvector structure of the covariance matrices $\boldsymbol{\Sigma}_j$ in order to construct constrained EM algorithms which guarantee the monotonicity property of the unconstrained version. We also propose a new set of simple constraints which are weaker than (2).

2 An Algebraic Analysis of the Covariance Matrices

The EM algorithm generates a sequence of estimates $\{\boldsymbol{\psi}^{(m)}\}_m$, where $\boldsymbol{\psi}^{(0)}$ denotes the initial guess and $\boldsymbol{\psi}^{(m)} \in \boldsymbol{\Psi}$ for $m \in \mathbb{N}$, so that the corresponding sequence $\{\mathcal{L}(\boldsymbol{\psi}^{(m)})\}_m$, is not decreasing. The theory of the EM algorithm assures that $\mathcal{L}(\boldsymbol{\psi}^{(m+1)}) \geq \mathcal{L}(\boldsymbol{\psi}^{(m)})$. The E-step, on the $(m+1)$ iteration computes the quantities

$$u_{nj}^{(m+1)} = \frac{\alpha_j^{(m)} p(\mathbf{x}_n; \boldsymbol{\mu}_j^{(m)}, \boldsymbol{\Sigma}_j^{(m)})}{\sum_{h=1}^k \alpha_h^{(m)} p(\mathbf{x}_n; \boldsymbol{\mu}_h^{(m)}, \boldsymbol{\Sigma}_h^{(m)})} \qquad n = 1, \ldots, N, \quad j = 1, \ldots, k \,.$$

The M-step on the $(m+1)$ iteration requires the global maximization of the complete log-likelihood

$$Q(\boldsymbol{\psi}) = \sum_{j=1}^{k} \sum_{n=1}^{N} u_{nj}^{(m+1)} (\ln \alpha_j) + \sum_{j=1}^{k} q_j(\boldsymbol{\mu}_j, \boldsymbol{\Sigma}_j) \tag{4}$$

with respect to $\boldsymbol{\psi}$ over the parameter space $\boldsymbol{\Psi}$ to give the update estimate $\boldsymbol{\psi}^{(m+1)}$. To achieve this global maximization, let us first study the three separate maximizations with respect to $\boldsymbol{\alpha} = [\alpha_1, ..., \alpha_k]'$, $\boldsymbol{\mu}_j$ and $\boldsymbol{\Sigma}_j$ $(j = 1, 2, \ldots, k)$.

1) Maximization with respect to $\boldsymbol{\alpha}$. It can be easily shown that the complete log-likelihood (4) obtains a maximum with respect to $\boldsymbol{\alpha}$ by setting

$$\alpha_j = \frac{1}{N} u_{\cdot j}^{(m+1)} = \frac{1}{N} \sum_{n=1}^{N} u_{nj}^{(m+1)} \qquad j = 1, \ldots, k \,,$$

2) Maximization with respect to $\boldsymbol{\mu}_j$. In this case, the maximization of (4) can be split into k independent maximizations of the terms $q_j(\boldsymbol{\mu}_j, \boldsymbol{\Sigma}_j)$

$(j = 1, \ldots, k)$. Thus, let us write

$$q_j(\boldsymbol{\mu}_j, \boldsymbol{\Sigma}_j) = \sum_{n=1}^{N} u_{nj}^{(m+1)} \ln p(\mathbf{x}_n; \boldsymbol{\mu}_j, \boldsymbol{\Sigma}_j)$$

$$= \frac{1}{2} \sum_{n=1}^{N} u_{nj}^{(m+1)} \left[-q \ln(2\pi) - \ln |\boldsymbol{\Sigma}_j| - (\mathbf{x}_n - \boldsymbol{\mu}_j)' \boldsymbol{\Sigma}_j^{-1} (\mathbf{x}_n - \boldsymbol{\mu}_j)\right] \quad (5)$$

>From (5) easily follows that (4) obtains a maximum with respect to $\boldsymbol{\mu}_j$ when

$$\boldsymbol{\mu}_j = \frac{1}{u_{.j}^{(m+1)}} \sum_{n=1}^{N} u_{nj}^{(m+1)} \mathbf{x}_n \qquad j = 1, \ldots, k\,;$$

3) Maximization with respect to $\boldsymbol{\Sigma}_j$ which is the most relevant for our scope. Again the maximization of $Q(\boldsymbol{\psi})$ can be split into k independent maximizations of the terms $q_j(\boldsymbol{\mu}_j, \boldsymbol{\Sigma}_j)$ $(j = 1, \ldots, k)$. By noting that

$$(\mathbf{x}_n - \boldsymbol{\mu}_j)' \boldsymbol{\Sigma}_j^{-1} (\mathbf{x}_n - \boldsymbol{\mu}_j) = \operatorname{tr}((\mathbf{x}_n - \boldsymbol{\mu}_j)' \boldsymbol{\Sigma}_j^{-1} (\mathbf{x}_n - \boldsymbol{\mu}_j))$$
$$= \operatorname{tr}(\boldsymbol{\Sigma}^{-1} (\mathbf{x}_n - \boldsymbol{\mu}_j)(\mathbf{x}_n - \boldsymbol{\mu}_j)'),$$

the function (5) can be also written as

$$q_j(\boldsymbol{\mu}_j, \boldsymbol{\Sigma}_j) =$$
$$= \frac{1}{2} \sum_{n=1}^{N} u_{nj}^{(m+1)} \left[-q \ln(2\pi) - \ln |\boldsymbol{\Sigma}_j| - \operatorname{tr}(\boldsymbol{\Sigma}_j^{-1} (\mathbf{x}_n - \boldsymbol{\mu}_j)(\mathbf{x}_n - \boldsymbol{\mu}_j)')\right].$$

After some algebras we get

$$q_j(\boldsymbol{\mu}_j, \boldsymbol{\Sigma}_j) = \gamma_j^{(m+1)} - \frac{1}{2} u_{.j}^{(m+1)} \left[\ln |\boldsymbol{\Sigma}_j| + \operatorname{tr}\left(\boldsymbol{\Sigma}_j^{-1} \mathbf{S}_j\right)\right] \quad (6)$$

where for simplicity we set

$$\gamma_j = -\frac{q}{2} \ln(2\pi) u_{.j}^{(m+1)}, \mathbf{S}_j = \frac{1}{u_{.j}^{(m+1)}} \sum_{n=1}^{N} u_{nj}^{(m+1)} (\mathbf{x}_n - \boldsymbol{\mu}_j)(\mathbf{x}_n - \boldsymbol{\mu}_j)'.$$

The relation (6) shows that the maximization of $q_j(\boldsymbol{\mu}_j, \boldsymbol{\Sigma}_j)$ with respect to $\boldsymbol{\Sigma}_j$ amounts to the minimization of $\ln |\boldsymbol{\Sigma}_j| + \operatorname{tr}\left(\boldsymbol{\Sigma}_j^{-1} \mathbf{S}_j\right)$.

Let $\boldsymbol{\Sigma}_j = \boldsymbol{\Gamma}_j \boldsymbol{\Lambda}_j \boldsymbol{\Gamma}_j'$ be the spectral decomposition of $\boldsymbol{\Sigma}_j$, where $\boldsymbol{\Lambda}_j = \operatorname{diag}(\lambda_{1j}, \ldots, \lambda_{qj})$ is the diagonal matrix of the eigenvalues of $\boldsymbol{\Sigma}_j$ in non decreasing order, and $\boldsymbol{\Gamma}_j$ is an orthogonal matrix whose columns are the standardized eigenvectors of $\boldsymbol{\Sigma}_j$. It is well known that (see for example: Theobald, 1975)

$$\operatorname{tr}\left(\boldsymbol{\Sigma}_j^{-1} \mathbf{S}_j\right) \geq \operatorname{tr}\left(\boldsymbol{\Lambda}_j^{-1} \mathbf{L}_j\right) = \sum_{i=1}^{q} \lambda_{ij}^{-1} l_{ij} \quad (7)$$

where $\mathbf{L}_j = \mathrm{diag}(l_{1j}, \ldots, l_{qj})$ is the diagonal matrix of the eigenvalues, in non decreasing order, of $\mathbf{S}_j$. In particular, the equality in (7) holds if and only if $\boldsymbol{\Sigma}_j$ and $\mathbf{S}_j$ have the same eigenvectors which are ordered with respect both $\lambda_{1j}, \ldots, \lambda_{qj}$ and $l_{1j}, \ldots, l_{qj}$. This implies that the minimum can be reached if and only if $\boldsymbol{\Sigma}_j$ has the same eigenvectors of $\mathbf{S}_j$. Under this condition, since $\ln|\boldsymbol{\Sigma}_j| = \sum_i \ln \lambda_{ij}$, the minimization of $\ln|\boldsymbol{\Sigma}_j| + \mathrm{tr}\left(\boldsymbol{\Sigma}_j^{-1}\mathbf{S}_j\right)$ with respect to $\boldsymbol{\Sigma}_j$ amounts to the minimization of

$$\sum_{i=1}^{q} \ln \lambda_{ij} + \sum_{i=1}^{q} \lambda_{ij}^{-1} l_{ij} = \sum_{i=1}^{q} \left(\ln \lambda_{ij} + \lambda_{ij}^{-1} l_{ij}\right) \tag{8}$$

with respect to $\lambda_{1j}, \ldots, \lambda_{qj}$, which is equivalent to q independent minimizations of $\ln \lambda_{ij} + \lambda_{ij}^{-1} l_{ij}$ with respect to $\lambda_{1j}, \ldots, \lambda_{qj}$, which give $\lambda_{ij} = l_{ij}$. In conclusion, the optimal $\boldsymbol{\Sigma}_j$ is obtained first by setting its eigenvectors equal to the ones of $\mathbf{S}_j$ and then doing the same with the eigenvalues. This can be simply done by setting $\boldsymbol{\Sigma}_j = \mathbf{S}_j$.

On the basis of the previous results, it should be noted that only the maximization with respect to $\boldsymbol{\Sigma}_j$ depends on the current values of the other parameters. It follows that the M-step can be done by maximizing:

1. $Q(\boldsymbol{\alpha}, \boldsymbol{\mu}_1^{(m)}, ..., \boldsymbol{\mu}_k^{(m)}, \boldsymbol{\Sigma}_1^{(m)}, ..., \boldsymbol{\Sigma}_k^{(m)})$ with respect to $\boldsymbol{\alpha}$ to get
$$\alpha_j^{(m+1)} = \frac{1}{N} u_{\cdot j}^{(m)} \quad ;$$
2. $Q(\boldsymbol{\alpha}^{(m+1)}, \boldsymbol{\mu}_1, ..., \boldsymbol{\mu}_k, \boldsymbol{\Sigma}_1^{(m)}, ..., \boldsymbol{\Sigma}_k^{(m)})$ with respect to $\boldsymbol{\mu}_j$ $(j = 1, \ldots, k)$ to get
$$\boldsymbol{\mu}_j^{(m+1)} = \frac{1}{u_{\cdot j}^{(m+1)}} \sum_{n=1}^{N} u_{nj}^{(m+1)} \mathbf{x}_n \quad ;$$
3. $Q(\boldsymbol{\alpha}^{(m+1)}, \boldsymbol{\mu}_1^{(m+1)}, ..., \boldsymbol{\mu}_k^{(m+1)}, \boldsymbol{\Sigma}_1, ..., \boldsymbol{\Sigma}_k)$ with respect to $\boldsymbol{\Sigma}_j$ $(j = 1, \ldots, k)$ to get
$$\boldsymbol{\Sigma}_j^{(m+1)} = \mathbf{S}_j^{(m+1)} = \frac{1}{u_{\cdot j}^{(m+1)}} \sum_{n=1}^{N} u_{nj}^{(m+1)} (\mathbf{x}_n - \boldsymbol{\mu}_j^{(m+1)})(\mathbf{x}_n - \boldsymbol{\mu}_j^{(m+1)})' \quad .$$

The third step can be regarded as obtained according to the following three substeps:

i) set $\boldsymbol{\Gamma}_j^{(m+1)}$ equal to the orthonormal matrix whose columns are standardized eigenvectors of $\mathbf{S}_j^{(m+1)}$;
ii) set $\boldsymbol{\Lambda}_j^{(m+1)} \leftarrow \mathrm{diag}(l_{1j}^{(m+1)}, \ldots, l_{qj}^{(m+1)})$;
iii) compute $\boldsymbol{\Sigma}_j^{(m+1)} \leftarrow \boldsymbol{\Gamma}_j^{(m+1)} \boldsymbol{\Lambda}_j^{(m+1)} \boldsymbol{\Gamma}_j^{(m+1)\prime}$.

This split into three substeps is not convenient in the ordinary EM algorithm. However, in the next section it will be shown how this help to formulate monotone algorithms for the constrained case.

3 Constrained Monotone EM Algorithms

The reformulation of the update of the covariance matrices $\boldsymbol{\Sigma}_j$ $(j = 1, \ldots, k)$ presented in the previous section suggests some ideas for the the construction of EM algorithms such that the constraints (1) are satisfied while the monotonicity is preserved.

Approach A. The simplest approach is the following:

i) if $\lambda_{\min}(\mathbf{S}_j^{(m+1)})/\lambda_{\max}(\mathbf{S}_j^{(m+1)}) \geq c$ then set $\boldsymbol{\Sigma}_j^{(m+1)} \leftarrow \mathbf{S}_j^{(m+1)}$ otherwise set $\boldsymbol{\Sigma}_j^{(m+1)} \leftarrow \mathbf{S}_j^{(m)}$.

Approach B. A more refined strategy is:

i) set $\boldsymbol{\Gamma}_j^{(m+1)}$ equal to the orthogonal matrix whose columns are standardized eigenvectors of $\mathbf{S}_j^{(m+1)}$;
ii) if $\lambda_{\min}(\mathbf{S}_j^{(m+1)})/\lambda_{\max}(\mathbf{S}_j^{(m+1)}) \geq c$ then
set $\boldsymbol{\Lambda}_j^{(m+1)} \leftarrow \mathrm{diag}(l_{1j}^{(m+1)}, \ldots, l_{qj}^{(m+1)})$ otherwise set $\boldsymbol{\Lambda}_j^{(m+1)} \leftarrow \boldsymbol{\Lambda}_j^{(m)}$;
iii) compute the covariance matrix by $\boldsymbol{\Sigma}_j^{(m+1)} \leftarrow \boldsymbol{\Gamma}_j^{(m+1)}\boldsymbol{\Lambda}_j^{(m+1)}\boldsymbol{\Gamma}_j^{(m+1)'}$.

Approach C. Another approach consists in imposing the constraints to the eigenvalues, that is to find an update of $\boldsymbol{\Sigma}_j$ which maximizes (4) under the constraints (3), see Ingrassia (2004). According to the results given in the previous section, the optimal update is a symmetric matrix having the same eigenvectors as $\mathbf{S}_j^{(m+1)}$, and eigenvalues minimizing (8) under (2). It is easy to show that this can be achieved by setting

$$\lambda_{ij} = \begin{cases} a & \text{if } l_{ij}^{(m+1)} < a \\ l_{ij}^{(m+1)} & \text{if } a \leq l_{ij}^{(m+1)} \leq b \\ b & \text{if } l_{ij}^{(m+1)} > b\,. \end{cases}$$

We can summarize this strategy as follows:

i) set $\boldsymbol{\Gamma}_j^{(m+1)}$ equal to the orthogonal matrix whose columns are standardized eigenvectors of $\mathbf{S}_j^{(m+1)}$;
ii) afterward set

$$\lambda_{ij}^{(m+1)} \leftarrow \min\left(b, \max\left(a, l_{ij}^{(m+1)}\right)\right)\,. \tag{9}$$

iii) update the covariance matrix as $\boldsymbol{\Sigma}_j^{(m+1)} \leftarrow \boldsymbol{\Gamma}_j^{(m+1)}\boldsymbol{\Lambda}_j^{(m+1)}\boldsymbol{\Gamma}_j^{(m+1)'}$.

In this way the monotonicity is guaranteed once the initial guess $\boldsymbol{\Sigma}_j^{(0)}$ satisfies the constraints.

Approach D. A different kind of constraints on eigenvalues is here proposed by introducing a suitable parameterization for the covariance matrices of the mixture components. Let us rewrite $\boldsymbol{\Sigma}_j = \eta^2 \boldsymbol{\Omega}_j$ $(j = 1, \ldots, k)$, where the matrices $\boldsymbol{\Omega}_j$ are such that

$$\lambda_i(\boldsymbol{\Omega}_j) \leq \frac{1}{c}, \qquad \text{and} \qquad \min_{ij} \lambda_i(\boldsymbol{\Omega}_j) = 1 \tag{10}$$

for $i = 1, \ldots, q$ and $j = 1, \ldots, k$. They are weaker than (2), indeed if constraints (2) are satisfied and we set $\eta^2 = \min_{ij} \lambda_i(\boldsymbol{\Sigma}_j)$ and $\boldsymbol{\Omega}_j = \boldsymbol{\Sigma}_j / \eta^2$ then, by noting that $\lambda_i(\boldsymbol{\Sigma}_j) = \lambda_i(\boldsymbol{\Omega}_j)\eta^2$, we get (10) since

$$\lambda_i(\boldsymbol{\Sigma}_j) \leq b \Rightarrow \lambda_i(\boldsymbol{\Omega}_j) \leq \frac{b}{\eta^2} \leq \frac{b}{a} = \frac{1}{c} \quad \text{and} \quad \min_{ij} \lambda_i(\boldsymbol{\Omega}_j) = 1.$$

However they are stronger than (1), indeed if the constraints (10) hold then

$$\lambda_{\min}(\boldsymbol{\Sigma}_h \boldsymbol{\Sigma}_j^{-1}) \geq \frac{\lambda_{\min}(\boldsymbol{\Sigma}_h)}{\lambda_{\max}(\boldsymbol{\Sigma}_j)} = \frac{\lambda_{\min}(\boldsymbol{\Omega}_h)}{\lambda_{\max}(\boldsymbol{\Omega}_j)} \geq \frac{1}{1/c} = c, \qquad 1 \leq h \neq j \leq k$$

In order to implement in the EM algorithm the new set of constraints, only the last step must be changed. The update of η^2 and $\boldsymbol{\Omega}_j$ must maximize the complete log-likelihood, i.e., they have to maximize the function

$$\sum_{j=1}^{k} q_j(\boldsymbol{\mu}_j, \eta^2, \boldsymbol{\Omega}_j) = \gamma_j - \frac{1}{2} \sum_{j=1}^{k} u_{\cdot j}^{(m+1)} \left[\ln |\eta^2 \boldsymbol{\Omega}_j| + \operatorname{tr}\left(\eta^{-2} \boldsymbol{\Omega}_j^{-1} \mathbf{S}_j \right) \right].$$

It can be shown that the maximum with respect to η^2 is achieved by setting

$$\eta^2 = \frac{1}{Nq} \sum_{j=1}^{k} u_{\cdot j}^{(m+1)} \left[\operatorname{tr}\left(\boldsymbol{\Omega}_j^{-1} \mathbf{S}_j \right) \right], \tag{11}$$

while, on the basis of the results shown in this section and in the previous one, it can be easily shown that the maximum with respect to $\boldsymbol{\Omega}_j$ is obtained by setting its eigenvectors equal to the eigenvectors of $\mathbf{S}_j$ and the eigenvalues

$$\lambda_i(\boldsymbol{\Omega}_j) = \min\left(\frac{1}{c}, \max\left(1, \frac{l_{ij}}{\eta^2} \right) \right). \tag{12}$$

We can summarize this fourth strategy as follows:

i) set $\boldsymbol{\Gamma}_j^{(m+1)}$ equal to the orthogonal matrix whose columns are standardized eigenvectors of $\mathbf{S}_j^{(m+1)}$;

ii) update η^2 as

$$(\eta^{(m+1)})^2 = \frac{1}{Nq} \sum_{j=1}^{k} u_{\cdot j}^{(m+1)} \left[\operatorname{tr}\left((\boldsymbol{\Omega}_j^{(m)})^{-1} \mathbf{S}_j^{(m+1)} \right) \right];$$

iii) set

$$\lambda_{ij}^{(m+1)} \leftarrow (\eta^{(m+1)})^2 \min\left(\frac{1}{c}, \max\left(1, \frac{l_{ij}^{(m+1)}}{(\eta^{(m+1)})^2}\right)\right);$$

iv) update the covariance matrix as $\boldsymbol{\Sigma}_j^{(m+1)} \leftarrow \boldsymbol{\Gamma}_j^{(m+1)} \boldsymbol{\Lambda}_j^{(m+1)} \boldsymbol{\Gamma}_j^{(m+1)'}$.

It is important to note that in (12) the maximizer depends on the current value of η^2, while in (11) the maximizer depends on the current values of $\boldsymbol{\Omega}_j$ $(j = 1, \ldots, k)$. It follows that the sequential implementation of the above four steps leads to an increment of the complete log-likelihood but does not necessarily maximize it with respect to η^2 and $\boldsymbol{\Omega}_j$ $(j = 1, \ldots, k)$. This implies that the resulting algorithm is of the class ECM (Expectation Conditional Maximization) (see e.g. McLachlan & Krishnan, 1997) rather than EM. It is also important to note that the proposed algorithm does not necessarily gives a solution satisfying the constraint $\min_{ij} \lambda_i(\boldsymbol{\Omega}_j) = 1$ in (10); in this case, a correct solution can be obtained by setting

$$\lambda_i(\boldsymbol{\Omega}_j) \longleftarrow \frac{\lambda_i(\boldsymbol{\Omega}_j)}{\min_{ij} \lambda_i(\boldsymbol{\Omega}_j)} \quad \text{and} \quad \eta^2 \longleftarrow \eta^2 \min_{ij} \lambda_i(\boldsymbol{\Omega}_j) \tag{13}$$

and thus a new solution is obtained that satisfy the complete set of constraints by giving the same value of the log-likelihood. Also in this case the monotonicity is guaranteed once the initial guess $\boldsymbol{\Sigma}_j^{(0)}$ satisfies the constraints. Finally, it should be noted that strategies *A* and *B* do not necessarily maximize the complete log-likelihood at each iteration.

4 Numerical Results and Concluding Remarks

In this section we present some numerical results in order to evaluate the performance of approaches *C* and *D*, corresponding to the constraints (2) and (10). Further experiments have been carried out and they are presented in Ingrassia and Rocci (2006). We considered samples of size $N = 200$ generated from a mixture of three bi-variate normal distributions ($k = 3$ and $q = 2$) having the parameters $\boldsymbol{\psi} = (\boldsymbol{\alpha}, \boldsymbol{\mu}_1, \boldsymbol{\mu}_2, \boldsymbol{\mu}_3, \boldsymbol{\Sigma}_1, \boldsymbol{\Sigma}_2, \boldsymbol{\Sigma}_3)$ where

$$\boldsymbol{\alpha} = (0.3, 0.4, 0.3)' \qquad \boldsymbol{\mu}_1 = (0,3)' \qquad \boldsymbol{\mu}_2 = (1,5)' \qquad \boldsymbol{\mu}_3 = (-3,8)'$$
$$\boldsymbol{\Sigma}_1 = \begin{pmatrix} 1 & 0 \\ 0 & 2 \end{pmatrix} \qquad \boldsymbol{\Sigma}_2 = \begin{pmatrix} 1 & -1 \\ -1 & 2 \end{pmatrix} \qquad \boldsymbol{\Sigma}_3 = \begin{pmatrix} 2 & 1 \\ 1 & 2 \end{pmatrix},$$

and the eigenvalues of the covariance matrices $\boldsymbol{\Sigma}_1, \boldsymbol{\Sigma}_2$ and $\boldsymbol{\Sigma}_3$ are respectively: $\boldsymbol{\lambda}_1 = (1,2)'$, $\boldsymbol{\lambda}_2 = (0.382, 2.618)'$ and $\boldsymbol{\lambda}_3 = (1,3)'$. We generated 200 samples from this mixture. For each sample, we run the constrained EM algorithms following approaches *C* and *D*, starting from a set of points randomly chosen; the computation stopped when the difference between two consecutive log-likelihood values resulted less than 0.0001. The results, displayed in table 1, have been summarized by considering the mean of the sum of squared differences between the true parameters and the corresponding estimates (SSE), and the average number of iterations (# iter). On the same

Table 1. Mean values of the sum of squared errors of estimation and mean values for the number of iterations for constrained EM algorithms C and D

Strategy C			Strategy D		
a, b	SSE	# iter	c	SSE	# iter
0.38, 3	1.50	75	0.38/3	2.08	111
0.20, 4	2.02	79	0.20/4	4.42	129
0.10, 8	3.67	95	0.10/8	6.05	103
0.01,80	6.11	99	0.01/80	6.11	100
1.14, 9	3.89	161			

datasets we run also the unconstrained algorithm obtaining an average number of iterations equal to 99 and an average SSE equal to 6.11.

We can not draw general conclusions from this limited simulation study. However, we note that the two constrained algorithms outperforms always the unconstrained one. They are equivalent only when $(a, b) = (0.01, 80)$ because in this case the constraints are not active. We note also that the performances of the algorithms decreases when the constraints are less tight. The same consideration applies if we compare the two constrained algorithms when $a/b = c$: approach C is always better than D because it is the most constrained. Only in the last setting, D is better than C (note that $1.14/9 = 0.38/3$). This is due to the fact that the constraints are wrong for C. In conclusion, it seems that the choice between the two approaches depends on the information available on the eigenvalues: use C if the location is known, use D if only the ratio between the highest and the lowest is known.

References

HATHAWAY, R. J. (1985): A constrained formulation of maximum-likelihood estimation for normal mixture distributions. *The Annals of Statistics, 13, 795–800.*

INGRASSIA, S. (2004): A likelihood-based constrained algorithm for multivariate normal mixture models. *Statistical Methods & Applications, 13, 151–166.*

INGRASSIA, S. and ROCCI, R. (2006): On the geometry of the EM algorithm for multinormal mixture models and constrained monotone versions. Technical report.

McLACHLAN, G. J. and KRISHNAN, T. (1997): *The EM Algorithm and Extensions.* John Wiley & Sons, New York.

McLACHLAN, G. J. and PEEL, D. (2000): *Finite Mixture Models.* John Wiley & Sons, New York.

THEOBALD, C. M. (1975): An inequality with applications to multivariate analysis. *Biometrika, 62, 461–466.*

Visualizing Dependence of Bootstrap Confidence Intervals for Methods Yielding Spatial Configurations

Henk A.L. Kiers[1] and Patrick J.F. Groenen[2]

[1] University of Groningen Grote Kruisstraat 2/1, 9712 TS Groningen, The Netherlands, email: h.a.l.kiers@rug.nl
[2] Erasmus University Rotterdam P.O.Box 1738, 3000 DR Rotterdam, The Netherlands, email: groenen@few.eur.nl

Abstract. Several techniques (like MDS and PCA) exist for summarizing data by means of a graphical configuration of points in a low-dimensional space. Usually, such analyses are applied to data for a sample drawn from a population. To assess how accurate the sample based plot is as a representation for the population, confidence intervals or ellipsoids can be constructed around each plotted point, using the bootstrap procedure. However, such a procedure ignores the dependence of variation of different points across bootstrap samples. To display how the variations of different points depend on each other, we propose to visualize bootstrap configurations in a bootstrap movie.

1 Assessing Sampling Inaccuracy for Spatial Configurations

Several techniques exist for summarizing data by means of a graphical configuration of points in a low-dimensional space. The most common examples are multidimensional scaling (MDS) and principal component analysis (PCA). Usually, such analyses are applied to data for a sample drawn from a population, while the researcher hopes that the configuration (at least roughly) holds for the full population. For instance, when PCA on data from a sample is used to display the relations between the variables in a plot, it is hoped that this plot also holds (roughly) for the whole population. Likewise, it is hoped that the stimulus configuration obtained by an MDS on a sample proximity data set for a set of stimuli, gives a good approximation for the optimal stimulus configuration representing the population of proximity data. To assess how accurate the sample based plot is as a representation for the population, one can try to set up confidence intervals for each parameter defining the configuration. Thus, one would get confidence intervals for all coordinates separately. Clearly, however, it is not very insightful to give confidence intervals for each x- and y-coordinate of each point representing a variable (in PCA) or stimulus (in MDS) separately, as if these are independent. Rather, one would like to see the region that with a certain probability will cover the "population position" of the variable or stimulus at hand. An example of such

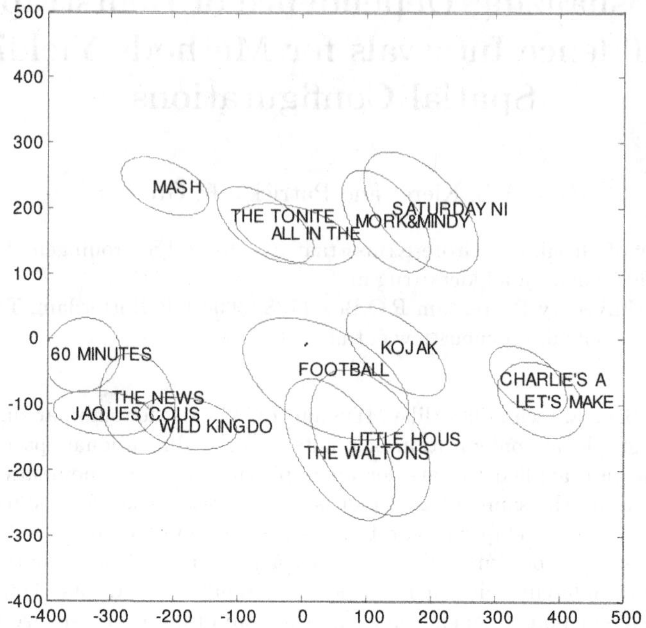

Fig. 1. Example of an MDS plot with 95% confidence regions.

a plot for an MDS analysis on the average (over 30 judges) proximity of 15 American TV shows (kindly made available by R.A. Harshman; see Lundy, Harshman and Kruskal, 1989) is given in Fig. 1. In this Figure, the labels indicate the positions of the stimuli in the MDS solution, and the ellipses around these positions indicate 95% confidence regions. These confidence regions are obtained by first applying the bootstrap procedure (Efron and Tibshirani, 1993) to the 30 similarity matrices we had available here (where each bootstrap sample consisted of the data for a random sample (with replacement) of 30 proximity matrices from the 30 proximity matrices given by the 30 different judges, while next on each bootstrap sample, an MDS was applied to the average across the proximity matrices); next all bootstrap solutions were optimally rotated towards the full sample solution by orthogonal Procrustes rotation (Cliff, 1966), and finally ellipses were computed that optimally cover 95% of the bootstrap solutions for each point at hand (using the procedure by Meulman and Heiser, 1983). Similar bootstrap procedures including Procrustes rotations have been proposed by Markus (1994) for correspondence analysis, Groenen, Commandeur and Meulman (1998) and Commandeur, Groenen and Meulman (1999) for special kinds of MDS called distance-based PCA, Linting, Groenen and Meulman (submitted for publication) for CatPCA, Kiers (2004) for three-way component analysis, Timmerman, Kiers and Smilde (2005) for principal components analysis.

2 Dependence of Points Across Bootstraps

Above it has been illustrated that the bootstrap procedure can be used to obtain confidence regions (or hypervolumes) for individual stimuli (or, in case of PCA, variables). By dealing with positions rather than coordinates, we implicitly have taken into account the mutual dependence among the different coordinates across the various bootstrap solutions. Indeed, this can be seen in Figure 1, where, for instance, the ellipse for Saturday Night live (left under) is fairly elongated. This indicates that, if one coordinate increases across bootstrap solutions, then, generally the other will decrease, etc., thus indicating a dependence among the x- and y-coordinates of this TV-show. This is important, because, if we would only use the confidence intervals, we would conclude that both the x- and the y-coordinate vary roughly from -300 to -100, and hence that the point could be anywhere in the block set out by these x and y values, whereas in reality it is restricted to a much smaller area. By using confidence regions, some dependence across bootstrap solutions can be taken into account, but the majority of dependences, namely those among different stimuli, has, to our knowledge always, been ignored. This is remarkable, since the configurations define mutual positions, and hence points in configurations are intrinsically dependent. It is important to study such dependences, because it gives information on the joint position of points (e.g., that it is likely that certain points in the population are close to each other (even if confidence ellipses do not overlap), or that they are relatively far from each other (even if the associated confidence ellipses touch). The reason for the ignorance of dependence may be that the dependence of locations of points across bootstraps alluded to in the present paper, is very hard to grasp. To indicate what is meant here, first consider how the dependence of two coordinates is visualized by means of a two-dimensional ellipse; would we have had a point in a three-dimensional space, the dependence could be visualized by a confidence ellipsoid, which is still imaginable. However, in the simple example of 15 TV-shows displayed in 2 dimensions being studied here, the mutual dependence of these points can only be fully visualized by means of confidence hyperellipsoids in 30-dimensional space.

3 Bootstrap Movies

Rather than trying to cover all the information in the many dimensions in one go, it is here proposed to use human memory and perception on the one hand, and human interaction on the other hand to allow for good insight into the dependence of points across bootstraps solutions. We propose here that the researcher inspects all bootstrap solutions, projected quickly after each other. However, simply projecting the configurations after each other does not give much insight, because it requires a memory of the previous positions of the points. Therefore, a procedure has been built in that shows

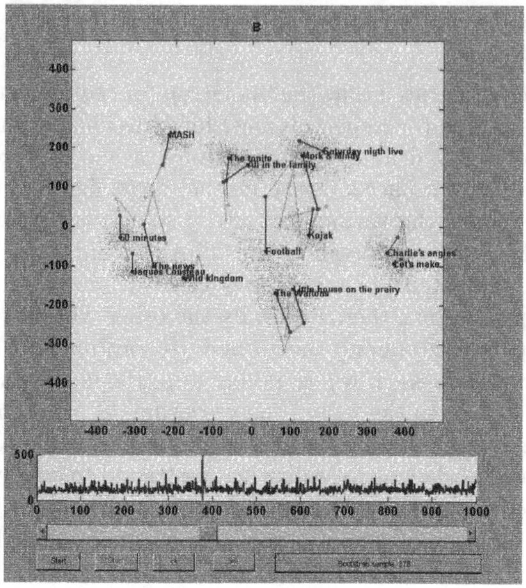

Fig. 2. Example of a Shot in a Bootstrap Movie for the MDS data.

the trail of the points linking the last two positions of the points. To smooth the transition between two points, we subdivide the line connecting the bootstrap points into four parts so that the trail moves in four steps to the next bootstrap point. In this way, at each moment we get an impression of the previous positions, and thereby of the direction of the latest movements. Thus, the subsequent plots actually give the impression of a movie. For this reason, we call this way of presenting the bootstrap results a bootstrap movie. Fig. 2 displays a "still" of the bootstrap movie for the same MDS data as were displayed in Fig. 1. The sample points are indicated with labels, and here also with black dots. Furthermore, in light grey dots the complete set of bootstrap solutions is given (as vague, overlapping nebula), and in heavy grey dots the current configuration in the movie is given. Furthermore, for each TV-show, a line connecting the current bootstrap point and the sample point is given, as well as a light line connecting it with the previous bootstrap points, thus giving the trailing effect, emphasizing the illusion of movement of points, as mentioned above. This is an important tool in order to recognize dependencies. In the graph right below the main window, we see the distances between subsequent configurations (as assessed by the square root of the sum of squared differences between the corresponding coordinates of all points in these configurations), so when the line is rather flat, movement is smooth, and when it is low, there is little movement. In this way, not only a bootstrap movie, but also insight into its smoothness is given. The movement displayed in a bootstrap movie, crucially depends on the order in

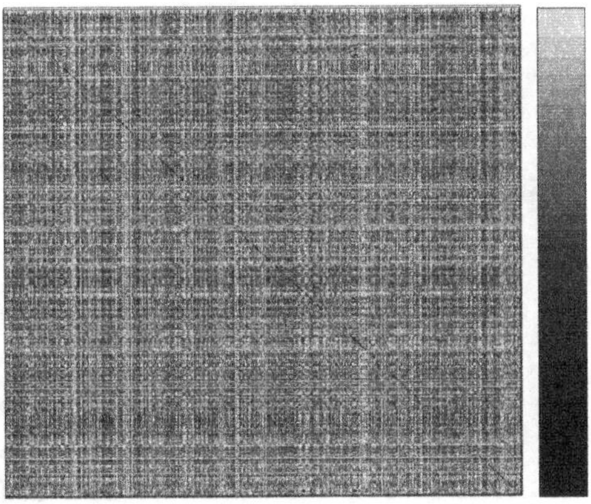

Fig. 3. Distances between randomly ordered bootstrap configurations.

which the bootstrap configurations are displayed. To give an impression of movement, we should see differences between the configurations, but, as in ordinary movements, these differences should not be too big, because too big differences would only give an impression of a chaotic sequence of images, and no smooth movie. Therefore, we search orders of configurations in which subsequent configurations differ relatively little. In fact, then, we search a path to "walk" through all bootstrap configurations in such a way that differences between subsequent configurations are minimal. Since configurations themselves are high-dimensional objects (in our example, each configuration is a 30-dimensional vector), it is no trivial task to find such a path. In the next section, we describe two suggestions for choosing such orders.

4 Ordering Bootstrap Configurations for Bootstrap Movies

To order bootstrap configurations in such a way that subsequent configurations are relatively similar, we here discuss two such suggestions. To assess the success of these procedures, we inspected the distances between all pairs of configurations. We started with a random order. For this randomly ordered set of configurations, we obtained the distances as displayed in Fig. 3 (where configurations are labelled by sequence number, and mutual distances by grey levels, according to the indicated grey level scale scale, with all distances larger than 300 displayed as white). It can be seen in Fig. 3 that there is no pattern in these distances, and that distances between subsequent configurations are not really smaller than those between arbitrary other configurations. The mean distance between subsequent configurations was 235.7, while the mean distance between configurations was 234.5. A first attempt to order the configurations in such a way that subsequent configurations are

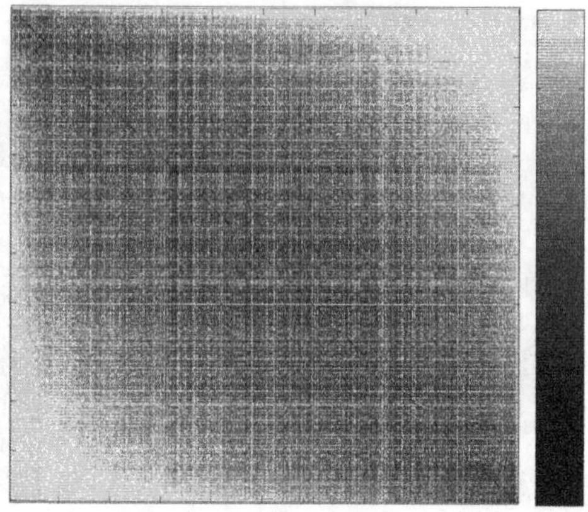

Fig. 4. Distances between bootstrap configurations ordered along first principal component of covariances between configurations.

relatively close to each other, consisted of first doing a Principal Component Analysis on the matrix of covariances between all 1000 configurations (where a covariance is computed simply as the covariance between all coordinates in one configuration with all corresponding coordinates in the other configuration), and next ordering the configurations on the basis of the loadings of the 1000 configurations on the first principal component. The rationale for this was that, in this way, we efficiently get a one dimensional MDS (seriation) representation of the 1000 configurations, which will roughly order configurations on the basis of their mutual distances. It turned out that now the mean distance between subsequent configurations had decreased considerably, to 190.6. In Fig. 4 we can clearly see that distance between configurations becomes larger as one moves away from the diagonal. A second attempt to order the configurations consisted of performing a hierarchical cluster analysis on all 1000 vectorized configurations, using Ward's linkage criterion, using its implementation in Matlab. With the dendrogram option in Matlab, we obtained a dendrogram of the 1000 configurations, and an ordering resulting from this dendrogram. The latter ordering is used here as an alternative ordering of the configurations. It turned out that for this ordering the mean distance between subsequent configurations was much smaller than those based on the first principal component: 129.9. Fig. 5 displays all mutual distances between configurations, and clearly reveals the clustering structure by dark blocks along the main diagonal, indicating groups with mutually small distances. It can also be said that, between such blocks, relatively big distances are found. For the movie this implies that within blocks movements are smooth, and between blocks sudden jumps will occur.

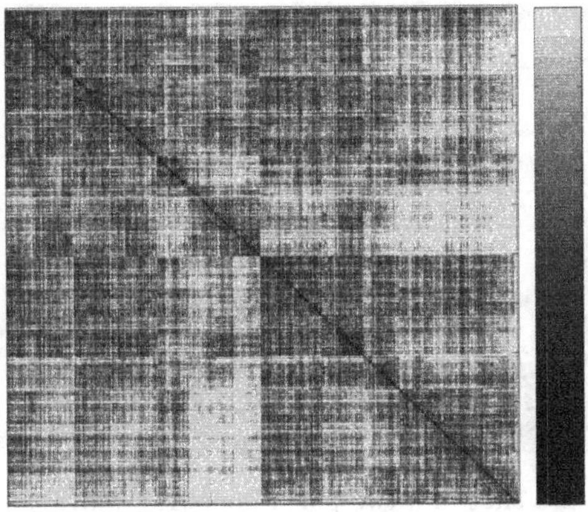

Fig. 5. Distances between bootstrap configurations ordered according to results of (Ward) hierarchical clustering of bootstrap configurations.

5 Two-Panel Movies

In the present paper, we have described the construction of a bootstrap movie for MDS solutions. A similar procedure can be used for other analysis results given in the form of spatial representations, like results of a Principal Components Analysis or Correspondence Analysis. Also three-way generalizations of PCA yield solutions that can be displayed by means of spatial representations, and procedures for computing bootstrap confidence intervals and regions have been proposed (see Kiers, 2004). When three-way principal component analysis is applied to, for instance, subjects, by variables, by situations data, then it will be interesting to study a display of the variables, and a display of the situations, and confidence regions for all points in those displays. However, these confidence regions do not only correlate with other confidence regions in the same configuration, but also with those for the other configuration. Therefore, to adequately display all joint variation, two movies should be played simultaneously, in a two-panel movie. Such a movie has indeed been developed, a shot of which is displayed in Fig. 6.

6 Discussion

The movies described here are a tool for visualizing joint variation across bootstrap configurations. They help the analyst in getting an impression of possible strong dependencies among certain points. Often, the analyst may want to focus specifically on such dependencies. Then, this could be done by means of a simple three-dimensional display featuring only such points, or by means of a dependence measure, which, then, however, still needs to be developed for such purposes. The bootstrap movies developed here could help

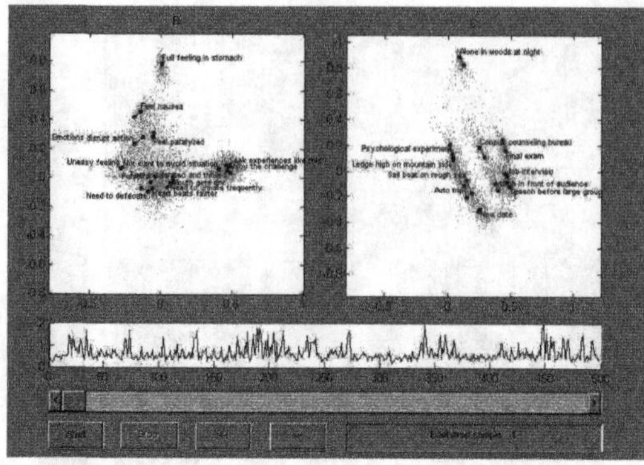

Fig. 6. Example of a shot in a two-panel bootstrap movie for three-way data.

in choosing and developing such measures. The Matlab programs for these movies are available from the authors upon request.

References

CLIFF, N. (1966): Orthogonal rotation to congruence. *Psychometrika, 31, 33–42.*

COMMANDEUR, J.J.F., GROENEN, P.J.F. and MEULMAN, J.J. (1999): A distance-based variety of nonlinear multivariate data analysis, including weights for object and variables. *Psychometrika, 64, 169–186.*

EFRON, B. and TIBSHIRANI, R.J. (1993): *An introduction to the bootstrap.* New York, Chapman and Hall.

GROENEN, P.J.F., COMMANDEUR, J.J.F. and MEULMAN, J.J. (1998): Distance analysis of large data sets of categorical variables using object weights. *British Journal of Mathematical and Statistical Psychology, 51, 217–232.*

KIERS, H.A.L. (2004): Bootstrap confidence intervals for three-way methods. *Journal of Chemometrics, 18, 22–36.*

LINTING, M., GROENEN, P.J.F. and MEULMAN, J.J. (2005): Stability of nonlinear principal components analysis by CatPCA: An empirical study. Submitted for publication.

LUNDY, M.E.,HARSHMAN, R.A. and KRUSKAL, J.B. (1989): A two-stage procedure incorporating good features of both trilinear and quadrilinear models. In R. Coppi and S. Bolasco (Eds.) *Multiway Data Analysis*, Amsterdam, Elsevier.

MARKUS, M.T. (1994): *Bootstrap confidence regions in nonlinear multivariate analysis.* Leiden, DSWO Press.

MEULMAN, J.J. and HEISER, W.J. (1983): The display of bootstrap solutions in multidimensional scaling. Unpublished technical report, University of Leiden, The Netherlands.

TIMMERMAN, M.E., KIERS, H.A.L., and SMILDE, A.K. (2005): Bootstrap confidence interval in principal component analysis. Manuscript submitted for publication.

Automatic Discount Selection for Exponential Family State-Space Models

Andrea Pastore

Dipartimento di Statistica
Università Ca' Foscari Venezia
andrea.pastore@unive.it

Abstract. In a previous paper (Pastore, 2004), a method for selecting the discount parameter in a gaussian state-space model was introduced. The method is based on a sequential optimization of a Bayes factor and is intended for on-line modelling purposes. In this paper, these results are extended to state-space models where the distribution of the observable variable belongs to the exponential family.

1 Introduction

In the dynamic linear model, the system equation represents a suitable law of evolution of the parameters vector, which is introduced to allow the model to be more flexible. In many parametric cases (essentially the class of generalized dynamic linear models), Bayesian inference and prediction is possible in close form (exactly or approximately) only conditioning on the variance of the system error (for instance: West and Harrison, 1997; Vidoni, 1999). This *hyperparameter*, unlike the system matrix and the measurement error variance, can be estimated only using some *non-sequential* methods, like maximum likelihood or MCMC techniques (Harvey and Fernandes, 1989; Lopez et al., 1999, among others). The *discounting* approach (Harrison and Stevens, 1976; Ameen and Harrison, 1985; West and Harrison, 1997) allows to solve the problem essentially by setting the variance of the system error such that the prior distribution of the state vector at time t has same location but greater variance than the posterior distribution at time $t-1$. The discount factor is the parameter which represents the ratio between these two variances. The discount factor is fixed, considering some suggested reference values. Bayesian inference and prediction is then possible conditioning on the discount factor.

In a previous paper (Pastore, 2004), for gaussian dynamic models, a method was proposed in order to select the discount factor. The method is based on a sequential optimization of a Bayes factor (see Kass and Raferty, 1995, for a review) and allows sequential filtering and prediction for the model. In this paper, the results are extended to models where the distribution of the observations belongs to the exponental family. Examples for Poisson and Binomial distributions, and an application to a binomial time series (Smith, 1980) are considered.

2 Automatic Discount Selection

The *state-space* form of a dynamic model can be *represented* through the following system of conditional distributions:

$$p(y_t|\eta_t) \tag{1}$$

$$p(\eta_t|\eta_{t-1}) \tag{2}$$

where, for $t = 1, 2, \ldots$, y_t is an univariate observable random variable (r.v.) and η_t is the state parameter, with $\eta_t \in H_t \subseteq \mathrm{R}^k$. These two conditional probability distributions can depend also by other parameters. Denote with y_t^0 the observed value of y_t and set: $Y_t^0 = (Y_{t-1}^0, y_t^0)$ with $y_0^0 = \emptyset$. Inference for this model is made obtaining sequentially at each time t the posterior distribution of η_t, given Y_t^0, $p(\eta_t|Y_t^0)$ combining the posterior distribution $p(\eta_{t-1}|Y_{t-1}^0)$ of η_{t-1} at time $t-1$, and the observation y_t^0. Two steps are required. First, the prior distribution of η_t is obtained using the conditional distribution (2):

$$p(\eta_t|Y_{t-1}^0) = \int_{H_{t-1}} p(\eta_t|\eta_{t-1}) \cdot p(\eta_{t-1}|Y_{t-1}^0)\, d\eta_{t-1}. \tag{3}$$

Then, given y_t^0, the posterior distribution $p(\eta_t|Y_t^0)$ is obtained following the Bayes theorem, where the (one-step-ahead) predictive distribution $p(y_t|Y_{t-1}^0)$ is implicitly considered.

If the prior distribution $p(\eta_t|Y_{t-1}^0)$ is conjugate with $p(y_t|\eta_t)$ and if this property holds also for the posterior distribution $p(\eta_{t-1}|Y_{t-1}^0)$, then the inference is essentially a set of updating equations for the parameters of the involved distributions.

Usually, the conditional distribution $p(\eta_t|\eta_{t-1})$ is defined such that the expectation of the prior is a known function of the expectation of the posterior, that is:

$$E\left[\eta_t|Y_{t-1}^0\right] = h_E\left(E\left[\eta_{t-1}|Y_{t-1}^0\right]\right), \tag{4}$$

where $h_E(\cdot)$ is a known non-random function, and

$$var\left[\eta_t|Y_{t-1}^0\right] = h_v\left(var\left[\eta_{t-1}|Y_{t-1}^0\right]\right) \cdot \frac{1}{\delta_t} \tag{5}$$

where δ_t is the *discount factor* (Harrison and Stevens, 1976; Ameen and Harrison, 1985; West and Harrison, 1997), with $0 < \delta_t \leq 1$, and $h_v(\cdot)$ is a known non-random function such that, if M is a variance matrix, $h_v(M)$ is also a variance matrix. Usually it is assumed that δ_t is known and is a constant with respect to t. In this case, for some model, it can be proved that the limit of $var(\eta_t|Y_t^0)$, for $t \to +\infty$, depends on the discount factor.

The proposed solution allows to obtain $p(\eta_t|Y_{t-1}^0)$ from $p(\eta_{t-1}|Y_{t-1}^0)$ and is based on the Bayes Factor which compares models *with* and *without* discount.

Denote with $p_{\delta_t}(\eta_t|\eta_{t-1})$ the system conditional distribution which depends on δ_t in the way described by conditions (4) and (5). Moreover, denote with $p_{\delta_t}(\eta_t|Y_{t-1}^0)$ the prior distribution which is obtained by means of equation (3) with the conditional distribution $p_{\delta_t}(\eta_t|\eta_{t-1})$ and with $p_{\delta_t}(y_t|Y_{t-1}^0)$ the related predictive distribution. When $\delta_t = 1$, the system conditional distribution has not a random effect. In this case, the system conditional distribution will be denoted with $p_0(\eta_t|\eta_{t-1})$, the prior distribution with $p_0(\eta_t|Y_{t-1}^0)$, and the predictive distribution with $p_0(y_t|Y_{t-1}^0)$.

The automatic discount selection need the following steps to be performed:

a. obtain the predictive distribution $p_0(y_t|Y_{t-1}^0)$ and the one-step ahead prediction

b. given the observed value y_t^0 of y_t, define the Bayes Factor:

$$BF_t(\delta_t) = \frac{p_{\delta_t}(y_t^0|Y_{t-1}^0)}{p_0(y_t^0|Y_{t-1}^0)} \tag{6}$$

c. solve: $\hat{\delta}_t = \arg\max_{\delta_t} BF_t(\delta_t)$, with $0 < \hat{\delta}_t < 1$ and $BF_t(\hat{\delta}_t) > 1$

d. if $\hat{\delta}_t$ exists then set: $p(\eta_t|Y_{t-1}^0) = p_{\hat{\delta}_t}(\eta_t|Y_{t-1}^0)$; otherwise set: $p(\eta_t|Y_{t-1}^0) = p_0(\eta_t|Y_{t-1}^0)$

3 Bayes Factor for Exponential Family State Space Models

Consider a model where the conditional distribution of the observation can be represented through the distribution:

$$p(y_t|\eta_t, \phi) = \exp\{\phi [y_t\eta_t - a(\eta_t)]\} \cdot b(y_t, \phi) \tag{7}$$

where η_t is the natural parameters, ϕ is the scale parameter, $a(\cdot)$ and $b(\cdot, \cdot)$ are suitable functions. In this case (Gelman et al., 1995) the conjugate prior for η_t can be represented in the canonical form:

$$p(\eta_t|Y_{t-1}^0) = c(r_t, s_t) \cdot \exp\{r_t\eta_t - s_t a(\eta_t)\} \tag{8}$$

where, under some mild regularity conditions, r_t/s_t is the location parameter and $1/s_t$ is a scale parameter. Moreover, the predictive distribution is:

$$p(y_t|Y_{t-1}^0) = \frac{c(r_t, s_t)}{c(r_t + \phi y_t, s_t + \phi)} \cdot b(y_t, \phi) \tag{9}$$

If the posterior distribution for η_{t-1}, $p(\eta_{t-1}|Y_{t-1}^0)$ is also conjugate:

$$p(\eta_{t-1}|Y_{t-1}^0) = c(r_{t-1}, s_{t-1}) \cdot \exp\{r_{t-1}\eta_{t-1} - s_{t-1}a(\eta_{t-1})\} \tag{10}$$

a discounted prior distribution $p_{\delta_t}(\eta_t|Y^0_{t-1})$ can be obtained using the *power discount* (Smith, 1979), setting:

$$p_{\delta_t}(\eta_t|Y^0_{t-1}) \propto p(\eta_{t-1}|Y^0_{t-1})^{\delta_t}$$

where δ_t is the discount factor at time t. The power discounting of the posterior density (10) gives the conjugate prior distribution:

$$p_{\delta_t}(\eta_t|Y^0_{t-1}) = c(\delta_t r_{t-1}, \delta_t s_{t-1}) \cdot \exp\{\delta_t r_{t-1}\eta_t - \delta_t s_{t-1} a(\eta_t)\} \tag{11}$$

with the same value of the location parameter and a scale parameter discounted by δ_t.

The predictive distribution, obtained with the prior (11) will be:

$$p_{\delta_t}(y_t|Y^0_{t-1}) = \frac{c(\delta_t r_{t-1}, \delta_t s_{t-1})}{c(\delta_t r_{t-1} + \phi_t y_t, \delta_t s_{t-1} + \phi_t)} \cdot b(y_t, \phi_t) \tag{12}$$

and the Bayes factor (6), as a function of the discount factor δ_t will be:

$$BF(\delta_t) = \frac{c(\delta_t r_{t-1}, \delta_t s_{t-1})}{c(r_{t-1}, s_{t-1})} \cdot \frac{c(r_{t-1} + \phi_t y^0_t, s_{t-1} + \phi_t)}{c(\delta_t r_{t-1} + \phi_t y^0_t, \delta_t s_{t-1} + \phi_t)} \tag{13}$$

Then steps **c** and **d** of the procedure for the optimal discount selection can be performed from the Bayes Factor (13). The optimal discount $\hat{\delta}_t$ is then obtained via the maximization of the Bayes Factor (13) by means of standard numerical algorithms.

3.1 Example: Poisson Model

The Poisson distribution:

$$p(y_t|\theta_t) = (y_t!)^{-1}\theta_t^{y_t} \cdot e^{-\theta_t}$$

can be expressed as the distribution (7) with:

$$\phi = 1, \quad b(y_t, \phi) = (y_t!)^{-1}, \quad \eta_t = \log(\theta_t), \quad a(\eta_t) = e^{\eta_t}.$$

The conjugate posterior for θ_{t-1} is a Gamma distribution:

$$p(\theta_{t-1}|Y^0_{t-1}) = \frac{\beta_{t-1}^{\alpha_{t-1}}}{\Gamma(\alpha_{t-1})}\theta_{t-1}^{\alpha_{t-1}} \cdot e^{-\beta_{t-1}\theta_{t-1}},$$

with parameters α_{t-1}, β_{t-1}, that can be written in the canonical form (10) with: $r_{t-1} = \alpha_{t-1}$, $s_{t-1} = \beta_{t-1}$ and:

$$c(r_{t-1}, s_{t-1}) = \frac{s_{t-1}^{r_{t-1}}}{\Gamma(r_{t-1})}.$$

The Bayes factor (13) will be:

$$BF(\delta_t) = \frac{(\delta_t s_{t-1})^{\delta_t r_{t-1}}}{\Gamma(\delta_t r_{t-1})} \cdot \frac{\Gamma(r_{t-1})}{s_{t-1}^{r_{t-1}}} \cdot \frac{(s_{t-1}+1)^{r_{t-1}+y^0_t}}{\Gamma(r_{t-1}+y^0_t)} \cdot \frac{\Gamma(\delta_t r_{t-1}+y^0_t)}{(\delta_t s_{t-1}+1)^{\delta_t r_{t-1}+y^0_t}}$$

3.2 Example: Binomial Model

The Binomial distribution

$$p(y_t|\theta_t) = \binom{n_t}{y_t} \theta_t^{y_t} (1-\theta_t)^{n_t - y_t} \tag{14}$$

can be represented through the expression (7) with:

$$\phi = 1, \quad b(y_t, \phi) = \binom{n_t}{y_t}, \quad \eta_t = \log(\frac{\theta_t}{1-\theta_t}),$$

and $a(\theta_t) = -n_t \log(1-\theta_t) = n_t \log(1+\exp(\theta_t))$. The conjugate posterior for θ_{t-1} is a Beta distribution:

$$p(\theta_{t-1}|Y_{t-1}^0) = [\mathrm{B}(\alpha_{t-1}, \beta_{t-1})]^{-1} \cdot \theta_{t-1}^{\alpha_{t-1}-1} \cdot (1-\theta_{t-1})^{\beta_{t-1}-1}$$

where $\mathrm{B}(\alpha, \beta)$ denotes the Beta function. This distribution can be written in the canonical form (10) with $r_{t-1} = \alpha_{t-1}$, $s_{t-1} = (\alpha_{t-1} + \beta_{t-1})/n_{t-1}$ and: $c(r_{t-1}, s_{t-1}) = [\mathrm{B}(r_{t-1}, s_{t-1} \cdot n_{t-1} - r_{t-1})]^{-1}$. The Bayes factor (13), as a function of the parameters α_{t-1} e β_{t-1} is:

$$BF(\delta_t) = \frac{\mathrm{B}(\delta_t \alpha_{t-1} + y_t^0, \delta_t \beta_{t-1} + n_t - y_t^0)}{\mathrm{B}(\delta_t \alpha_{t-1}, \delta_t \beta_{t-1})} \cdot \frac{\mathrm{B}(\alpha_{t-1}, \beta_{t-1})}{\mathrm{B}(\alpha_{t-1} + y_t^0, \beta_{t-1} + n_t - y_t^0)}$$

4 An Application

A simple illustration is proposed on a dataset of a binomial time series from Smith (1980). Here the problem is to identify change points in the series as modelled by changes in η_t. The conclusion of Smith, obtained with a non-sequential procedure, is that evidence suggests two consecutive changes at $t = 5$ and $t = 6$. The results obtained with the automatic discount selection procedure is compared with ones obtained with the Bayesian model monitoring proposed by West (1986). This is a sequential procedure where a standard model (a model without discount) is compared with an alternative model (a model with a fixed discount) using the Bayes Factor. The Bayesian model monitoring requires to set a value for the discount factor and a threshold value for the Bayes Factor and requires also a manual intervention. For this reason, when dealing with high frequency data, the approach is ineffective. The results are summarized in Table 4, where $\hat{y}_{t|t-1}(\mathrm{a})$ and $\hat{y}_{t|t-1}(\mathrm{w})$ denote one-step-ahead predictions respectively by automatic discount and Bayesian model monitoring (West, 1986). The prior distribution of θ_t at $t = 1$ has been set accordingly to West (1986), in order to compare the results. Figure 1 displays the same results. In the first two panels the one-step-ahead predictions and the related errors are respectively compared. It is worth to note that prediction error of the automatic discount selection are less or equal than the ones of the Bayesian model monitoring for most of the observations. The third and the fourth panels of Figure 1 show also the values of the Bayes Factor and of the optimal discount factor obtained with the automatic discount selection procedure. Particularly, Notice how the change-points at $t = 5$ and $t = 6$ are correctly identified by the values of $BF(\hat{\delta}_t)$, Even if the main purpose of the automatic discount selecion is not the change point detection, but the system error definition via the discount parameter. The results of the automatic discount are quite good, considering that it is a full automatic procedure, while the Bayesian model monitoring requires manual intervention and depends on some control parameters.

t	1	2	3	4	5	6	7	8	9	10	11	12	13
n_t	21	36	44	30	52	45	48	57	48	22	20	21	20
y_t	12	26	31	24	28	34	38	46	41	19	17	17	16
$\hat{y}_{t\|t-1}$(a)	10.5	20.3	30.0	20.8	38.8	25.3	35.0	43.3	37.3	17.9	16.5	17.4	16.5
$\hat{y}_{t\|t-1}$(w)	10.5	20.2	29.0	21.0	36.9	29.7	32.6	39.9	34.6	16.3	14.8	15.7	15.0
$\hat{\delta}_t$	1.00	0.53	1.00	0.34	0.11	0.12	1.00	1.00	0.28	1.00	1.00	1.00	1.00
$BF(\hat{\delta}_t)$	1.00	1.06	1.00	1.04	3.50	2.51	1.00	1.00	1.06	1.00	1.00	1.00	1.00

Table 1. Results for a binomial time series (West, 1986)

5 A Simulation Experiment

We wanted to test the performance of the automatic discount selection with a more extensive simulation experiment. To this purpose, a Binomial time series model was choosen with a change point situation similar to the one of the previously considered data set. A binomial time series $y_t, t = 1, \ldots, T$ was then considered with probability function $p(y_t|n_t, \theta_t)$ defined as in equation (14) with $\theta_t = \lambda + \omega \cdot I(t \geq k)$ where $I(\cdot)$ denotes the indicator function and $k \in \{1, \ldots, T\}$. Here k denotes the time where the probability parameter of the binomial distribution change from a value λ to a value $\lambda + \omega$. For the experiment, series of length $T = 20$ were simulated, and the change point was set $k = 5$. The probability λ was set equal to 0.20, while four different values of ω were considered: $\omega = 0.10, 0.20, 0.30, 0.40$ For each value of ω, 5000 series were generated. The automatic discount has been applied with an uniform prior distribution for θ_1. For each value of ω, we considered the percentages for classes of values of the optimal Bayes factor $BF(\hat{\delta}_t)$ for $t = k$. The classes have been defined following the guidelines specified by Kass and Raferty (1995). These results are reported in Table 5. The delay of detection of the change point has also been considered, from $t = k$ (no delay) to $t = k + 3$. These results are reported in Table 5

The algorithm gave satisfactory results. The performance is better when the values of the perturbation ω are higher. In order to evaluate these results,

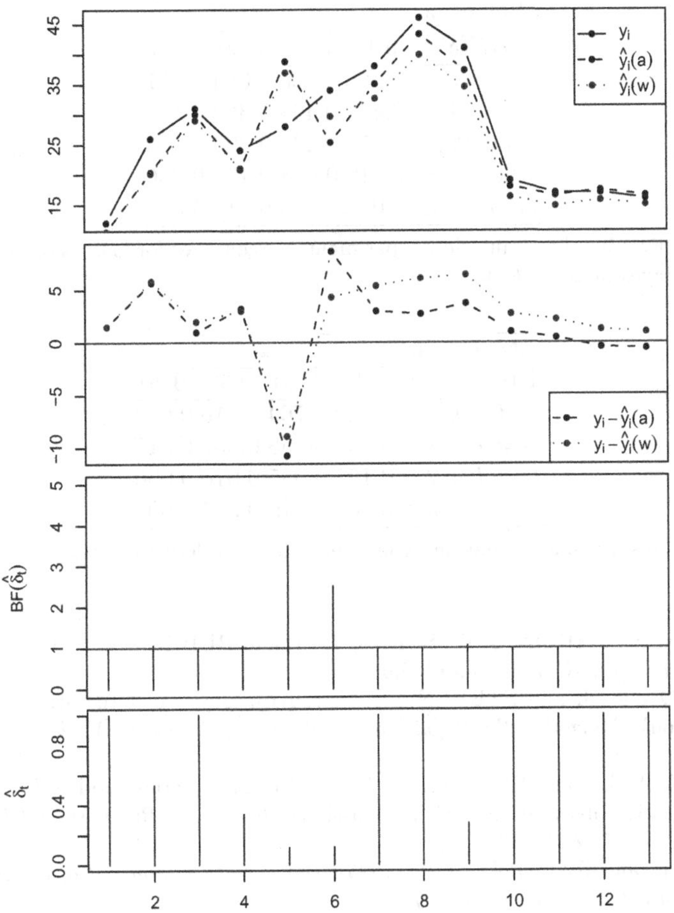

Fig. 1. One-step-ahead predictions errors for the automatic discount selection (dashed lines) and Bayesian model monitoring (dotted lines); prediction errors errors for the automatic discount selection (dashed lines) and Bayesian model monitoring (dotted lines); values of the Bayes Factor and values of the optimal discount for the automatic discount selection

we have to consider that if the perturbation is small, the algorithm tends to identificate more subsequent small discount factor rather than an high one.

References

AMEEN, J.R.M., HARRISON, P.J. (1985): Normal Discount Bayesian Models (with discussion). In: J.M Bernardo, M.H. DeGroot, D.V. Lindley and A.F.M. Smith (Eds.): *Bayesian Statistics 2*, Elsevier Science Publisher, 271–298.

Values of $BF(\hat{\delta})$	ω 0.10	0.20	0.30	0.40
(0, 1]	0.480	0.236	0.084	0.017
(0, 3]	0.421	0.431	0.322	0.128
(3, 20]	0.077	0.230	0.305	0.253
(20, 150]	0.019	0.078	0.173	0.266
(150, $+\infty$)	0.003	0.025	0.116	0.336

Table 2. Results of a simulation experiment: percentages for classes of value of the optimal Bayes factor $BF(\hat{\delta}_t)$

Change point detected at:	ω 0.10	0.20	0.30	0.40
$t = k$	0.520	0.764	0.916	0.983
$t = k + 1$	0.191	0.158	0.067	0.017
$t = k + 2$	0.125	0.055	0.010	0.000
$t = k + 3$	0.062	0.019	0.004	0.000

Table 3. Results of a simulation experiment: delay of detection of the change point

GELMAN, A., CARLIN, J. B., STERN, H. S. and RUBIN, D.B. (1995): *Bayesian data analysis.* Chapman and Hall.

HARRISON, P. J. and STEVENS, C. F. (1976): Bayesian forecasting (with discussion). *Journal of the Royal Statistical Society, Series B, Methodological, 38, 205–247.*

HARVEY, A.C. and FERNANDES, C. (1989): Time series models for count or qualitative observations (with discussion). *Journal of Business and Economic Statistics, 7, 407–422.*

KASS, R.E. and RAFERTY, A.E. (1995): Bayes factors. *Journal of the American Statistical Association, 90, 773–795.*

LOPES H.F, MOREIRA, A.R.B., SCHMIDT, A.M. (1999): Hyperparameter estimation in forecast models. *Computational Statistics and Data Analysis, 29, 387–410.*

PASTORE, A. (2004): Optimal discount selection in Gaussian state-space models via Bayes Factor. *Università Ca' Foscari - Dipartimento di Statistica, Redazioni provvisorie*

SMITH, A.F.M. (1980): Change-point problems: Approaches and applications. In: J. M. Bernardo, M. H. DeGroot, D. V. Lindley and A. F. M. Smith (Eds.): *Bayesian Statistics*, Valencia University Press, 83–98.

SMITH, J.Q. (1979): A generalization of the Bayesian steady forecasting model. *Journal of the Royal Statistical Society, Series B, Methodological, 41, 375–387.*

VIDONI, P. (1999): Exponential family state space models based on a conjugate latent process. *Journal of the Royal Statistical Society, Series B, Methodological, 61, 213–221.*

WEST, M. (1986): Bayesian model monitoring. *Journal of the Royal Statistical Society, Series B, Methodological, 48,70–78.*

WEST, M. and HARRISON, P.J. (1997): *Bayesian forecasting and dynamic models.* Springer-Verlag.

A Generalization of the Polychoric Correlation Coefficient

Annarita Roscino and Alessio Pollice

Dipartimento di Scienze Statistiche,
Università degli Studi di Bari, Italy
{aroscino;apollice}@dss.uniba.it

Abstract. The polychoric correlation coefficient is a measure of association between two ordinal variables. It is based on the assumption that two latent bivariate normally distributed random variables generate couples of ordinal scores. Categories of the two ordinal variables correspond to intervals of the corresponding continuous variables. Thus, measuring the association between ordinal variables means estimating the product moment correlation between the underlying normal variables (Olsonn, 1979). When the hypothesis of latent bivariate normality is empirically or theoretically implausible, other distributional assumptions can be made. In this paper a new and more flexible polychoric correlation coefficient is proposed assuming that the underlying variables are skew-normally distributed (Roscino, 2005). The skew normal (Azzalini and Dalla Valle, 1996) is a family of distributions which includes the normal distribution as a special case, but with an extra parameter to regulate the skewness. As for the original polychoric correlation coefficient, the new coefficient was estimated by the maximization of the log-likelihood function with respect to the thresholds of the continuous variables, the skewness and the correlation parameters. The new coefficient was then tested on samples from simulated populations differing in the number of ordinal categories and the distribution of the underlying variables. The results were compared with those of the original polychoric correlation coefficient.

1 Introduction

Data in the social and medical sciences are often based on ordinal measurements and represented in contingency tables. A first approach to the analysis of this kind of variables is to measure their association in order to know if some relationship exists and to quantify its strength. To achieve this purpose, it is possible either to estimate the concordance between the scores of each ordinal variable or to assume that those variables derive from the categorization of some continuous variables.
The first type of measure includes Kendall's τ, Somers' e, Goodman and Kruskal's γ and many others more (Agresti, 2004). They estimate the association between ordinal variables comparing the frequencies of each category without any distributional assumption. The polychoric correlation coefficient, instead, is based on the assumption that the ordinal variables derive from partitioning the range of some continuous normally distributed variables into

categories. Consequently, it does not compare two sets of scores, but rather estimates the correlation between two unobserved continuous variables underlying the two ordinal variables assuming a bivariate normal distribution with means zero and variances one.
Some studies have been carried out in order to compare the most important measures of association. Joreskog and Sorbon (1988) performed an experiment based on a bivariate normal distribution for the underlying variables and showed that, under this condition, the polychoric correlation coefficient is always closer to the real correlation than all measures evaluated in the same study. Moreover, the matrix of the polychoric correlation coefficients is largely used to replace the covariance matrix in order to estimate the parameters of structural equation models when the observed variables are ordinal. On the other hand, experience with empirical data (Aish and Joreskog, 1990) shows that the assumption of underlying bivariate normality seldom holds. It is also believed that this assumption is too strong for most ordinal variables used in the social sciences (Quiroga, 1991). Therefore, there is a need to find a shape of the underlying variables more plausibly compatible with the real data.
Many studies performed in order to analyse the distributions of underlying variables showed that asymmetric distributions are very frequent. Muthen (1984) proved that the distributions of underlying variables can be highly skewed, causing lack of convergence and/or negative standard errors when estimating structural equation model parameters. Moreover, Muthen and Kaplan (1985) noticed that the presence of asymmetric latent distributions can bias the results of chi square tests used to assess the goodness of fit of structural equation models. The former studies suggest a need to find a distribution that takes into account the potential asymmetry of the underlying variables.
In this paper a new polychoric correlation coefficient is proposed, based on the hypothesis that underlying variables have a bivariate skew normal distribution (Roscino, 2005). The bivariate skew normal distribution (Azzalini and Dalla Valle, 1996) belongs to a family of distributions which includes the normal distribution as a special case, but with two extra parameters to regulate the skewness. As for the polychoric correlation coefficient, maximum likelihood was used in order to estimate the new polychoric correlation coefficient under the assumption of underlying skew normally distributed variables. A simulation study was then carried out in order to compare the performance of the new coefficient with that of the original polychoric correlation coefficient.
In the first section of this paper, the generalised polychoric correlation coefficient is defined and estimated. In the second section the simulation study is presented and in the third section the results of the simulation study are shown and the efficacy of the new polychoric correlation coefficient is discussed.

2 An Extension of the Polychoric Correlation Coefficient

The polychoric correlation coefficient (Olsonn, 1979) is based on the assumption that underlying a pair of ordinal variables there is a couple of continuous latent variables which have a bivariate normal distribution. Ordinal variables X and Y, with I and J categories each, are thus assumed to be related to underlying continuous variables Z_1 and Z_2 by

$$\begin{cases} X = i \text{ if } a_{i-1} \leq Z_1 \leq a_i, \; i = 1, 2, \ldots, I \\ Y = j \text{ if } b_{i-1} \leq Z_2 \leq b_i, \; j = 1, 2, \ldots, J \end{cases} \tag{1}$$

where Z_1 and Z_2 have a bivariate normal distribution with correlation coefficient ρ and a_i and b_j are referred to as thresholds. Measuring the polychoric correlation means estimating the product moment correlation ρ between underlying normal variables. This correlation is estimated by the maximum likelihood method, assuming a multinomial distribution of the cell frequencies in the contingency table. If n_{ij} is the number of observations in cell (i, j), and K is a constant, the likelihood of the sample is given by

$$L = K \prod_{i=1}^{I} \prod_{j=1}^{J} P_{ij}^{n_{ij}}, \tag{2}$$

where

$$P_{ij} = \Phi_2(a_i, b_j) - \Phi_2(a_{i-1}, b_j) - \Phi_2(a_i, b_{j-1}) + \Phi_2(a_{i-1}, b_{j-1}) \tag{3}$$

and Φ_2 is the bivariate normal distribution function with unknown correlation coefficient ρ. The estimator of the polychoric correlation coefficient between variables X and Y corresponds to the value of ρ which maximizes equation 2, where the choice of the number of categories I and J has a crucial influence on the dimensionality of the likelihood function.

A problem with the polychoric correlation coefficient concerns the robustness of the method to departures from symmetric distributional assumptions. Quiroga (1991) carried out a Monte Carlo study in order to analyze the effects of the departure from the normal assumption on the estimation of the polychoric correlation coefficient. The author simulated samples from underlying distributions affected by asymmetry and showed that the polychoric correlation underestimates the association between ordinal variables, particularly when the sample size is large and the categories are few. Such results reveal that there could be an advantage in considering a latent distribution more compatible with real data. As discussed in the Introduction, the underlying variables are often asymmetric (Muthen, 1984), therefore the bivariate skew normal distribution was chosen.

A random variable $Z = (Z_1, Z_2)$ is said to be distributed according to a bivariate skew normal $SN(\alpha_1, \alpha_2, \omega)$ if its density function is given by

$$g(z_1, z_2) = 2\phi(z_1, z_2; \omega)\Phi(\alpha_1 z_1 + \alpha_2 z_2), \tag{4}$$

where $\phi(\cdot;\omega)$ is the bivariate normal distribution with null mean, unit variance and correlation ω and $\Phi(\cdot)$ is the univariate standard normal distribution function. Skewness α_1 and α_2 can vary in $(-\infty,\infty)$ and imply bivariate normality when they are both null. The correlation coefficient associated to the bivariate skew normal distribution is given by:

$$\rho_{SN} = \frac{\omega - 2\pi^{-1}\delta_1\delta_2}{\{(1-2\pi^{-1}\delta_1^2)(1-2\pi^{-1}\delta_2^2)\}^{1/2}}, \tag{5}$$

where δ_1 and δ_2 are linked to α_1 α_2 and ω by expressions:

$$\alpha_1 = \frac{\delta_1 - \delta_2\omega}{\{(1-\omega^2)(1-\omega^2-\delta_1^2-\delta_2^2+2\delta_1\delta_2\omega)\}^{1/2}}$$
$$\alpha_2 = \frac{\delta_2 - \delta_1\omega}{\{(1-\omega^2)(1-\omega^2-\delta_1^2-\delta_2^2+2\delta_1\delta_2\omega)\}^{1/2}}, \tag{6}$$

with δ_1 and δ_2 in $[-1,1]$.

Under the new assumption, the joint distribution of the underlying variables Z_1 and Z_2 is bivariate skew normal, as given in (4). Thus the product moment correlation of Z_1 and Z_2, ρ_{SN} estimates the polychoric correlation coefficient between X and Y.

As for the original polychoric correlation coefficient, the new coefficient is estimated by maximization of the log-likelihood function L (see 2) with respect to the thresholds, the skewness and the correlation parameters, where the new expression of the probability P_{ij} in the likelihood of the sample is equal to:

$$P_{ij} = P[X = i \wedge Y = j] = 2\int_{a_{i-1}}^{a_i}\int_{b_{j-1}}^{b_j} \phi(z_1, z_2, \Omega)\Phi(\alpha_1 z_1 + \alpha_2 z_2) dz_1 dz_2. \tag{7}$$

In order to work with standardized parameters, a different parametrization of the skew normal distribution was considered (Azzalini and Dalla Valle, 1996). The correlation parameter ω was replaced by ψ, where

$$\psi = (\omega - \delta_1\delta_2)[(1-\delta_1)(1-\delta_2)]^{-1/2} \tag{8}$$

and the skewness parameters $\alpha_1$and α_2 were replaced by δ_1 and δ_2 (see 6).

The function `sn.polychor` (Roscino, 2005) was written in R to perform the maximization of the log-likelihood function using a numerical optimization method, according to Nelder and Mead (1965). This method works reasonably well for non-differentiable functions as it uses only function values and does not require to evaluate the gradient of the log-likelihood.

The function `sn.polychor` first computes the maximum likelihood estimates of ψ, δ_1 and δ_2 and their standard errors. Then, after replacing ψ, δ_1 and δ_2 with their estimated values in 8, it calculates $\widehat{\omega}$ and $\widehat{\rho}_{SN}$ (see 5).

The function `sn.polychor` is available on request by emailing the first author.

3 The Simulation Study

The new polychoric correlation coefficient $\widehat{\rho}_{SN}$ was calculated for 360 samples from simulated bivariate populations differing in the number of ordinal categories, correlation and skewness parameters of the underlying distributions, as shown in Table 1. One sample was considered for each combination of the parameters ψ, δ_1, δ_2, I, J and n. The sampling distribution of $\widehat{\rho}_{SN}$ and $\widehat{\rho}$ was analysed only for the combination ($\psi = 0.5, \delta_1 = 0.7, \delta_2 = -0.7, I = 3, J = 3, n = 400$) where 100 samples were extracted and means and standard errors of $\widehat{\rho}_{SN}$, $\widehat{\rho}$, $\widehat{\delta_1}$, $\widehat{\delta_2}$ were computed. The analysis of the sampling distributions associated with the remaining combinations of parameters is currently being undertaken and the results will be presented in the near future.
The `R` library `MASS` was used to produce samples from bivariate normal distributions, while a new function called `sn.simul` (Roscino, 2005) was implemented in order to generate samples from bivariate skew normal distributions. The generated samples of the underlying variables (Z_1, Z_2) were grouped according to intervals and each interval was associated with a category of the corresponding ordinal variable. The values of ψ, δ_1 and δ_2 were chosen to

ψ	0.3	0.5	0.8			
n	250	400	600	800		
(I, J)	(2,2)	(2,3)	(3,3)	(3,5)	(5,5)	
(δ_1, δ_2)	(0,0)	(0,0.7)	(0.7,0.7)	(0.7,-0.7)	(0.4,0.4)	(0.4,-0.4)

Table 1. Parameters of the simulated distributions

include as many different shapes of the distributions of underlying variables as possible. In particular, when δ_1 and δ_2 are equal to zero, the underlying variables have bivariate normal distribution with correlation coefficient equal to ψ. For all the other cases, the simulated distributions are bivariate skew normals and the associated values of the polychoric correlation coefficient can be found in Table 2.

The `R` functions `sn.polychor` and `polychor` (Johnson, 2004) were used to compute $\widehat{\rho}_{SN}$ and $\widehat{\rho}$ respectively. While the output of `polychor` consists of the estimators of ρ_{SN}, a_i, b_j (for $i = 1, ..., I$ and $j = 1, ..., J$) with their standard errors, the function `sn.polychor` estimates the additional parameters ψ, δ_1 and δ_2 and their standard errors, together with ρ_{SN}, a_i, b_j (for $i = 1, ..., I$ and $j = 1, ..., J$) and their standard errors.
The values of $\widehat{\rho}_{SN}$ and $\widehat{\rho}$ were compared with the true value of the polychoric correlation coefficient for each of the simulated samples. The performance of both estimators with respect to the value of the polychoric correlation coefficient in the underlying population (as the absolute value of the difference)was

(δ_1, δ_2)	ψ 0.3	0.5	0.8
(0, 0)	0.3	0.5	0.8
(0, 0.7)	0.2582	0.4304	0.6887
(0.7, 0.7)	0.4811	0.6293	0.8517
(0.7, -0.7)	-0.0364	0.1118	0.3341
(0.4, 0.4)	0.3453	0.5323	0.8129
(0.4, -0.4)	0.2158	0.4028	0.6834

Table 2. Values of ρ_{SN}

evaluated only for the combination ($\psi = 0.5, \delta_1 = 0.7, \delta_2 = -0.7, I = 3, J = 3, n = 400$).

4 Some Results

In this section some results of the simulations are summarized. It is clear that a complete evaluation of the performance of the estimator would need a more extensive simulation study which is currently being undertaken.
The simulations involved one sample for each combination of parameters and showed that $\widehat{\rho}_{SN}$ is always closer to the real correlation than $\hat{\rho}$ when:

1. The number of categories of ordinal variables is small, ie. less than or equal to 3 (See Figure 1a, where the solid line represents the real polychoric correlation coefficient while the dashed and the dotted lines are respectively $\widehat{\rho}$ and $\widehat{\rho}_{SN}$) or
2. The sample size is large - 400 units or above, or
3. The skewness parameters are discordant, that is when they have opposite signs (See Figure 1b).

Furthermore, under these conditions the estimators of the skewness parameters are always very close to their values in the population.
These results are confirmed by the analysis of the sampling distribution of $\widehat{\rho}_{SN}$ for the combination of parameters ($\psi = 0.5, \delta_1 = 0.7, \delta_2 = -0.7, I = 3, J = 3, n = 400$). The mean and standard deviation of $\widehat{\rho}_{SN}$ were equal to 0.1173 and 0.0072 respectively while the mean and standard deviation of $\widehat{\rho}$ were 0.1003 and 0.0651. The mean of $\widehat{\rho}_{SN}$ is closer to ρ than the mean of $\widehat{\rho}$ (see Table 2) and the standard deviation is lower than the standard deviation of $\widehat{\rho}$ by a factor of almost ten.
On the other side, the polychoric correlation coefficient is closer to the real correlation when:

1. The sample size is small, or
2. The number of categories is large.

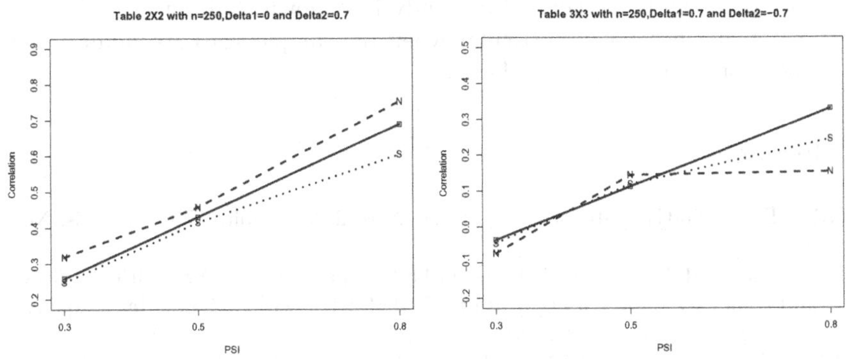

Fig. 1. Value of the estimated correlation coefficients for each value of ψ.

For $(I, J) = (5, 5)$, the results showed a high degree of variability and were therefore of limited use. This case is currently being studied to improve the quality of the results.
When the sample size is small, the poor results of the generalised polychoric correlation coefficient could be determined by an irregularity in the likelihood function of the bivariate skew normal distribution. Azzalini and Capitanio (1999) showed that for small sample sizes, the maximum likelihood estimators of the parameters of a skew normal distribution can overestimate their real values. This is due to the analytical expression of the likelihood function and cannot be modified using a different parametrisation.
The conclusions of the paper of Joreskog and Sorbon (1988) hold when the number of categories of the ordinal variables is large. The authors compared six measures of ordinal association and found that the polychoric correlation coefficient is more robust to departures from normality in the presence of ordinal variables with a large number of categories.

5 Conclusions

In this paper we propose a new polychoric correlation coefficient based on the assumption that the underlying continuous variables are skew normally distributed. By definition, ρ_{SN} is equal to ρ when the underlying variables are normally distributed, but it is more flexible than ρ as it takes into account the potential skewness of the underlying variables.
An `R` function was written in order to compute $\widehat{\rho}_{SN}$ and and 360 samples were generated with the aim of comparing $\widehat{\rho}_{SN}$ and $\widehat{\rho}$.
The examples presented in the simulation study indicates that $\widehat{\rho}_{SN}$ is more appropriate than $\widehat{\rho}$ when the sample size is large or the number of categories of ordinal variables is small or the skewness parameters have opposite signs.

On the other hand, the simulation study has shown that some further developments are needed in particular when the sample sizes are small or the number of ordinal categories is large.

References

AGRESTI, A. (2004): Categorical Data Analysis, 2nd ed. John Wiley & Sons. New York

AISH, A. M. and JORESKOG, K. G. (1990): A panel model for political efficacy and responsiveness: An application of LISREL 7 with weighted least squares. *Quality and Quantity, 24, 405–426.*

AZZALINI, A. and CAPITANIO, A. (1999): Statistical applications of the multivariate skew normal distribution. *Journal of the Royal Statistical Society, Series B, 61, 579–602.*

AZZALINI, A. and DALLA VALLE, A. (1996): A multivariate skew-normal distribution. *Biometrika, 83, 715–726.*

JOHNSON, T.R. (2004): *Polychor function for* R. http://www.webpages.uidaho.edu/ trjohns/polychor.R.

JORESKOG, K. G. and SORBON, D. (1988): A *program for multivariate data screening and data summarization.* A *processor for* LISREL. Scientific Software Inc. Mooresville, Indiana.

MUTHEN, B. (1984): A general structural equation model with dichotomous, ordered categorical and continuous latent variable indicators. *Psychometrika, 49, 115–132.*

MUTHEN, B. and KAPLAN, D. (1985): A comparison of some methodologies for the factor analysis of non-normal Likert variables. *British Journal of Mathematical Statistical Psychology, 38, 171–189.*

NELDER, J. A. and MEAD, R. (1965): A simplex method for function minimization. *Computer Journal, 7, 308–317.*

OLSONN, U. (1979): Maximum likelihood estimation of the polychoric correlation coefficient. *Psychometrika, 44, 443–460.*

QUIROGA, A. M. (1991): Studies of the Polychoric Correlation and other Correlation Measures for Ordinal Data. *Uppsala University, Department of Statistics. PhD Thesis.*

ROSCINO, A. (2005): Una generalizzazione del coefficiente di correlazione policorica. *Università degli Studi di Chieti - Pescara. Tesi di Dottorato di Ricerca.*

The Effects of MEP Distributed Random Effects on Variance Component Estimation in Multilevel Models

Nadia Solaro[1] and Pier Alda Ferrari[2]

[1] Dipartimento di Statistica
Università di Milano-Bicocca, Italy
nadia.solaro@unimib.it
[2] Dipartimento di Scienze Economiche, Aziendali e Statistiche
Università di Milano, Italy
pieralda.ferrari@unimi.it

Abstract. An in-depth investigation on maximum likelihood estimators for variance components is proposed, where the reference is a multilevel model with misspecifications on random effect distribution. The multivariate distributions here introduced for the random effects belong to the family of the Multivariate Exponential Power (MEP) distributions. Our primary interest is devoted to the variability of such estimators, since the MEPs have a noteworthy influence upon it.

1 Introduction

Multilevel models represent a comprehensive approach for the analysis of data organized in a nesting (or hierarchical) structure: they are defined by a set of multiple regression equations by which the variation present at different levels, namely within-groups and between-groups variability, is explicitly modelled (Goldstein (1995), Snijders and Bosker (1999)).

As usual, the maximum likelihood approach for estimating parameters (basically, fixed effects and variance components) relies on the assumption of normality of the error terms (level-1 errors and higher level errors called "random effects") included in the equations at each level. Empirical practice indicates however that the assumption of normally distributed random effects, even approximatively, is not always satisfied. For instance, this might happen when the dependent variable is not a directly measured variable, but rather a latent dimension extract from a set of qualitative variables through application of techniques such as multidimensional scaling or multiple correspondence analysis (Gori and Vittadini (1999)). The question arising, therefore, relates to what extent maximum likelihood estimation procedures may be considered robust against departures from normality.

To examine the performance of maximum likelihood estimators under misspecifications of random effects, in a recent study dealing with a two-level model (Solaro and Ferrari (2004)) we introduced a specific class of multivariate distributions. It consisted of the family of Multivariate Exponential

Power distributions (MEP) (Gómez et al. (1998)), which belongs to the class of symmetric and elliptical multivariate, Kotz type distributions (Fang et al. (1990)). Since it also includes the multivariate normal as a special case, a family of this kind reveals that it is fairly appealing, especially in studies relating to robustness.

To examine the consequences of MEP distributed random effects on the estimator performance, both REML and ML type estimates were obtained as if the "traditional" normal assumptions on both random effects and level-1 model residuals would be satisfied. The complexity of the problem urged us to undertake the study by applying Monte Carlo (MC) simulation procedures under specific experimental conditions, defined by different combinations of number of groups, group size, variance–covariance structure of random effects. Results achieved are described in detail in Solaro and Ferrari (2004).

Here it is worth briefly remarking that MEP distributed random effects appeared to have no significant influence on the bias of all estimators. The same may be said apropos of the variability of estimators for, respectively, fixed effects and level-1 variance. On the other hand, a significant MEP effect was observed on the variability of level-2 variance components.

Taking stock of these results, in this work we specifically focus on the level-2 variance component estimation problem. In particular, the role of MEP distributed random effects on the variability of such estimators will be deeper investigated. In fact, depending on the type of MEP, these estimators appear to be systematically more or less variable than those obtained under the normal distribution. Another point at issue involves the performance of model-based standard errors, namely standard errors estimated through the inverse of Fisher's information matrix. In fact, what they reveal is that they are very inaccurate estimates of the *true* variability of the estimators. It would seem that this situation cannot be solved by increasing sample size.

This paper is set out as follows: Section 2 contains brief remarks on the standard Gaussian multilevel model as well as on the main estimation methods. A multilevel model with MEP distributed random effects is further introduced. In Section 3 the method used for investigation is described. In Section 4 the main results are presented and briefly discussed. Finally, Section 5 is giving over to some concluding remarks.

Our work is part of the context of studies related to robustness issues in multilevel models. For instance, studies on robustness in presence of misspecifications of random effect distribution can be found in Verbeke and Lesaffre (1997) and Maas and Hox (2004).

2 Two Multilevel Model Definitions

Throughout this work a two-level hierarchical, balanced data structure is assumed, with J groups (level-2 units) and n units within group (level-1 units or elementary units) ($j = 1, \ldots, J$). The linear multilevel model may be defined through Laird and Ware's matrix formulation (Laird and Ware (1982)):

$$\mathbf{Y}_j = \mathbf{Z}_j\boldsymbol{\gamma} + \mathbf{X}_j\mathbf{U}_j + \boldsymbol{\varepsilon}_j, \tag{1}$$

being $\mathbf{Y}_j$ a r.v. (random vector) of quantitative Y_{ij} variables ($i = 1, \ldots, n$); $\mathbf{Z}_j$ and $\mathbf{X}_j$ the so called design matrices; $\boldsymbol{\gamma}$ a vector of p fixed effects; $\mathbf{U}_j$ a r.v. of q level-2 residuals, also said random effects; finally, $\boldsymbol{\varepsilon}_j$ a r.v. of level-1 residuals, ($j = 1, \ldots, J$).

Fixed effects along with variance components, that is the variances and the covariances of the random part of the model, are the unknown parameters to be estimated. The main methods are represented by the maximum likelihood approach (full -ML- and restricted maximum likelihood -REML-; see e.g. Pinheiro and Bates (2000)) or by suitable extensions of generalized least squares (IGLS and RIGLS; see Goldstein (1995)). In particular, REML and RIGLS methods are conceived as modifications of, respectively, ML and IGLS methods to allow for the correction of bias of variance component estimators. In all these methods the estimates cannot be obtained in a closed form, but rather through iterative numerical procedures. Whenever the maximum likelihood approach is involved, a so called Gaussian multilevel model is usually required. In this work, in addition to this we propose another definition which assumes MEP distributions on random effects.

1) Gaussian multilevel model. In standard theory the multilevel model is specified by introducing normality assumption on errors at each level. Formally: $\mathbf{U}_j \sim N_q(\mathbf{0}, \mathbf{T})$, $\boldsymbol{\varepsilon}_j \sim N_n(\mathbf{0}, \sigma^2\mathbf{I}_{(n)})$, $\mathbf{U}_j$ and $\boldsymbol{\varepsilon}_j$ independent ($j = 1, \ldots, J$), where the elements τ_{hm} of $\mathbf{T}$ ($h, m = 0, 1, \ldots, q-1$) and σ^2 are, respectively, level-2 and level-1 variance components.

Likewise the standard result in multiple linear regression, under normality of all error terms IGLS/RIGLS estimates coincide with ML/REML ones, so that IGLS/RIGLS methods may be viewed as the algorithm by which computing ML/REML estimates (Goldstein (1995)). Moreover, under the Gaussian multilevel model maximum likelihood estimators for, respectively, fixed effects and variance components are asymptotically uncorrelated and asymptotically normally distributed, with the variance-covariance matrix represented by the inverse of Fisher's information matrix (see e.g. Pinheiro and Bates (2000)).

2) Multilevel model with MEP distributed random effects. As previously mentioned, we assume, however, MEP distributed random effects $\mathbf{U}_j$ for all j, along with normally distributed residuals $\boldsymbol{\varepsilon}_j$. Formally, an n-dimensional r.v. $\boldsymbol{S}$ is MEP distributed with parameters $\boldsymbol{\mu}$, $\boldsymbol{\Sigma}$ and κ (shortly, $\boldsymbol{S} \sim \text{MEP}_n(\boldsymbol{\mu}, \boldsymbol{\Sigma}, \kappa)$), if its p.d.f. may be expressed in this form:

$$f(\mathbf{s}; \boldsymbol{\mu}, \boldsymbol{\Sigma}, \kappa) = \frac{n\Gamma(n/2)\exp\left\{-\frac{1}{2}\left[(\mathbf{s}-\boldsymbol{\mu})^T\boldsymbol{\Sigma}^{-1}(\mathbf{s}-\boldsymbol{\mu})\right]^{\frac{\kappa}{2}}\right\}}{\pi^{n/2}\Gamma\left(1+\frac{n}{\kappa}\right)2^{1+\frac{n}{\kappa}}|\boldsymbol{\Sigma}|^{1/2}}, \tag{2}$$

where $\boldsymbol{\mu} \in \mathbf{R}^n$, $\kappa > 0$ is called the non–normality parameter and $\boldsymbol{\Sigma}$ ("characteristic matrix") is positive-definite; mean vector is: $\text{E}(\boldsymbol{S}) = \boldsymbol{\mu}$ and variance-covariance matrix is: $\mathrm{V}(\boldsymbol{S}) = c(\kappa, n)\boldsymbol{\Sigma}$, being $c(\kappa, n) = 2^{2/\kappa}\Gamma((n+2)/\kappa)/(n\Gamma(n/\kappa))$. Depending on the value of κ, either a leptokurtic ($\kappa < 2$) or a platikurtic ($\kappa > 2$) distribution may be obtained; the multivariate normal ($\kappa = 2$) is also included.

Our assumption on $\mathbf{U}_j$ in model (1) is so specified: $\mathbf{U}_j \sim \mathrm{MEP}_q(\mathbf{0}, \boldsymbol{\Phi}, \kappa)$. In such a case: $\mathrm{V}(\mathbf{U}_j) = \mathbf{T} = \boldsymbol{\Phi} \cdot 2^{2/\kappa}\Gamma\left(\frac{q+2}{\kappa}\right)/(q\Gamma(\frac{q}{\kappa}))$. As pointed out elsewhere (Solaro and Ferrari (2004)), in this case the likelihood function is defined by a somewhat complex formulation, that suggests we should rule out any analytical treatment in exact form, but for $\kappa = 2$. This matter would require a more in-depth examination.

3 Method

In our simulation settings a specific definition of model (1) was considered, which includes one explanatory variable at both levels (i.e. level-1 X and level-2 W), $p = 4$ fixed effects and $q = 2$ random effects. Substantially, this is a random intercept and slope model, which also includes a cross-level interaction term XW. Datasets were artificially constructed under different experimental conditions, shown in Table 3 together with target values and the other quantities set up for the study.

Pseudo-observations $\mathbf{y}_j$ were obtained through random variate generation from MEP distributed random effects $\mathbf{U}_j$ (with $\mathrm{E}\left([U_{0j}, U_{1j}]^T\right) = \mathbf{0}$ and $\mathrm{V}\left([U_{0j}, U_{1j}]^T\right) = \mathbf{T}$, $\mathbf{T}$ fixed) and normal distributed residuals $\boldsymbol{\varepsilon}_j$, with σ^2 fixed, for all j. As for MEPs, both leptokurtic ($\kappa = 1; 1.5$) and platikurtic ($\kappa = 8; 14$) distributions were considered; the normal case ($\kappa = 2$) was also included in the study as the reference distribution.

The generated data were then employed in order to estimate the four fixed effects: γ_{00} (general intercept), γ_{10} (coeff. of X), γ_{01} (coeff. of W), γ_{11} (coeff. of XW), and the four variance components: $\tau_{hh} = \mathrm{Var}(U_{hj})$, $(h = 0, 1)$, $\tau_{01} = \mathrm{Cov}(U_{0j}, U_{1j})$ and σ^2. In order to examine if the maximum likelihood method is resistant against MEP-type misspecifications, we computed ML and REML estimates as if the Gaussian multilevel model were valid.

Subsequently, we dealt with the main features of the distribution of the "simulated" estimators, such as bias and variability. Particular attention was given to the level-2 variance component estimators and to their variability. Approximate confidence intervals were also calculated through application of asymptotic results, which involve the inverse of Fisher's information matrix, and ultimately the so called model-based standard errors. To assess the significance of the MEP effect, in addition to the other factors involved, upon the main estimator features, Friedman's and Page's nonparametric tests were performed (see e.g. Hollander and Wolfe (1973)). Let ϑ_g denote the effect of the g-th experimental condition or the g-th value for κ ($g = 1, \ldots, G$). Then by these two tests the null hypothesis: $\mathrm{H}_0 : \vartheta_1 = \ldots = \vartheta_G$ may be evaluated against the general alternative that not all the ϑ's are equal (Friedman's test)

Experimental conditions	
Number of groups (NG)	$J = 5; 10; 20; 50$
Group size (GS)	$n = 5; 10; 15; 20; 25; 30$
Non-normality parameter	$\kappa = 1; 1.5; 2; 8; 14$
Target values	
Fixed effects	$\gamma_{00} = 20;\ \gamma_{10} = 1;\ \gamma_{01} = -0.5;\ \gamma_{11} = 8$
Level-2 variance components	$\tau_{00} = 2;\ \tau_{11} = 10;\ \tau_{01} = 0.8944;\ \rho_{01} = 0.2$*
Level-1 variance	$\sigma^2 = 10$
Explanatory variables	
Level-2 variable	J values from: $W \sim N(0.45, 9)$
Level-1 variable	Jn values from: $X \sim N(3, 16)$
Number of simulation runs	$K = 1,000$

**Note.* ρ_{01} is the correlation coefficient of U_{0j} and U_{1j}, for all j.

Table 1. Experimental conditions and other quantities fixed for the study

or against an ordered alternative (Page's test). Specifically, in order to evaluate the MEP effect upon the estimator variability the ordered alternative of Page's test is: $\mathrm{H}_1 : \vartheta_{\kappa=14} \leq \vartheta_{\kappa=8} \leq \vartheta_{\kappa=2} \leq \vartheta_{\kappa=1.5} \leq \vartheta_{\kappa=1}$, where at least one of those inequalities is strict. Similarly, in order to test the group size effect and the group number effect the ordered alternatives are, respectively: $\mathrm{H}_1 : \vartheta_{n=30} \leq \ldots \leq \vartheta_{n=5}$ and $\mathrm{H}_1 : \vartheta_{J=50} \leq \ldots \leq \vartheta_{J=5}$.

With comparative purposes in mind we examined whether the discrepancy between the model-based standard errors of, respectively, the REML and the ML estimators, which is well known for $\kappa = 2$, might be observed also for $\kappa \neq 2$. Therefore, we compared the estimated densities of these standard errors under each experimental condition and for each κ. Density estimation was performed through the kernel approach and by using the normal kernel function (Bowman and Azzalini (1997)). The procedure applied in order to compare the density estimates was the permutation test described in Bowman and Azzalini (1997, pp.107–112), by means of which the hypothesis: $\mathrm{H}_0 : \phi(x) = \psi(x)$ for all x is tested against: $\mathrm{H}_1 : \phi(x) \neq \psi(x)$ for some x, where x represents a model-based standard error. According to the authors' remarks, when yielding the density estimates for each estimation method a common smoothing parameter was used, which is given by the geometric mean of the correspondent normal optimal smoothing parameters (Bowman and Azzalini (1997), pp.31–37).

All point and interval estimates were determined by means of the S-plus library `nlme3` by Pinheiro and Bates (2000). As regards variance component estimation, this library deals with an unconstrained parameterization which is termed "matrix logarithm", the main purposes being to avoid the "negative-variance" problem and to speed up numerical routines (Pinheiro and Bates

(1996)). However, when building approximate confidence intervals on level-2 variance components another type of parameterization, so called "natural", is used. It simply uses $\log(\sqrt{\tau_{hh}})$ for variances τ_{hh} and the generalized logit of the correlations of random effects, that is $\log[(1+\rho_{hm})/(1-\rho_{hm})]$. In this manner it is possible to transform the confidence intervals computed for the natural parameters back to the original ones (Pinheiro and Bates (2000)). Finally, both estimating and comparing the standard error densities of the REML and ML estimators were carried out through the S-plus library sm, version 2, by Bowman and Azzalini (1997).

4 Results and Discussion

Here the main syntheses concerning simulation results on level-2 variance component estimators are presented. Both REML and ML estimates will be considered.

The first remark concerns the variability of the empirical distributions of the estimators, which is synthetized by the Monte Carlo (MC) standard error. What may be perceived is that both REML and ML estimators for level-2 variance components tend to have greater MC standard errors under leptokurtic MEPs than those obtained under normality, while they have smaller MC standard errors under platikurtic MEPs than under normality. This feature finds strong empirical evidence from Page's test results. The first part of Table 4 displays the results achieved for the MEP effect in the REML case, which are very similar to those ones obtained in the ML case (here omitted).

Effect	*param.*	*Number of groups*			
		$J=5$	$J=10$	$J=20$	$J=50$
MEP	τ_{00}	322**	322**	326**	326**
– net of GS –	τ_{01}	324**	323**	329**	327**
(Page)	τ_{11}	326**	329**	329**	328**
GS	τ_{00}	19.29*	24.54**	24.54**	23.63**
– net of MEP –	τ_{01}	17.91*	23.17**	23.63**	24.08**
(Fried.)	τ_{11}	5.11	11.97+	6.26	8.54

Note. **: $p < 0.001$; *: $p < 0.01$; +: $p < 0.05$.

Table 2. Page's (L statistic) and Friedman's (large sample approx. χ^2) test results on MC standard errors of level-2 variance component REML estimators.

The role of the group size seems to be less clear than MEPs: in the case of the REML and ML estimators for τ_{00} and τ_{01} the group size effect turns out to be significant in both Friedman's and Page's tests. Specifically, the latter indicates that, for each J and net of the type of MEP, the estimator variability tends to diminish when group size increases. On the other hand, as

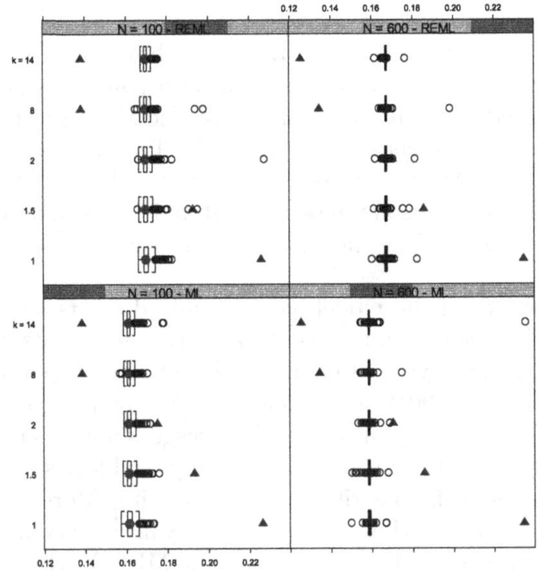

Fig. 1. REML and ML estimators for variance τ_{11} (natural param.): boxplots of the standard errors, for $J = 20$, $\kappa = 1, 1.5, 2, 8, 14$, $n = 5$ (1st column of panels) and $n = 30$ (2nd column). MC standard error is depicted as ▲.

for the REML estimator for τ_{11} there are no significant results, but for $J = 10$ ($p < 0.05$) (second part of Table 4). This is also a common feature in Page's test results. In addition, as far as the ML estimator for τ_{11} is concerned, the role of the group size turns out to be unclear. In Friedman's test its effect results significant for each J considered, but in Page's test the ordered alternative finds no empirical support.

As concerns the group number effect, Page's test, performed for each κ and net of the group size, provides strong empirical evidence in this direction upon all REML and ML estimators. In other words, under the same conditions, MC standard errors tend to become smaller as the number of groups increases. These results are here omitted.

A second remark concerns the model-based standard errors of the "natural" variance component estimators. In this case, a systematic under- or over-estimation of the *true* estimator variability is observed depending on the type of MEP. Specifically, in the case of leptokurtic MEPs there is a tendency to systematically underestimate the true variability, while platikurtic MEPs are involved with its systematic overestimation.

For a clearer understanding, it is worth taking account of a specific case. As regards the REML estimator for the "natural" variance $\log(\sqrt{\tau_{11}})$, the two panels in the first row of Figure 1 display boxplots of the model-based standard errors, for $J = 20$ and, respectively, $n = 5$ and $n = 30$, when random effects are distributed according to leptokurtic MEPs ($\kappa = 1, 1.5$), to platikurtic MEPs ($\kappa = 8, 14$) and to the normal ($\kappa = 2$). To allow for simple comparisons, MC standard errors, computed with respect to the natural parameterization, are also depicted. Both the above mentioned features appear evident. As it may be clearly seen, in both panels MC standard errors are underestimated in the presence of leptokurtic MEPs; viceversa, they are overestimated in the presence of platikurtic MEPs. And this occurs for each sample size (J, n) here considered. In a similar fashion, the two panels on the bottom of Figure 1 display model-based and MC standard errors of the ML estimator for $\log(\sqrt{\tau_{11}})$. Once again, the features just described may be clearly noticed.

The performance of the model-based standard errors has an immediate effect on the inference regarding level-2 variance components, which is traditionally based on the asymptotical normality of the estimator distributions (Pinheiro and Bates (2000)). In particular, as for confidence intervals, under our experimental conditions the nominal coverage is not generally guaranteed when random effects are MEP distributed: empirical levels tend to approach the nominal one for both κ nearing 2 and J enlarging. More precisely, empirical coverages smaller than those under normality are observed in the presence of underestimated standard errors (leptokurtic MEPs), while greater empirical coverages are obtained in the presence of overestimated standard errors (platikurtic MEPs).

MEP	Number of groups							
	$J = 10$				$J = 50$			
κ	$n = 15$	$n = 20$	$n = 25$	$n = 30$	$n = 15$	$n = 20$	$n = 25$	$n = 30$
1	0.859	0.836	0.856	0.865	0.861	0.851	0.849	0.865
1.5	0.909	0.890	0.911	0.913	0.919	0.918	0.919	0.905
2	0.928	0.918	0.932	0.923	0.953	0.944	0.940	0.942
8	0.961	0.964	0.964	0.969	0.991	0.979	0.992	0.982
14	0.973	0.976	0.964	0.964	0.990	0.983	0.991	0.985
κ	P: $L = 327.5$, appr. $p = 0$				P: $L = 328$, appr. $p = 0$			
n	F: $\chi^2_5 = 3.56$, appr. $p = 0.615$				F: $\chi^2_5 = 6.31$, appr. $p = 0.227$			

**Note.* The two tests are performed on the complete two-way layout tables, which include also coverages for the group sizes $n = 5$ and $n = 10$.

Table 3. REML estimator for $\sqrt{\tau_{11}}$: coverages (nominal confidence level: 0.95). Page's (P) and Friedman's (F) significance tests*.

For instance, Table 4 displays coverages relating to confidence intervals on $\sqrt{\tau_{11}}$ obtained by the 0.95-nominal level and w.r.t. 10 and 50 groups. A few facts may be focused on: 1) for $\kappa = 2$ and for each n the empirical coverages tend to approach the nominal level with J passing from 10 to 50; 2) for $\kappa > 2$ and for each n the observed over-coverages clearly tend to widen further when J enlarges; 3) for J and n fixed, the coverages become wider as κ increases. Further, what may be noted is that the group size effect is actually negligible.

The significance of the effects of the factors involved upon the coverages has been as well assessed. In particular, from Page's test the MEP effect results significant for each J, whereas from Friedman's test group size effect is not significant at all (Table 4). On the other hand, the group number effect finds support in the 0.05-level significant results of Page's test yielded for each group size. In other terms, net of MEPs and with the group size kept fixed, the coverages tend to widen as J increases.

As an example of the comparison between REML and ML, Figure 2 displays the empirical coverages achieved for the REML and ML estimator of $\sqrt{\tau_{11}}$ for each κ and n, with $J = 10$ and $J = 50$. The MEP effect is clearly noticeable in both cases.

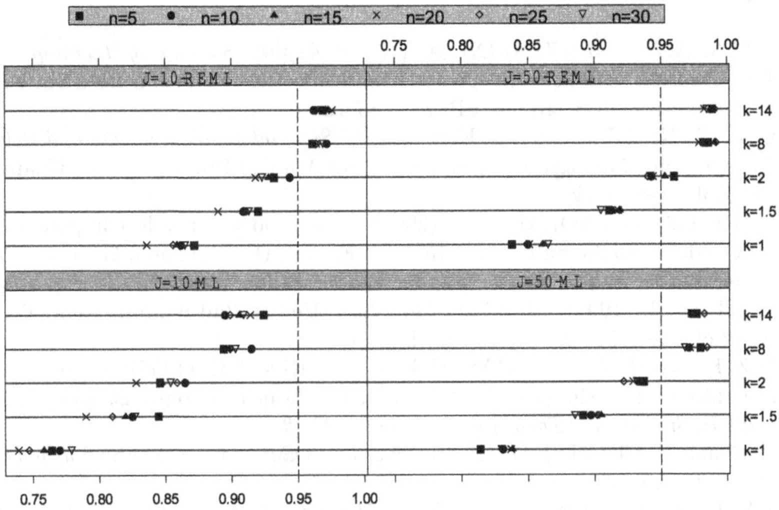

Fig. 2. REML and ML estimators for $\sqrt{\tau_{11}}$: dotplots of the coverages for each κ and n, with $J = 10$ (1st column of panels) and $J = 50$ (2nd column). The dash-line represents the 0.95-nominal level.

Finally, the application of the permutation test reveals that the discrepancy between the model-based standard errors of, respectively, the REML and the ML estimators holds also for $\kappa > 2$ and $\kappa < 2$. In the pairwise comparison the estimated densities for the REML and the ML estimators differ significantly. In addition, from the analysis of the p-values it would seem that such a difference does not strengthen or weaken as κ varies.

5 Concluding Remarks

Our work has shown that MEP-type misspecifications on random effect distribution have a noticeable influence upon the variability of level-2 variance component (REML and ML estimators). Further, model-based standard errors are far from being accurate estimates of the true variability of these estimators. Hence corrections of the estimated standard errors are absolutely necessary. In fact, neither by increasing the number of groups nor by increasing the group size would seem to resolve this situation.

In literature a type of correction often recommended is represented by the so-called "sandwich estimator", in the Huber-White sense (see, e.g., Goldstein (1995), Verbeke and Lesaffre (1997)). Sandwich-type corrections are implemented in most of the softwares for multilevel model analysis (e.g., the software MLwiN by the Multilevel Project (Rasbash et al. (2000)), but they are not yet in S-Plus. This is one of the objectives of our future endeavours in this area.

References

BOWMAN, A.W. and AZZALINI, A. (1997): *Applied Smoothing Techniques for Data Analysis. The kernel approach with S-Plus illustrations.* Oxford Statistical Science Series, 18, Clarendon Press, Oxford.

FANG, K.-T., KOTZ, S. and NG, K.W. (1990): *Symmetric Multivariate and Related Distributions.* Monograph on Statistics and Applied Probability, 36, Chapman and Hall, New York.

FERRARI, P.A. and SOLARO, N. (2002): Una proposta per le componenti erratiche del modello multilivello. In: B.V. Frosini, U. Magagnoli, G. Boari: *Studi in onore di Angelo Zanella.* V&P Università, Milano, 273–291

GOLDSTEIN, H. (1995): *Multilevel Statistical Models.* 2nd edition, Kendall's Library of Statistics, 3, Arnold Ed., London

GÓMEZ, E., GÓMEZ-VILLEGAS, M.A. and MARÍN, J.M. (1998): A multivariate generalization of the power exponential family of distributions. *Communications in Statistics – Theory and Methods, 27, 3, 589–600*

GORI, E. and VITTADINI, G. (1999): *Qualità e valutazione nei servizi di pubblica utilità.* ETAS, Milano.

HOLLANDER, M. and WOLFE, D.A. (1973): *Nonparametric Statistical Methods.* Wiley, New York

LAIRD, N.M. and WARE, J.H. (1982): Random-effects models for longitudinal data. *Biometrics, 38, 963–974*

MAAS, C.J.M. and HOX, J. (2004): The influence of violations of assumptions on multilevel parameter estimates and their standard errors. *Computational Statistics and Data Analysis, 46, 427–440*

PINHEIRO, C.J. and BATES, D.M. (1996): Unconstrained parametrizations for variance-covariance matrix. *Statistics and Computing, 6, 289–296*

PINHEIRO, C.J. and BATES, D.M. (2000): *Mixed-Effects Models in S and S-Plus.* Statistics and Computing, Springer-Verlag, New York

RASBASH, J., BROWNE, W., GOLDSTEIN, H. and YANG, M. (2000): *A user's guide to MLwiN.* 2nd edition, London, Institute of Education. `http://www.mlwin.com`

SNIJDERS, T. and BOSKER R. (1999): *Multilevel Analysis – An Introduction to Basic and Advanced Multilevel Modelling.* Sage Pubblications, London

SOLARO, N. and FERRARI, P.A. (2004): Robustness of parameter estimation procedures in multilevel models when random effects are MEP distributed. *Submitted.* Preprint: `http://service.bepress.com/unimi/`

VERBEKE, G. and LESAFFRE, E. (1997): The effect of misspecifying the random-effects distribution in linear mixed models for longitudinal data. *Computational Statistics and Data Analysis, 23, 541–556*

Calibration Confidence Regions Using Empirical Likelihood

Diego Zappa

Institute of Statistics
Cattolica University of Milan, Italy
diego.zappa@unicatt.it

Abstract. The literature on multivariate calibration shows an increasing interest in non-parametric or semiparametric methods. Using Empirical Likelihood (EL), we present a semiparametric approach to find multivariate calibration confidence regions and we show how a unique optimum calibration point may be found weighting the EL profile function. In addition, a freeware VBA for Excel© program has been implemented to solve the many relevant computational problems. An example taken from a process of a semiconductor industry is presented.

1 Introduction

Statistical calibration deals with the inference on the unknown values of explanatory variables given a vector of response variables. It results particularly useful when it is much simpler (or much more economically convenient) to measure (or to fix) the responses of an experiment and then to find the levels of the explanatory variables that may have produced them. Examples may be found in chemistry, biometrics, where the interest is on the causes of the experiments, or in engineering where targets must be suitably chosen to let the process work.
In Section 2 we briefly describe the multivariate calibration problem, stressing some of the difficulties in the construction of confidence regions in a parametric context, and some recent non parametric proposals. In Section 3, using a semi-parametric approach, a new methodology for calibration is proposed, based on empirical likelihood. Then in Section 4 an application of the proposal to a semiconductor process is presented.

2 Multivariate Calibration: Object, Problems, Recent Proposals

A brief introduction to multivariate calibration is deemed necessary for the remainder of the paper. Many references can be found e.g. in Brown (1993), Naes et al. (2002) and Zappa, Salini (2005). According to most of the literature on this topic, we consider two steps.

1) *The calibration step.* An experiment is run. n observations on q response variables $\boldsymbol{Y}_1 = \{Y_1, Y_2, ..., Y_q\}$ and p explanatory variables $\boldsymbol{X} =$

$\{X_1, X_2, ..., X_p\}$, are collected in the matrices $\mathbf{Y}_1$ and $\mathbf{X}$. A vector $\mathbf{g}(\cdot)$ $(1 \times q)$ of transfer functions is used to link the two sets of variables. $\mathbf{g}(\cdot)$ is usually supposed to be composed of linear models. Let $\boldsymbol{E}_1$ be a $(1 \times q)$ vector of random variables (r.v.). This vector will represent measurement errors and we will suppose, for the sake of simplicity, that it is additive with respect to $\mathbf{g}(\cdot)$. Let $\boldsymbol{\Theta}$ be a $(p \times q)$ matrix of unknown parameters. The calibration model is defined as:

$$\boldsymbol{Y}_1 = \mathbf{g}(\boldsymbol{\Theta}, \boldsymbol{X}) + \boldsymbol{E}_1 \tag{1}$$

2) *The prediction step.* Analogously to the previous step, suppose that a $\mathbf{Y}_2$ $(m \times q)$ set of response data is available, where m represents the replications of the experiment in unknown homogeneous conditions. Then the prediction model is

$$\mathbf{Y}_2 = \mathbf{g}(\widehat{\boldsymbol{\Theta}}, \mathbf{1} \cdot \xi) + \mathbf{E}_2 \tag{2}$$

where $\widehat{\boldsymbol{\Theta}}$ has been estimated at step 1, $\mathbf{E}_2$ is a $(m \times q)$ matrix of measurement errors, $\mathbf{1}$ is a $(m \times 1)$ vector of ones and we are interested in the unknown $(1 \times p)$ levels ξ of the $\boldsymbol{X}$ variables. Independently from the properties of $\mathbf{g}(\cdot)$, some problems may occur when $q > p$. Using maximum likelihood based on multinormal distributional assumptions, it can be shown that the confidence region for ξ depends on a quantity usually called R, which is a measure of the unreliability of the $\mathbf{Y}_2$ sample set to calibrate ξ. The main problem, among others, is that the confidence region has an anomalous behavior with respect to R: it increases as R decreases and decreases as R increases. When $q < p$ the problem may have no solution (see Brown (1993) for details), mainly because the space spanned by ξ is greater than the one spanned by $\boldsymbol{Y}_1$. Thus no uniqueness in the solution arises. Different additional problems, mostly related to the maximum likelihood procedure, are due to the nature of $\boldsymbol{X}$: if it is supposed to be fixed we will talk of controlled calibration; if it is random, calibration will be said to be uncontrolled.

Zappa and Salini (2005) present a semiparametric solution to some of the above problems using a data depth procedure. This proposal requires no distributional assumptions, nor the choice of a linear transfer function, as well as the application of no multivariate technique in order to reduce the complexity of the problem. Furthermore, a non-empty, non-infinity confidence region is found for any combination of q and p. This proposal has two major properties:1) all the information included in the set of variables is used, and 2) a preliminary (non-parametric) test is run to verify if the hypothesis of linear relationship between the set of variables may be accepted. The counterpart of this approach (like most of the statistics that use a depth function) is the relevant computational effort needed. At present no sufficiently powerful (fast and reliable) software has been prepared and most of the available algorithms (as the one specifically implemented by the authors) have been programmed for research reasons. Additionally, it may be used only for uncontrolled calibration experiments and it does not apply when the dataset is small.

All these considerations will support the following new proposal.

3 Nonparametric Calibration Regions Based on Empirical Likelihood

Many are the references on Empirical Likelihood (EL). The most relevant is certainly Owen (2001), where historical background, main results, applications in very different fields and computational details may be found.

Briefly on EL: let F_Z be the distribution function of a continuous r.v. Z and Z_n be an iid sample from Z. Supposing that w_i is the weight that F_Z places on observation z_i, define the EL for the whole sample as $\prod w_i$; the maximum is achieved for $w_i = 1/n \ \forall i$ and the empirical likelihood ratio will be $R(F_Z) = \prod nw_i$. Starting from these preliminaries, we may define the profile likelihood ratio of a vector of parameters θ and its corresponding EL confidence region. Supposing that $\boldsymbol{T}(F_Z) = \sum w_i \mathbf{h}(Z_n)$ is an estimator of θ, where $\mathbf{h}(\cdot)$ is a vector of functions that may itself be dependent on F_Z, then, on the domain of θ, the empirical profile likelihood will be defined as

$$R(\theta) = \max\left(\prod nw_i | \boldsymbol{T}(F_Z) = \theta, w_i \geq 0, \sum w_i = 1\right) \tag{3}$$

and the corresponding EL confidence region will be

$$\{\theta | R(\theta) \geq r_0\} = \left(\boldsymbol{T}(F_Z) | \prod nw_i \geq r_0, w_i \geq 0, \sum w_i = 1\right) \tag{4}$$

where r_0 must be chosen such that the coverage probability equals a chosen $(1 - \alpha)$ probability level. Theorem 3.6 in Owen (2001) shows that, in distribution,

$$-2 \log R(\xi) \rightarrow \chi_p^2 \qquad \text{as} \qquad n \rightarrow \infty \tag{5}$$

which is a result analogous to the Wilk's theorem. Then r_0 will be, at least asymptotically, the $(1-\alpha)$ percentile of χ_p^2. More generally the constrains in (3) may be substituted by what Owen calls "estimating equations" that is a vector of functions, $\mathbf{m}(\cdot)$, such that $\sum\sum w_i m_j(\mathbf{z}_i, \theta_j) = 0$.

To solve (3) some routines are actually freely available e.g. in *R*, *Gauss* softwares and very probably in other program language. One of the contribution of this work is a routine freely available from the Author to solve (3) in VBA-Excel©. Details of the algorithm are in Zappa (2005).

The application of EL to calibration problems may present some difficulties and at least two problems.

The difficulties concern the implementation of the constrains. In the prediction step stacking equations (1) and (2) and assuming to keep the estimate of $\boldsymbol{\Theta}$ fixed, the unknown ξ must be found subject to the constraints needed to find the solution to the unknowns in (1). As it is customary, supposing that a standard least squares method is used in (1) and supposing that $\mathbf{g}(\cdot)$ is at least twice differentiable, we should include in (3) at least the following set of $(qp \times q)$ constraints (the estimating equations) corresponding to the matrix of gradients equated to zero:

$$\left(\frac{\partial \mathbf{g}\left(\boldsymbol{\Theta}, \begin{bmatrix} \mathbf{1}\xi \\ \mathbf{X} \end{bmatrix}\right)}{\partial \boldsymbol{\Theta}} \right)^T_{\boldsymbol{\Theta}=\hat{\boldsymbol{\Theta}}} \left\{ \left(\begin{bmatrix} \mathbf{Y}_2 \\ \mathbf{Y}_1 \end{bmatrix} - \mathbf{g}\left(\hat{\boldsymbol{\Theta}}, \begin{bmatrix} \mathbf{1}\xi \\ \mathbf{X} \end{bmatrix}\right) \right) \odot \underset{(m+n)\times q}{\mathbf{W}} \right\} = \underset{(q\cdot p)\times q}{\mathbf{0}} \tag{6}$$

where the first argument is a $(q \cdot p)\times(m+n)$ matrix, $\mathbf{1}$ is a $(m \times 1)$ vector of ones and $\odot$ is the Hadamard product. More generally if equations (1) and (2) were a set of generalized linear models with random structure belonging to the exponential family, then with some little additional complications we should have constrains like (6) (for the application of EL to GLM see Kolaczyk (1994)). Note also that the matrix $\mathbf{W}$ may be thought of either q different columns $[w_1, w_2, \dots w_q]$ of vectors of weights or $\mathbf{W} = \mathbf{w} \cdot \mathbf{1}^T$, where $\mathbf{1}^T$ is $(1 \times q)$. In the former case we assume that the q models in (1) should be estimated independently, that is q different experiments have to be run. Then the maximum in (3) becomes $\prod_{i=1}^{n+m} \prod_{j=1}^{q} (n+m) w_{ij}$. In the latter case, as it is much more frequent, we assume that only one experiment is run and then the q models are jointly studied and estimated. Now, supposing that (1) and (2) are models linear in the parameters, with obvious notation we may write

$$\begin{aligned} \boldsymbol{Y}_1 &= \alpha + \boldsymbol{X}\,\mathbf{B} + \boldsymbol{E}_1 && \text{(a)} \\ \mathbf{Y}_2 &= \mathbf{1}\hat{\alpha} + \mathbf{1}\,\xi\,\hat{\mathbf{B}} + \mathbf{E}_2 && \text{(b)} \end{aligned} \tag{7}$$

Using (6) we have the following $q(p+1) \times q$ constraints

$$\left(\begin{bmatrix} 1 & \mathbf{1}\xi \\ \underset{(n\times 1)}{\mathbf{1}} & \mathbf{X} \end{bmatrix} \otimes \underset{(q\times 1)}{\mathbf{1}^T} \right)^T \left\{ \left(\begin{bmatrix} \mathbf{Y}_2 \\ \mathbf{Y}_1 \end{bmatrix} - \underset{(m+n)\times 1}{\mathbf{1}} \hat{\alpha} - \begin{bmatrix} \mathbf{1}\xi \\ \mathbf{X} \end{bmatrix} \hat{\mathbf{B}} \right) \odot \underset{(m+n)\times q}{\mathbf{W}} \right\} = \mathbf{0} \tag{8}$$

where $\otimes$ stands for the Kronecker product. If q=1 and $\mathbf{W} = \mathbf{w} \cdot \mathbf{1}^T$ then (8) corresponds to:

$$\begin{cases} \sum_{j=1}^{m} e_{\xi j} w_{\xi j} + \sum_{i=1}^{n} e_i w_i = 0 \\ \sum_{j=1}^{m} \xi_1 e_{\xi j} w_{\xi j} + \sum_{i=1}^{n} x_{i1} e_i w_i = 0 \\ \dots \\ \sum_{j=1}^{m} \xi_p e_{\xi j} w_{\xi j} + \sum_{i=1}^{n} x_{ip} e_i w_i = 0 \end{cases} \quad \text{where} \quad \begin{aligned} e_{\xi j} &= (y_{2j} - \hat{\alpha} - \sum_{s=1}^{p} \hat{\beta}_s \xi_s) \\ e_i &= (y_{1i} - \hat{\alpha} - \sum_{s=1}^{p} \hat{\beta}_s \xi_{is}) \end{aligned} \tag{9}$$

and $\mathbf{y}_1, \mathbf{y}_2$ are, respectively, the data coming from the calibration and the prediction step, and $w_{\xi j}$ is the weight of $e_{\xi j}$.

Computationally, the problem of finding the maximum of the log of (3) with the constrains in (9) can be more efficiently solved by implementing its dual problem. (9) may be written in a compact form as $\mathbf{X}_e^T \cdot \mathbf{w} = \mathbf{0}$, where $\mathbf{X}_e$ represents the matrix of residuals. Let $\mathbf{A}^* = \{a_{ij}^{-1}\}$. Using the hints of Owen (2001) pag. 60, the peculiar solution for the calibration problem, is $\mathbf{w} = \frac{1}{n+m}\mathbf{x}^*_{\xi_\lambda}$ where $\mathbf{x}_{\xi_\lambda} = \mathbf{1} + \mathbf{X}_e \cdot \lambda$ and λ is a $(p \times 1)$ Lagrange multiplier that satisfies the $p+1$ equations

$$\frac{1}{n+m}[(\mathbf{X}_e \odot (\mathbf{x}^*_{\xi_\lambda} \underset{1\times p}{\mathbf{1}}))^T \cdot \underset{(n+m)\times 1}{\mathbf{1}}] = \mathbf{0}$$

By substituting $\mathbf{w}$ in (3) we have $\log R(\xi_\lambda) = -\left[\log \mathbf{x}_{\xi_\lambda}\right]^T \cdot \mathbf{1}$, which must be minimized over λ subject to the constraint that the argument of log is positive and $\sum w_i = 1$.

Additional constraints, e.g. on the deviance of the residuals, may be added (see Zappa (2005) for details). Without loss of generality, in the rest of this paragraph we will make use only of the constrains in (9).

The first problem is that, given $\mathbf{y}_2$, not unique solution to (7b) may exists and then all the ξ that belong to the calibration model and that satisfy the constrains in (9) will yield $R(\xi) = 1$. Notwithstanding this, we may show that the calibration region is closed at least in probability measure. In fact, suppose for simplicity and without loss of generality that $q = m = 1$. Suppose that the variables have been centred (this is often the case especially in controlled calibration): then $\hat{\alpha} = \mathbf{0}$. Let $d = \|\xi - \mathbf{X}\|_2$ be the distance of ξ from the observed set $\mathbf{X}$. As $\mathbf{X}$ is fixed, it is sufficient that, e.g., the *j-th* element of ξ diverges, so that $d \to \infty$. It follows that $\lim_{|\xi_j| \to \infty} \xi_j e_{\xi j} = \infty$: that is, if the whole vector ξ does not simultaneously change as ξ_j increases, the contribution to the estimating equations in (9) goes to infinity. In order to fulfil the constraints to zero in (9) the weight $w_\xi \to 0$ and then $R(\xi) \to 0$. But even for small fixed $\mathbf{e}_\xi$, that is even if we choose the whole vector ξ systematically close to the model, the previous limit will be ∞ and then $R(\xi) \to 0$. These considerations support the conjecture that ξs too far from $\mathbf{X}$ are not good calibrating solutions and this will be exploited in the following. Then as $d \to \infty$ we may have only $R(\xi) \cong 0$ or $R(\xi) = 1$ respectively, if ξ does or does not belong to the model (that is for a subset of measure zero). In the latter situation the contribution to the estimating equations in (9) is zero, as it is shown in the next equation

$$\lim_{\substack{|\xi_j| \to \infty \\ |e_{\xi j}| \to 0}} \xi_j (y_{2j} - \hat{\beta}^T \xi) = \lim_{\substack{|\xi_j| \to \infty \\ |e_{\xi j}| \to 0}} [-\hat{\beta}_j / (y_{2j} - \hat{\beta}^T \xi)^2]^{-1} = 0 \qquad (10)$$

and then $w_i = 1/(n+1)\ \forall i$. Obviously for small or moderate d the contribution to the equations in (9) is at least moderate and the weights will not be necessarily all zeros. Once the problem of the closure of the confidence region at least in probability has been solved, we may apply (5) to find operatively the confidence region.

The second problem is that, if e.g. $p > q$, then the estimates satisfying $R(\xi) = 1$ may be infinite and the choice of a unique calibration point appears indeterminate. We propose the introduction of a penalty function that will downweight those ξ which will increase the overall entropy to the sets $(\mathbf{X}, \mathbf{Y}_1)$ collected in the calibration step. Consider that this step is usually run in a very accurate and precise conditions. Then it seems reasonable that any solution "far" from the set $(\mathbf{X}, \mathbf{Y}_1)$ may be considered not a good calibrating solution. This proposal introduces a measure coherent with the previous considerations based on d and similar to the inconsistency measure in the classical approach: the greater is the entropy the less is the prediction ability of the $\mathbf{Y}_2$ set with respect to $\mathbf{Y}_1$.

Let $(\mathbf{X}_\xi, \mathbf{Y})=\begin{bmatrix} 1\xi \ \mathbf{Y}_2 \\ \mathbf{X} \ \mathbf{Y}_1 \end{bmatrix}$. To measure entropy, we have chosen the global variability of the second order

$$\Delta_2(\mathbf{X}_\xi, \mathbf{Y}) = \left\{ \sum_{i,j}^{n+m} \left[\left((\mathbf{x}_\xi, \mathbf{y})_i - (\mathbf{x}_\xi, \mathbf{y})_j \right)^2 w_i w_j \right] \right\}^{0.5} \tag{11}$$

where $(\mathbf{x}_\xi, \mathbf{y})_i$, for $i(j) = 1, 2, ..., m + n$, is the ith (jth) row of the matrix $(\mathbf{X}_\xi, \mathbf{Y})$, because it does not need the estimate of any additional parameter and it is based on Euclidean distances. Note that the distance between the ith and the jth row is weighted by the estimated $\mathbf{w}$. Then it balances the increase in d and the relative importance of the new ξ with respect to the set $\mathbf{X}$. Moreover a property of (11) is that, for any ξ such that $w_\xi = 0$, $\Delta_2(\mathbf{X}_\xi, \mathbf{Y}) = const$ that is any solution far from the calibration set is to be considered indifferent for Δ_2. This proposal may be applied also to the Maximum likelihood (ML) approach, but in this case no weights will be used in (11).

4 An Example: The Critical Dimension (CD) Process

In microelectronics, (I wish to thank Gabriella Garatti of STMicroelectronics in Agrate -Italy- for having proposed me the CD problem, provided the dataset and helped me in giving a simple technical explanation of the problem) calibration is one of the key methodology to guarantee the highest precision and, where possible, a reduction of costs. Many are the steps requested to produce what is elementarily called a chipset. Among these very many steps the so called Critical Dimension (CD) process is considered (as its name suggests) very critical. This is also due to the progressive shrinking of CD. In very few words, CD concerns the control of a lithography process whose aim is to "print" on a wafer surface the map of geometric structures and to control that the distance or the shape of structures, such as channels, boxes, holes etc., has a particular profile.

To control this process key factors are: image contrast, image focus and exposure dose. Visual test based methods have played significant roles in both production and development environments. For example, the use of checking completely developed and cleared photoresist patterns from a dose-focus matrix is very common in semiconductor industry. While visual tests are easy to implement, they are not easy to automate. Scanning electron microscopes (SEM) can measure patterned features. However they are very expensive, and can be either time-consuming or destructive, and thus not suitable for run-to-run monitoring. Moreover, electrical measurements can provide information on final CD linewidths, but cannot provide reliable resist profile information. Besides the airborne base contamination of chemically amplified photoresist is a yield-limiting factor in deep UV lithography and will remain so as device features continue to shrink.

Because of all the above reasons, the availability of a reliable calibration algorithm is needed in order to find the best treatment for pre-chosen CDs.

The variables and the data used to calibrate the experiment are in Zappa (2005). In particular, the parameters n, p, q, m result to be 9,2,1,1 respectively. Note that n is very small and that $p > q$, that is the two problems described in § 3 are present. According to the standard literature for this process, a linear combination of variables has been preferred. We have chosen for Y, the CD response variable, the target value of 0. The output of the VBA program is in Fig. 1. Looking at $R(\xi)$ it turns out that exists, as expected, a locus of maximum corresponding to the estimated model, plus an unexpected local suboptimum area a bit far from the estimated linear model and in the neighbor of {0,0}. Such a region is absolutely absent in the ML profile. The Δ_2 surface in EL is evidently different from the analogous one in ML where no weights are attributed to the data. From the contour plot of $R(\xi)/\Delta_2$ we have found that the optimum calibration point is $(-1.58, -1.63)$ for EL and $(-1.51, -1.52)$ using ML. From the profile surfaces in Fig.1, it appears intuitive that applying (5) the resulting confidence region will be closed at least asymptotically using $R(\xi)$ but in no way this will happen using the ML approach.

5 Conclusions

Empirical likelihood seems to be very promising for a variety of problems. Its main property is the possibility to exploit some standard results of the parametric likelihood theory in a non parametric context. Certainly the main limit to a wide application of EL is the availability of a computational tool. To solve this limit we hope the freeware spreadsheet and the connected VBA program we have implemented will be a useful tool in making EL wider known, especially, as it has been in our experience, in those communities where small dataset, very high experimental costs and the difficulty to test parametric assumptions are very frequent in the daily labour. At present it has been specifically prepared for the calibration problem but it can be easily adjustable to many other contexts.

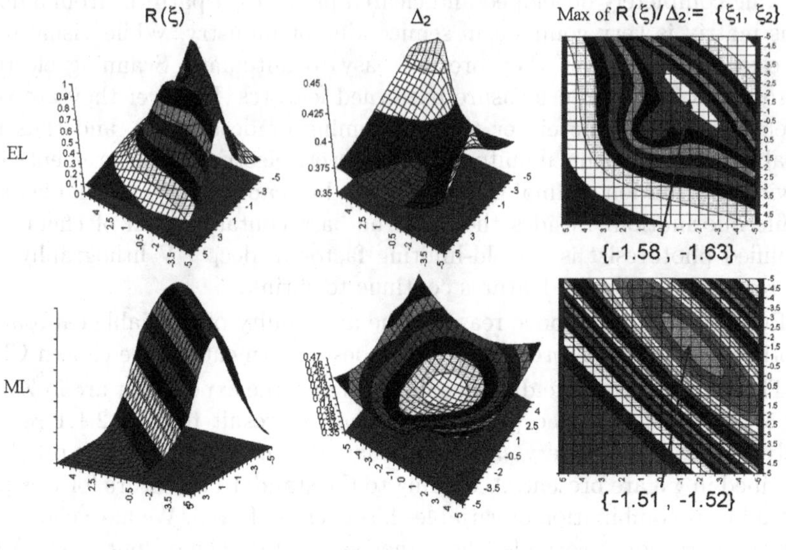

Fig. 1. Profile EL and ML, Δ_2, contour plot of $R(\xi)/\Delta_2$ and optimum ξ_1, ξ_2

References

BROWN, P.J. (1993): *Measurement, Regression and Calibration.* Oxford University Press, Oxford.

KOLACZYK E.D. (1994): Empirical likelihood for generalized linear models. *Statistica Sinica, 4, 199-218.*

NAES T., ISAKSSON T., FEARN T., DAVIES T. (2002): *Multivariate Calibration and Classification.*NIR, Chichester

OWEN A.B. (2001): *Empirical Likelihood.* Chapman and Hall, London

ZAPPA D., SALINI S. (2005): Confidence Regions for Multivariate Calibration: a proposal. In: M. Vichi et al. (Eds.): *New Developments in Classification and Data Analysis.* Springer, Berlin, 225–233.

ZAPPA D. (2005): Calibration confidence regions using Empirical Likelihood and the VBA-Excel tool. *Working paper series - Institute of Statistics, Cattolica University of Milan, n.123.*

Part III

Robust Methods and the Forward Search

Random Start Forward Searches with Envelopes for Detecting Clusters in Multivariate Data

Anthony Atkinson[1], Marco Riani[2], and Andrea Cerioli[2]

[1] Department of Statistics
London School of Economics
a.c.atkinson@lse.ac.uk

[2] Department of Economics
University of Parma, Italy
mriani@unipr.it, statecl@ipruniv.cce.unipr.it

Abstract. During a forward search the plot of minimum Mahalanobis distances of observations not in the subset provides a test for outliers. However, if clusters are present in the data, their simple identification requires that there are searches that initially include a preponderance of observations from each of the unknown clusters. We use random starts to provide such searches, combined with simulation envelopes for precise inference about clustering.

1 Introduction

The forward search is a powerful general method for detecting unidentified subsets and multiple masked outliers and for determining their effect on models fitted to the data. The search for multivariate data is given book length treatment by Atkinson et al. (2004). To detect clusters they use forward searches starting from subsets of observations in tentatively identified clusters. The purpose of this paper is to demonstrate the use of randomly selected starting subsets for cluster detection that avoid any preliminary data analysis. The goal is a more automatic method of cluster identification.

2 Mahalanobis Distances and the Forward Search

The main tools that we use are plots of various Mahalanobis distances. The squared distances for the sample are defined as

$$d_i^2 = \{y_i - \hat{\mu}\}^T \hat{\Sigma}^{-1} \{y_i - \hat{\mu}\}, \tag{1}$$

where $\hat{\mu}$ and $\hat{\Sigma}$ are estimates of the mean and covariance matrix of the n observations.

In the forward search the parameters μ and Σ are estimated by maximum likelihood applied to a subset of m observations, yielding estimates $\hat{\mu}(m)$ and $\hat{\Sigma}(m)$. From this subset we obtain n squared Mahalanobis distances

$$d_i^2(m) = \{y_i - \hat{\mu}(m)\}^T \hat{\Sigma}^{-1}(m) \{y_i - \hat{\mu}(m)\}, \qquad i = 1, \ldots, n. \tag{2}$$

We start with a subset of m_0 observations which grows in size during the search. When a subset $S(m)$ of m observations is used in fitting, we order the squared distances and take the observations corresponding to the $m+1$ smallest as the new subset $S(m+1)$. Usually this process augments the subset by one observation, but sometimes two or more observations enter as one or more leave.

In our examples we look at forward plots of quantities derived from the distances $d_i(m)$. These distances tend to decrease as n increases. If interest is in the latter part of the search we may use **scaled** distances

$$d_i^{\mathrm{sc}}(m) = d_i(m) \times \left(|\hat{\Sigma}(m)|/|\hat{\Sigma}(n)|\right)^{1/2v}, \tag{3}$$

where v is the dimension of the observations y and $\hat{\Sigma}(n)$ is the estimate of Σ at the end of the search.

To detect outliers we examine the minimum Mahalanobis distance amongst observations not in the subset

$$d_{\min}(m) = \min d_i(m) \quad i \notin S(m), \tag{4}$$

or its scaled version $d_{\min}(m)^{\mathrm{sc}}(m)$. In either case let this be observation $i_{\min}$. If observation $i_{\min}$ is an outlier relative to the other m observations, the distance (4) will be large compared to the maximum Mahalanobis distance of the m observations in the subset.

3 Minimum and Ordered Mahalanobis Distances

Now consider the ordered Mahalanobis distances with $d_{[k]}(m)$ the kth largest distance when estimation is based on the subset $S(m)$. In many, but not necessarily all, steps of the search

$$d_{[m+1]}(m) = d_{\min}(m). \tag{5}$$

Instead of using $d_{\min}(m)$ as an outlier test, we could use the value of $d_{[m+1]}(m)$. In this section we describe when the difference in the two distances can arise and what the lack of equality tells us about the presence of outliers or clusters in the data. We then use simulation to compare the null distribution of tests in the forward search based on the two distances.

Lack of equality in (5) can arise because the observations in $S(m)$ come from ordering the n distances $d_i(m-1)$ based on $S(m-1)$ not on $S(m)$. The effect is most easily understood by considering the case when the observation added in going from $S(m-1)$ to $S(m)$ is the first in a cluster of outliers. In that case the parameter estimates $\hat{\mu}(m)$ and $\hat{\Sigma}(m)$ may be sufficiently different from $\hat{\mu}(m-1)$ and $\hat{\Sigma}(m-1)$ that the other observations in the cluster will seem less remote. Indeed, some may have smaller distances than some of those in the subset. More formally, we will have

$$d_{\min}(m) < d_{[k]}(m) \quad k \leq m, \tag{6}$$

for one or more values of k. Then the difference

$$g_1(m) = d_{\text{min}}(m) - d_{[m]}(m) \tag{7}$$

will be negative, whereas when (5) holds, which it typically does in the absence of outliers,

$$g_2(m) = d_{[m+1]}(m) - d_{[m]}(m) = d_{\text{min}}(m) - d_{[m]}(m) \tag{8}$$

is positive. The forward plot of $g_1(m)$ and $g_2(m)$ is called a gap plot, appreciable differences between the two curves indicating the entry of a group of outliers or of a new cluster of observations into the subset. At such moments interchanges may occur when one or more of the observations in $S(m)$ leave the subset as two or more enter to form $S(m+1)$. A more detailed discussion of the ordering of observations within and without $S(m)$ is on pp. 68-9 of Atkinson et al. (2004). The gap plot for the Swiss banknote data, which contains two clusters, is on p.118.

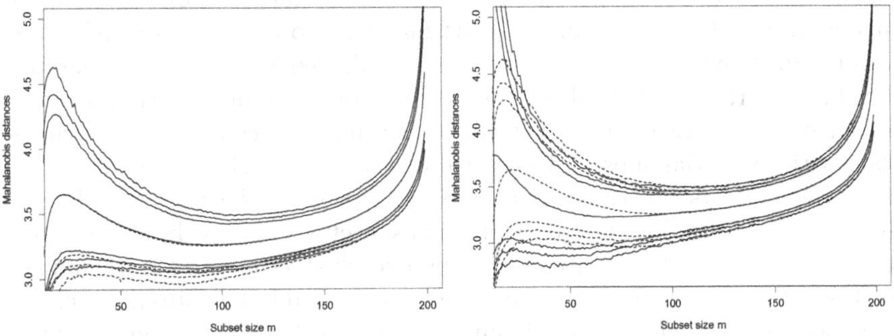

Fig. 1. Envelopes from 10,000 simulations of forward searches with multivariate normal data when n = 200 and v = 6. Left-hand panel - elliptical starts: continuous lines, the order statistic $d_{[m+1]}(m)$; dotted lines, $d_{\text{min}}(m)$, the minimum distance amongst observations not in the subset. Right-hand panel - plots of $d_{\text{min}}(m)$: dotted lines, elliptical starts as in the left-hand panel; continuous lines, random starts. 1, 2.5, 5, 50, 95, 97.5 and 99 % envelopes

The above argument suggests that, for a single multivariate population with no outliers, (5) will hold in most steps of the search and that use of $d_{[m+1]}(m)$ or of $d_{\text{min}}(m)$ as an outlier test will give identical results. To demonstrate this we show in the left-hand panel of Figure 3 forward plots of simulated percentage points of the empirical distribution of the unscaled versions of the two quantities from 10,000 simulations of 200 observations from a six-dimensional normal distribution. The continuous curves are for $d_{[m+1]}(m)$, whereas $d_{\text{min}}(m)$ is represented by dotted lines. There is no discernible difference over the whole search in the median and upper percentage points of the distribution. There is some difference in the lower percentage points where the average values of $d_{\min}(m)$ are slightly lower. This is explained because, in the earlier stages of the search there are a few samples in which the observations are not well ordered and the subset is unstable, so that condition (6) holds. That the difference between the two distributions is only in the lowest tails shows that such behaviour is comparatively rare. Since we use the upper tails of the distribution for detection of outliers, the figure confirms that the test is indifferent to the use of $d_{[m+1]}(m)$ or of $d_{\min}(m)$. In the remainder of this paper we only consider the minimum distances $d_{\min}(m)$.

4 Elliptical and Random Starts

When the observations come from a single multivariate normal population with some outliers, these outlying observations enter at the end of the search. To start the search under these conditions Atkinson et al. (2004) use the robust bivariate boxplots of Zani et al. (1998) to pick a starting set $S^*(m_0)$ that excludes any two-dimensional outliers. The boxplots have elliptical contours, so we refer to this method as the elliptical start. However, if there are clusters in the data, the elliptical start may lead to a search in which observations from several clusters enter the subset in sequence in such a way that the clusters are not revealed. Searches from more than one starting point are in fact needed to reveal the clustering structure. Typically it is necessary to start with an initial subset of observations from each cluster in turn, when the other clusters are revealed as outliers. An example using the data on Swiss banknotes is in Chapter 1 of Atkinson et al. (2004). In this example finding initial subsets in only one of the two clusters requires a preliminary analysis of the data. Such a procedure is not suitable for automatic cluster detection. We therefore instead run many forward searches from randomly selected starting points, monitoring the evolution of the values of $d_{\min}(m)$ as the searches progress.

In order to interpret the results of such plots we again need simulation envelopes. The right-hand panel of Figure 3 repeats, in the form of dotted lines, the envelopes for $d_{\min}(m)$ from the left-hand panel, that is with elliptical starts. The continuous lines in the figure are for the values of $d_{\min}(m)$ from random starts. At the start of the search the random start produces some very large distances. But, almost immediately, the distances for the random start are smaller, over the whole distribution, than those from the elliptical start. This is because the elliptical start leads to the early establishment of subsets $S(m)$ from the centre of the distribution. But, on the other hand, the subsets $S_{\mathrm{R}}(m)$ from the random start may contain some observations not from the centre of the distribution. As a consequence, the estimate of variance will be larger than that from the elliptical start and the distances to all units will be smaller. As the search progresses, this effect decreases as the $S_{\mathrm{R}}(m)$ for individual searches converge to the $S(m)$ from the elliptical start. As the figure shows, from just below $m = 100$ there is no difference between the envelopes from the two searches. Further, for appreciably smaller values of m inferences about outliers from either envelope will be similar.

The results of this section lead to two important simplifications in the use of envelopes in the analysis of multivariate normal data. One is that procedures based on either $d_{[m+1]}(m)$ or on $d_{\min}(m)$ are practically indistinguishable. The other is that the same envelopes can be used, except in the very early stages of the search, whether we use random or elliptical starts. If we are looking for a few outliers, we will be looking at the end of the search. If we are detecting clusters, their confirmation involves searches of only the cluster members so that, as we see in §6, we are again looking only at the end of the search.

5 Swiss Banknotes and Swiss Heads

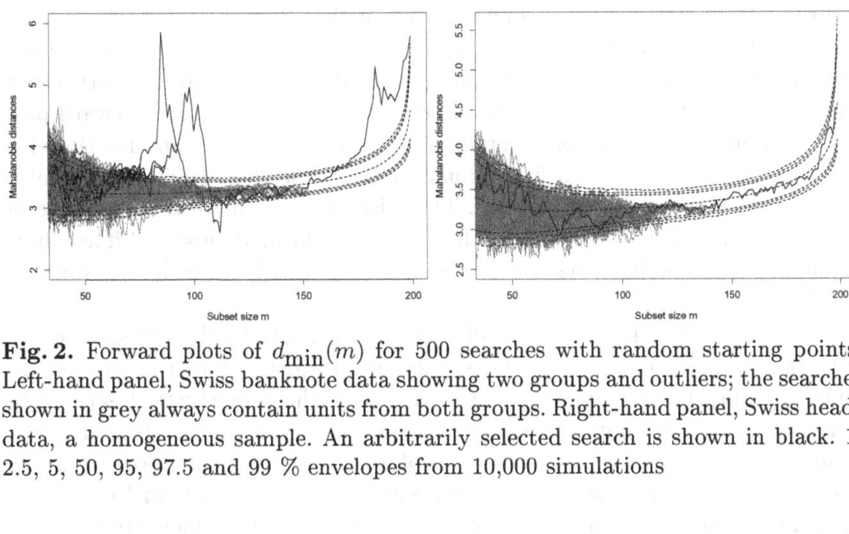

Fig. 2. Forward plots of $d_{\min}(m)$ for 500 searches with random starting points. Left-hand panel, Swiss banknote data showing two groups and outliers; the searches shown in grey always contain units from both groups. Right-hand panel, Swiss heads data, a homogeneous sample. An arbitrarily selected search is shown in black. 1, 2.5, 5, 50, 95, 97.5 and 99 % envelopes from 10,000 simulations

There are two hundred observations in the Swiss banknote data. The notes have been withdrawn from circulation and contain 100 notes believed to be genuine and 100 probable forgeries, on each of which six measurements were made. The left-hand panel of Figure 5 contains the results of 500 forward searches from randomly selected starting subsets with $m_0 = 10$. For each search we have plotted the outlier test $d_{\min}(m)$, the minimum unscaled Mahalanobis distance amongst observations not in the subset. Also included in the plot are 1, 2.5, 5, 50, 95, 97.5 and 99 % simulation envelopes for $d_{\min}(m)$ when the observations come from a single six-dimensional normal distribution.

The first feature of the plot is that, from m around 150, all searches follow the same trajectory, regardless of starting point. This is empirical justification of the assertion of Atkinson et al. (2004) that the starting point is not of consequence in the latter part of the search. The end of the search shows a group of 20 outliers, most of which, in fact, come from Group 2, the forgeries (there seem to have been two forgers at work). The peak around m = 98 is for searches containing only units from Group 1. At these values of m the outliers from Group 1 and observations from Group 2 are all remote and have large distances. Because of the larger number of outliers from Group 2, the peak for this cluster comes earlier, around $m = 85$. The searches that do not give rise to either peak always contain units from both clusters and are non-informative about cluster structure. They are shown in grey in the figure.

This plot shows the clear information that can be obtained by looking at the data from more than one viewpoint. It also shows how quickly the search settles down: the first peak contains 70 searches and the second 62. Fewer searches than this will have started purely in one cluster; because of the way in which units are included and excluded from the subset, the searches tend to produce subsets located in one or other of the clusters.

The left-hand panel of the figure can indeed be interpreted as revealing the clusters. But we also need to demonstrate that we are not finding structure where none exists. The right-hand panel of the figure is again a forward plot of the minimum distance of observations not in the subset, but this time for the 200 observations of six-dimensional data on the size of Swiss heads also analysed by Atkinson et al. (2004). This plot shows none of the structure of clustering that we have found in the banknote data. It however does show again how the search settles down in the last one third, regardless of starting point.

The plot in the left-hand panel of Figure 5 leads to the division of the data into two clusters, the units in the subsets just before the two peaks. Once the data have been dissected in this way, the procedures described in Atkinson et al. (2004) can be used to explore and confirm the structure. For example, their Figure 3.30 is a forward plot of all 200 Mahalanobis distances when the search starts with 20 observations on genuine notes; in Figure 3.35 the search starts with 20 forgeries. In both these plots, which are far from identical, the structure of two groups and some outliers is evident. However, in their Figure 3.28, in which the search starts with a subset of units from both groups, there is no suggestion of the group structure.

6 Bridge Data

In their §7.5 Atkinson et al. (2004) introduce the "bridge" data; 170 two-dimensional observations that consist of a dispersed cluster of 80 observations, a separate tight cluster of 60 observations and an intermediate bridge joining the two groups consisting of 30 observations. The data are plotted in their Figure 7.18. An important feature of these data without the bridge is that the

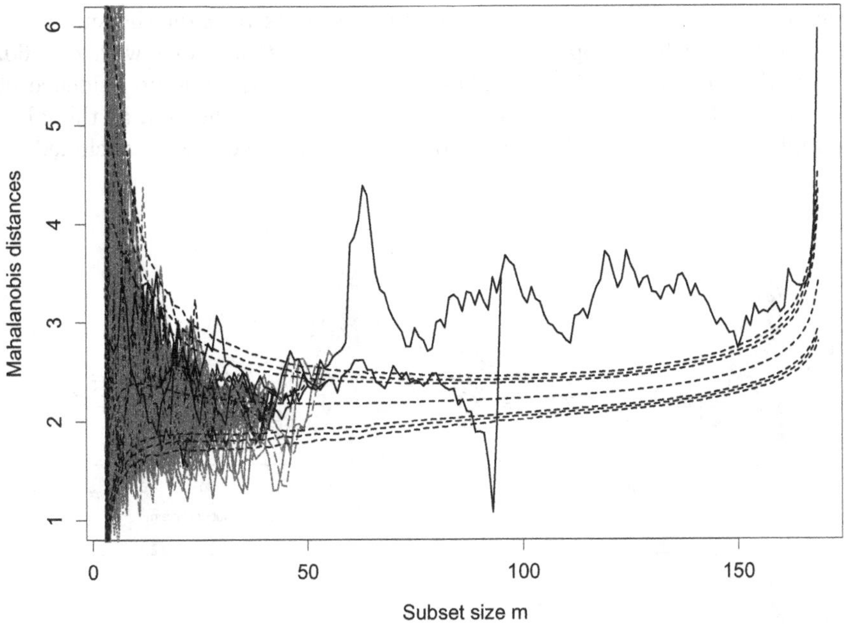

Fig. 3. "Bridge" data: $d_{\text{min}}(m)$ for 500 searches with random starting points. The peak at m = 61 comes from searches that contain only units in the compact group. The other main trajectory is for searches based on the dispersed group

cluster structure is not detected by very robust statistical methods which fit to a subsample coming from both clusters and so fail to reveal the clustered nature of the data. We first use these data as a second example of the power of random starts to indicate clustering. We then show how the repeated use of envelopes for varying sample sizes n can lead to the virtually exact determination of cluster membership.

Figure 5 shows plots of $d_{\text{min}}(m)$ for 500 searches with random starting points. The general structure is that, from around $m = 50$, there are two trajectories. The upper one, which at this point contains units from the compact group of 60 observations, has a peak at $m = 61$. The lower trajectory initially contains units from the dispersed group of 80 observations and then, for larger m, neighbouring units from the bridge are included. There follows a large interchange of units when most of those from the dispersed group are removed from the subset and, from $m = 95$, both trajectories are the same; the subset subsequently grows by inclusion of units from the dispersed group.

We now consider a careful analysis of the trajectory of $d_{\text{min}}(m)$ using subsets of the data of increasing sizes identified from Figure 5 as giving the upper trajectory, that is seemingly coming from the compact group. To discuss individual observations we use the ordering imposed by the forward search, notated as observation $[i]$. The top-left panel of Figure 6 is for 500 searches with random starts using the first 60 units to enter the subset, so the envelopes, which stop at $m = 59$ are found by simulations with $n = 60$. The trajectory lies within the simulated envelopes; there is no evidence of any outlier. In fact the trajectory is a little too flat at the end, as though a large, but not outlying, observation or two has been incorrectly excluded.

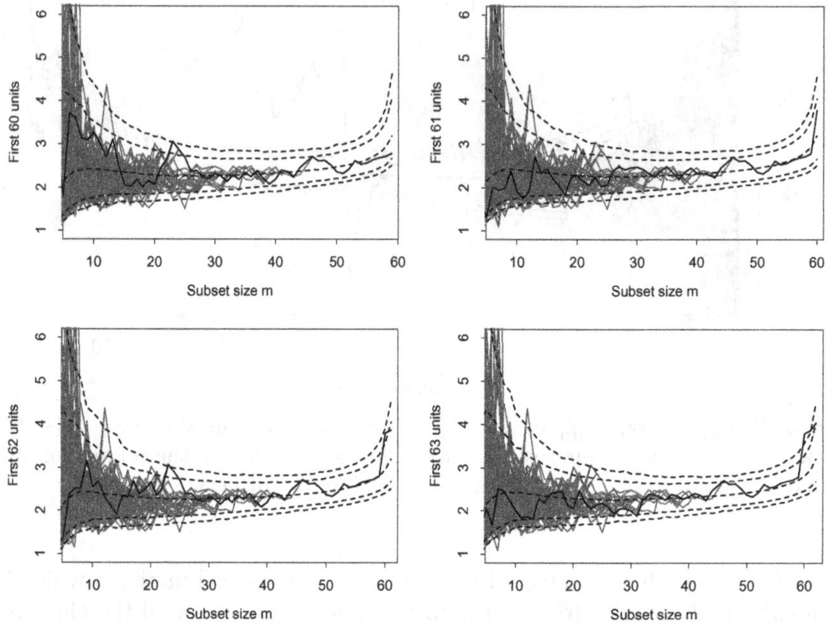

Fig. 4. "Bridge" data: $d_{\text{min}}(m)$ for 500 searches with random starting points for n = 60, 61, 62 and 63 observations giving the upper trajectory in Figure 5. The first 60 observations are shown to belong to a homogenous group. 1, 5, 50, 95, and 99 % envelopes

In fact, the first 60 observations in the search consist of 59 from Group 1 and one from the bridge. Of course, because the data are simulated with random error, group membership can be overlapping. The upper-right panel in the figure is for $n = 61$, with new simulation envelopes from this value of n. It is important in the exact detection of outliers that the upwards curve towards the end of the search is sensitive to sample size. Addition of observation [61] (observation 118 in Table A.15 of Atkinson et al., 2004) causes an upwards jump in the trajectory, although not a sufficiently large jump to take the trajectory outside the envelopes.

Observation [61] is the last from Group 1. Addition of observation [62] (161), shown in the lower-left panel of the figure, takes observation [61] partially outside the envelopes, although observation [62] remains inside. Finally, the plot for the first 63 observations shows observation [61] well outside all envelopes, with the trajectory returning inside the envelope.

The behaviour of the trajectory in the lower plots is typical of the effect of adding a cluster of different observations from those in the subset earlier in the search, which was discussed in §3. What these plots do show is that observation [61] is indeed an outlier and that the first 60 units form a homogeneous group. The next stage in the analysis would be to remove these sixty observations and to run further series of searches with random starts to identify any remaining structure.

7 The Importance of Envelopes

The analysis in this paper has depended crucially on the use of simulation envelopes in the forward search, a feature missing from our books Atkinson and Riani (2000) and Atkinson et al. (2004). The envelopes used here are similar to those described by Riani and Atkinson (2006) for testing for outliers in multivariate normal data. Since they are looking for outliers from a single population, they only use elliptical starts, rather than the random starts we use here to detect clusters. For large samples in high dimensions, the repeated simulation of envelopes for increasing sample sizes, as in §6, can be excessively time consuming. Riani and Atkinson (2006) describe methods for numerical approximation to the envelopes, particularly for the scaled distances (3). They also give theoretical results, based on order statistics from scaled F distributions, that give excellent approximations to the envelopes, even for moderate n and v.

References

ATKINSON, A.C. and RIANI, M (2000): *Robust Diagnostic Regression Analysis*, Springer–Verlag, New York.

ATKINSON, A.C., RIANI, M. and CERIOLI, A. (2004): *Exploring Multivariate Data with the Forward Search*, Springer–Verlag, New York.

RIANI, M and ATKINSON, A.C. (2006): Finding an unknown number of multivariate outliers in larger data sets, (Submitted).

ZANI, S., RIANI, M and CORBELLINI, A. (1998): Robust bivariate boxplots and multiple outlier detection, *Computational Statistics and Data Analysis, 28, 257–270.*

Robust Transformation of Proportions Using the Forward Search

Matilde Bini and Bruno Bertaccini

Dipartimento di Statistica "G. Parenti"
Università di Firenze, Italy
bini@ds.unifi.it; brunob@ds.unifi.it

Abstract. The aim of this work is to detect the best transformation parameters to normality when data are proportions. To this purpose we extend the forward search algorithm introduced by Atkinson and Riani (2000), and Atkinson *et al.* (2004) to the transformation proposed by Aranda-Ordaz (1981). The procedure, implemented by authors with R package, is applied to the analysis of a particular characteristic of Tuscany industries. The data used derive from the Italian industrial census conducted in the year 2001 by the Italian National Statistical Institute (ISTAT).

1 Introduction

Most applications carried out with multivariate methods yield results that assume normality assumptions. At present few methods are available to assess the stability of the results. Often the assumption of multivariate normality is approximately true only after an appropriate transformation of the data. The most widely used methods of transformation have been proposed by Box and Cox in 1964, and by Aranda-Ordaz in 1981 when data are respectively continuous and proportions. Unfortunately, when we manage multivariate data it is very difficult to find out, test and validate a particular transformation due to the well known masking and swamping problems (Atkinson, 1985; Velilla, 1995). The lack of a proper choice of the most appropriate transformation may lead to overestimate (or underestimate) the importance of particular variables (Riani and Atkinson, 2001), and to a wrong interpretation of the results from the methods used. The difficulties and intricacies of the choice of the best multivariate transformation usually lead the analyst to apply statistical methods to untransformed data.

The forward search (FS) is a powerful general method for detecting multiple masked outliers and for determining their effect on models fitted to data. Atkinson and Riani (2000) and Atkinson *et al.* (2004) describe its use in generalized linear models, and in multivariate methods.

The aim of this work is to show how the FS algorithm can be extended to multivariate normality transformations where data are proportions, with the aim to find the best transformation parameters. The implementation of this new approach to the transformation of proportions is carried out in a library of R package which is available on request by authors. To show how our proposal works, we perform an example using a data set concerning characteristics of the Italian industries obtained from the census of industries conducted in the year 2001 by the Italian National Statistical Institute. In particular, we consider a subset of three different variables concerning the economic sectors Commerce, Transport and Communication, and Finance, observed on 287 municipalities of the Tuscany region. In particular, the variables y_j are the percentage of firms with small-medium size (with ≤ 250 employees) to the total number of firms in each sector.

2 The Use of Forward Search in Multivariate Analyses

In the FS procedure subsamples of outlier free observations are found by starting from small subsets and incrementing them with observations which have small Mahalanobis distances, and so are unlikely to be outliers. More precisely, the FS is made up of three steps (Riani and Bini, 2002; Atkinson *et al.*, 2004): choice of the initial subset, adding observations during the FS and monitoring the search.

- **Step 1: Choice of the initial subset**
  We find an initial subset of moderate size by robust analysis of the matrix of bivariate scatter plots. If the data are composed by v variables, the initial subset of r observations (with $r \geq v$) consists of those observations which are not outlying on any scatter plot, found as the intersection of all points lying within a robust contour (a fitted B-spline) containing a specified portion of the data (Riani *et al.*, 1988). Since the evidence for transformations is provided by the extreme observations, such a robust subset will provide a good start to the search for many values of the transformation parameter.
- **Step 2: Adding observations during the FS**
  In every step of the forward search, given a subset $S_*^{(m)}$ of size m ($m = r, \ldots, n-1$), we move to a subset of size $(m+1)$ by selecting the $(m+1)$ units with the smallest Mahalanobis distances:

$$d^2_{i,S_*^{(m)}} = \left(y_i - \hat{\mu}_{i,S_*^{(m)}}\right)^T \hat{\Sigma}^{-1}_{S_*^{(m)}} \left(y_i - \hat{\mu}_{i,S_*^{(m)}}\right) \qquad i = 1, \ldots, n$$

  where $\hat{\mu}_{S_*^{(m)}}$ is the centroid of $S_*^{(m)}$ and $\hat{\Sigma}_{S_*^{(m)}}$ is the $S_*^{(m)}$ sub-sample covariance matrix. The procedure ends when all the n observation enter the analysis.
  In moving from $S_*^{(m)}$ to $S_*^{(m+1)}$, usually just one new unit joins the subset. It may also happen that two or more units join $S_*^{(m+1)}$ as one or more leave. However, our experience is that such an event is quite unusual, only occurring when the search includes one unit which belongs to a cluster of outliers. At the next step the remaining outliers in the cluster seem less

outlying and so several may be included at once. Of course, several other units then have to leave the subset.

- **Step 3: Monitoring the search**
  In multivariate analysis we first try to find a set of transformation parameters to reach approximate normality. In each step of the search as m goes from r to n we initially monitor the evolution of the Mahalanobis distances, of the maximum likelihood estimates (MLE) and of the likelihood ratio test for transformations, using the procedure described in the section below. The changes which occur, will be associated with the introduction of particular observations into the subset m used for fitting. We plot all n Mahalanobis distances $d_{i,S_*^{(m)}}$ for each value of m. The trajectories of this plot are informative about the behavior of each unit throughout the search.

3 The Aranda-Ordaz Transformation and Tests for Transformation in Multivariate Analysis

Let Y be a sample data matrix of dimension $n \times v$ where y_{ij} is the ith observation on variable j with $j = 1, ..., v$, and let $p_{ij} = y_{ij}/(1 - y_{ij})$ the corresponding odds. In the extension of the Aranda-Ordaz (1981) family to multivariate responses the normalized transformation of $u_{AO} = 2(p^\lambda - 1)/\lambda(p^\lambda + 1)$ is

$$z_{AO}(\lambda) = \begin{cases} \frac{2(p^\lambda-1)}{\lambda(p^\lambda+1)} \{G_{AO}(\lambda)\}^{-1}, (\lambda \neq 0) \\ \log\{y/(1-y)\} G\{y(1-y)\}, (\lambda = 0) \end{cases}$$

where G is the geometric mean function and $G_{AO}(\lambda) = G\left(\frac{4p^{\lambda-1}(1+p)^2}{(p^\lambda+1)^2}\right)$. Note that the first expression of $z_{AO}(\lambda)$ yields the second one with $\lambda \to 0$.

Figure 1 shows the values of the proportions obtained under transformation, using some common values of λ $(0, 0.5, 1, 1.5, 2, 2.5, 3)$. The figure includes two panels which show the $z_{AO}(\lambda)$ shapes respectively for the transformation and normalized transformation of y_{ij}. Similarly to Box-Cox transformation, these shapes highlight the effects of restriction and expansion of the proportions values respectively when $\lambda > 1$ and $\lambda < 1$ (Zani, 2000). Moreover, the effect of the λ values on the transformation is even more higher as y_{ij} values leave from 0.5. Since in the present application we manage data having very small values (almost all Y_{ij} are less than 0.30) it is significant the choice of the best transformation parameter. To do this we make use of FS procedure.

We recall the main features of the forward search analysis for the Aranda-Ordaz transformation to normality:

Step 1. Run a forward search through the untransformed data, ordering the observations at each m step by Mahalanobis distances calculated for the whole data set. After that we derive the MLE of λ at each step; the collection

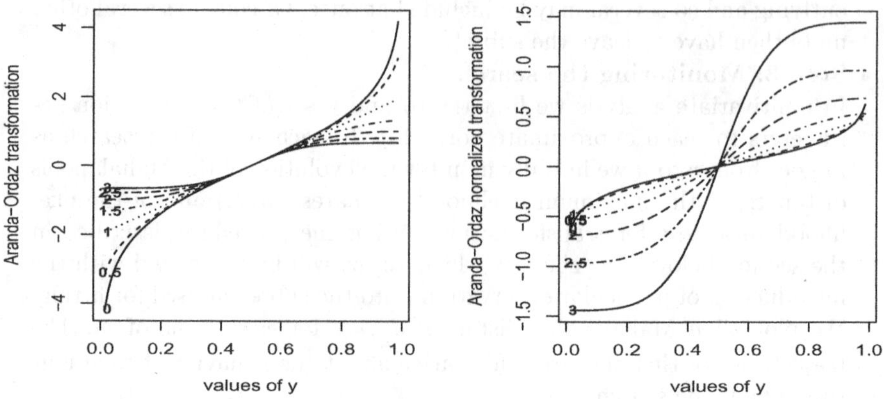

Fig. 1. Aranda-Ordaz transformation functions ($z_{AO}(\lambda)$) for some common values of λ.

of $\hat{\lambda}_m$ is analyzed in order to check for the presence of anomalous observations and to select a set of transformation parameters that are not affected by their presence.

Step 2. Rerun the forward search using distances calculated with the parameters selected in the previous step, again estimating λ for each m. If some change is suggested in $\hat{\lambda}_m$, repeat this step until a reasonable set of transformations has been found. Let this be $\hat{\lambda}_R$.

Step 3. Test the suggested transformation. In general, the likelihood ratio test to validate the hypothesis $\lambda = \lambda_0$, can be expressed as the ratio of the determinants of two covariance matrices (Atkinson, 2002): $T_{LR} = n \log \left\{ \left| \widehat{\Sigma}(\lambda_0) \right| / \left| \widehat{\Sigma}(\widehat{\lambda}) \right| \right\}$ where $\widehat{\lambda}$ is the vector of v parameter estimates found by numerical search maximizing the transformed normal log likelihood, $\widehat{\Sigma}(\lambda_0)$ and $\widehat{\Sigma}(\widehat{\lambda})$ are the covariance matrices calculated respectively when $\lambda = \lambda_0$ and $\lambda = \widehat{\lambda}$. The value of T_{LR} must be compared with the distribution χ^2 on v degrees of freedom. In this particular case, for each variable we expand each transformation parameter in turn around some values of the found estimate, using the values of λ_R for transforming the other variables. In this way, we turn a multivariate problem into a series of univariate ones. In each search we can test the transformation by comparing the likelihood ratio test χ^2 with on 1 degree of freedom. But we prefer to use the signed square root of the likelihood ratio in order to learn whether lower or higher values of λ are indicated.

In order to find the most appropriate transformation we use both the MLE and the values of the test statistic. However, it is better to look at the value of the test rather than parameter estimates, because if the likelihood is flat, the estimates can vary widely without conveying any useful information about the transformation.

4 Forward Transformation of Industries Census Data

We performed, according to the Aranda-Ordaz proposal, the forward search and tested for normalized transformation to normality on the industry census data. We decided to use the normalized transformation of observations since it allows to use a simpler form of the likelihood than in the other case. We started with a default value of 1 (i.e. $H_0 : \lambda_j = 1, j = 1, ..., v$). First, we found the initial subset as the intersection of robust bivariate contours superimposed in each bivariate scatter diagram as is showed in Figure 2. The boxplots show the skewed distributions of the three sectors y_j (with $j = 1, 2, 3$), Commerce, Transport and Communication, and Finance (also labelled Sector 7, Sector 9 and Sector 10).

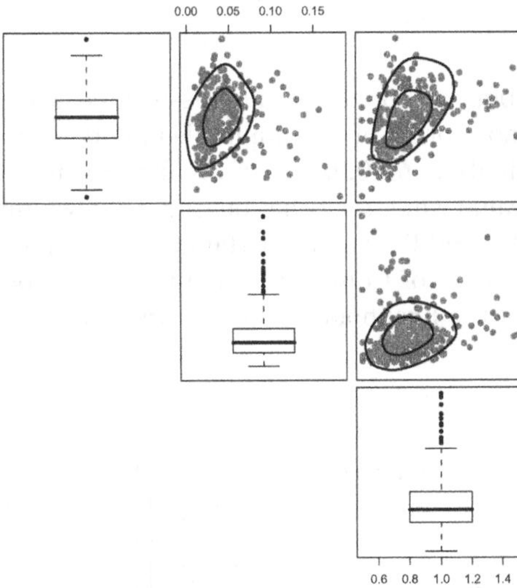

Fig. 2. Bivariate scatterplot matrix of the three variables observed on 287 municipalities with two superimposed robust contours containing respectively 50% and 95% of the data.

The starting forward search concerns the monitoring of the 3 maximum likelihood estimates $\widehat{\lambda}$. The resulting forward plots are shown in Figure 3.

The left panel of this figure shows the graphical output of monitoring of maximum likelihood estimates, while the right panel shows the monitoring of the likelihood ratio test statistic, for the null hypothesis of no transformation, that is $T_{LR(m)} = m \log \left\{ \left| \widehat{\Sigma}(\lambda_0) \right| \Big/ \left| \widehat{\Sigma}(\widehat{\lambda}) \right| \right\}$, $m = r, ..., n$.

Given that the values of the likelihood ratio test must be compared with a χ^2_3 it is clear that the data must be transformed. In fact, the right panel of Figure 3 shows that the values of test increase steadily throughout the search,

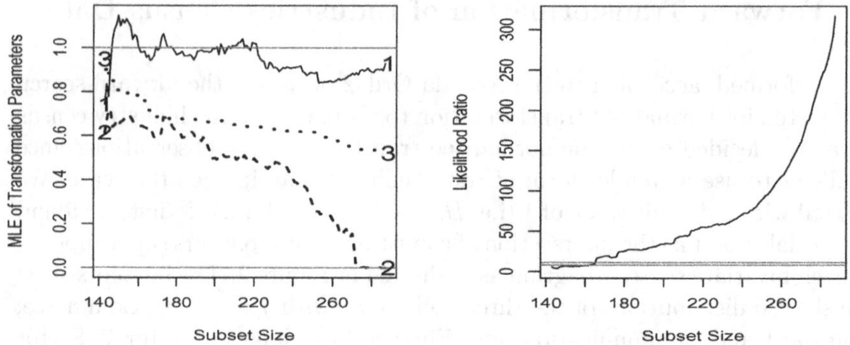

Fig. 3. Monitoring of maximum likelihood estimates (left panel) and likelihood ratio test (right panel) of $\widehat{\lambda}$, during the forward search.

and are very far from the horizontal lines associated with the 95% and 99% points of χ_3^2. The monitoring of maximum likelihood estimates leads to select the tentative combination λ_0 $(0.9, 0, 0.5)^T$. In fact, the first trajectory remains stable around 0.9 value while the other ones trend down during all the search long, until to reach 0 and 0.5 values at the last steps. Then, we repeated the analysis using the values of from step 1. Figure 4 describes the forward plot of the monitoring of the likelihood ratio test statistic, for the null hypothesis of λ_0 $(0.9, 0, 0.5)^T$.

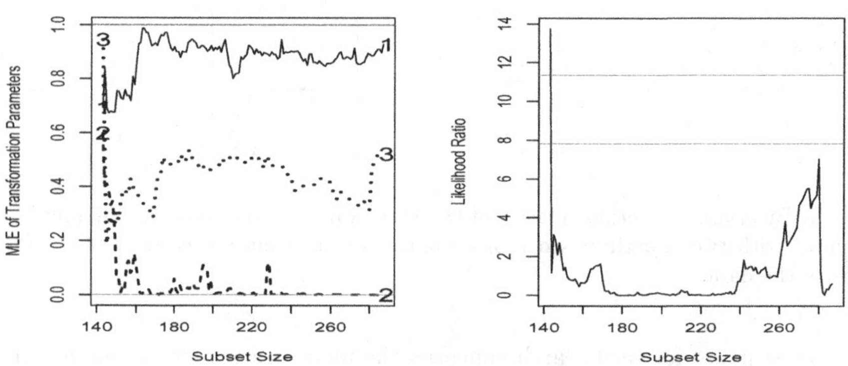

Fig. 4. Monitoring of maximum likelihood estimates (left panel) and likelihood ratio test (right panel) for $H_0 : \lambda_0$ $(0.9, 0, 0.5)^T$, during the forward search.

This figure shows that the values of the test lie inside the confidence bands. From step $m = 260$ until step $m = 280$ it is possible to notice an upward jump. This is certainly due to the presence of some outliers which affect the single variable y_3, as it is clearly shown for the same steps interval, in the left panel of the same figure. This means that using the FS, it is possible to detect the effect of each unit on the results of the statistic. Again, the trajectories of the estimates in the left panel show a stable behavior which confirms the choice of the three values of λ_0. An additional step of the analysis (not given here for lack of space) could be to investigate the characteristics of the units which are responsible of this sudden upward jump. Some more interesting forward plots can be done to have an idea of the best transformation of the parameters. Figure 5, for example, shows the profile likelihoods with asymptotic 95% confidence intervals for the three parameters. In all the these panels the shape of profiles, since it is not flat, confirm the values (corresponding to their maximum) obtained throughout the search. In particular, the likelihood surface of the third variable is sharply peaked.

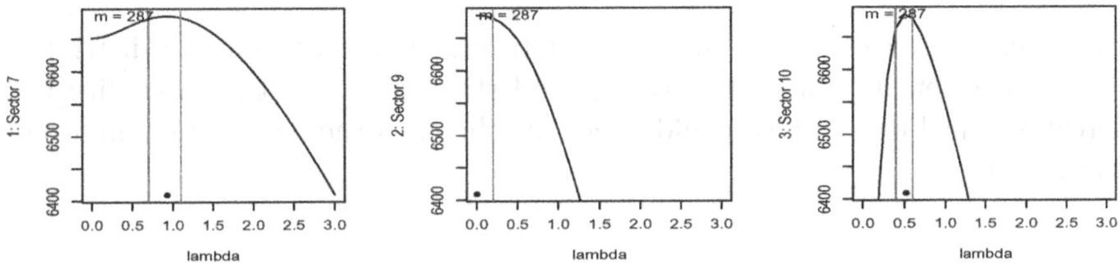

Fig. 5. Profile likelihoods for the three transformation parameters when $m = 287$.

Similar information as above, can be derived from the fan plots of the three variables, showed in Figure 6. In these plots we monitor all the possible values of the transformation parameters. Usually they are used to monitor some possible parameter values closer to that one which seems to be the best transformation, in order to confirm it. Any transformation value would be acceptable if it lies within the 99% interval for the asymptotic normal distribution throughout the searches. Here, the panels of this figure display and confirm the choice of 0.9, among possible values (0.5, 0.7, 0.9, 1.1, 1.3) for Sector 7, the choice of 0 among (0, 0.1, 0.2, 0.3, 0.5) for Sector 9 and the choice 0.5 among (0.1, 0.3, 0.5, 0.7, 1) for Sector 10. For example, from the right panel we can see how the presence of outliers affect the choice of λ_3: in fact, for $\lambda_3 = 0.1; 0.3$ the MLE estimates show an upward jump, while for λ_3 taking values around 0.5 the trajectory is stable. Instead, traces for $\lambda_3 = 0.7; 1$ decline from the first steps of the search, confirming $\lambda_3 = 0.5$ as the best choice of the transformation parameter.

5 Concluding Remarks

The FS procedure has been proposed as a powerful tool to detect and investigate single or groups of units which differ from the bulk of data and which can affect on model fitted to data. In this work we demonstrated that

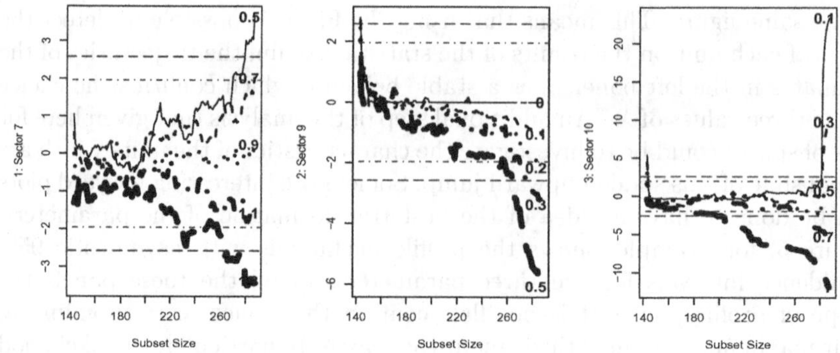

Fig. 6. Fan plots for individual transformations of λ.

its use can be extended to find the best transformation to normal distribution, also when data are proportions, and still preserving the capability to detect atypical units which could affect the choice of the best transformation parameter.

References

ARANDA-ORDAZ F. J. (1981): On two families of transformations to additivity for binary response data. *Biometrika, 89*, 357–363.

ATKINSON A. C. (1985): *Plots, Transformation and Regression.* Oxford University Press, Oxford.

ATKINSON A. C. (2002): Robust diagnostics for multivariate analysis and classification problems. *Atti della XLI Riunione Scientifica della Società Italiana di Statistica.* CLEUP, Milano, 283–294.

ATKINSON A. C., RIANI M. (2000): *Robust Diagnostic Regression Analysis.* Springer, New York.

ATKINSON A. C., RIANI M., CERIOLI A. (2004): *Exploring Multivariate Data with the Forward Search.* Springer, New York.

RIANI M., ATKINSON A. C. (2001): A Unified Approach to Outliers, Influence and Transformations in Discriminant Analysis. *Journal of Computational and Graphical Statistics, Vol. 10*, 513–544.

RIANI M., BINI M. (2002): Robust and Efficient Dimension Reduction. *Proceedings of the XLI Riunione Scientifica della Societá Italiana di Statistica.* CLEUP, Milano, 295–306.

RIANI M., ZANI S., and CORBELLINI A. (1998): Robust Bivariate Boxplots and Visualization Multivariate Data. In I. BALDEJAHN, R. MATAHAR and M. SCHADER (Eds.): *Classification data Analysis and Data Highways.* Springer, Berlin, 93–100.

VELILLA S. (1995): Diagnostics and robust estimation in multivariate data transformations. *Journal of the American Statistical Association, 90*, 945–951.

ZANI S. (2000): *Analisi dei Dati Statistici (Volume II).* Giuffrè, Milano.

The Forward Search Method Applied to Geodetic Transformations

Alessandro Carosio[1], Marco Piras[2], and Dante Salvini[1]

[1] IGP - Institute of Geodesy and Photogrammetry
Swiss Federal Institute of Technology, ETH Zurich, Switzerland
[2] DITAG - Land, Environment and Geo-Engineering Department
Politecnico di Torino, Italy

Abstract. In geodesy, one of the most frequent problems to solve is the coordinate transformation. This means that it is necessary to estimate the coefficients of the equations that transform the planimetric coordinates, defined in a reference system, to the corresponding ones in a second reference system. This operation, performed in a 2D-space is called planar transformation. The main problem is that if outliers are included in the data, adopting non-robust methods of adjustment to calculate the coefficients cause an arbitrary large change in the estimate. Traditional methods require a closer analysis, by the operator, of the computation progress and of the statistic indicator provided in order to identify possible outliers.
In this paper the application of the Forward Search in geodesy is discussed and the results are compared with those computed with traditional adjustment methods.

1 Introduction

Angular measures, distances and coordinates, are only some of the basic observations that can be found in a geodetic problem. The treatment of these observations can become complex, if not supported by the appropriate means. In geodesy, statistics is indispensable, in particular inference and other statistical operations are applied in order to gain the most realistic solution. Some statistical methods were initially developed to cope with geodetic issues.
Characteristical for most of geodetic problems, are low redundancy and precision differences within the observations due to the use of various instrumentations (GPS, theodolite, etc). The low redundancy of the data (in many cases the number of the observations available is at most twice the number of the unknown parameters) and the presence of possible outliers, request an intensive use of statistics. In fact, it is an imperative to discard the minor number of observations, identifying just the effectively wrong ones, in order to avoid singularity. Especially in those cases in which the precision and the accuracy are important factors (positioning of high precision, monitoring of deformations, etc.), it is necessary to pay particular attention to the problem of the outliers.
With the traditional methods used for geodetic adjustment, like the Least Squares, the identification of the error source is difficult, because these methods are based on the assumption of absence of errors in the treated data.

Robust statistics, on the other hand, are based on the opposite principle and offer a valid alternative for the solution of such type of problems.

2 Planar Transformations of Reference Systems

One of the issues studied in geodesy involves the reference systems and their transformations. A system of coordinates is defined through a tern of functions of points $x_i(P)$, with $i = 1, 2, 3$ that guarantee the univocal position of the point P. A coordinate system, defined by the opportune choice of a series of control points, measures and other compatible choices is called system of reference or *datum*.
In applied geodesy, it is often preferred to work with local systems of reference, instead of the global reference system. As already mentioned, a local system is defined by a cartesian tern with origin in a selected, well-known point. The local systems have validity in inverse proportion to the distance from the origin. In fact, above a certain distance from the origin, the precision of the position is out of tolerance and therefore can not be accepted. In this case it is necessary to define a new local system of reference, with its own origin. For this reason, considering a vast area, the various defined local systems of reference, must be interconnected, in order to provide the coordinates of the points relative to the global system of reference.
Mathematically, a planar transformation defines a univocal bidirectional relation between two sets of planimetric points (coordinates). The most frequently used transformation in geodesy is by far the similarity transformation. Essentially, this is defined by a rotation and two translations of the origin and an isotropic scale factor (independent from the orientation). The equations which describe the transformation from a local system of coordinates (O', X', Y') to global one (O, X, Y) are the following:

$$\begin{aligned} X_P &= T_X + \mu(X'_P \cos\alpha + Y'_P \sin\alpha) \\ Y_P &= T_Y + \mu(-X'_P \sin\alpha + Y'_P \cos\alpha) \end{aligned} \tag{1}$$

The common problem of every planar transformation is the determination of the unknown parameters; in our case these are: μ, α, T_X, T_Y.
The equations can be solved by including a sufficient number of points, of which coordinates are known in both reference systems in the computation. For every known point in both reference systems, it is possible to compile the equations (1). To determine univocally the four unknown parameters, at least four equations are necessary; that means two points (P_1 e P_2) known in both reference systems:

$$\begin{aligned} X_{P_1} &= T_X + \mu(X'_{P_1} \cos\alpha + Y'_{P_1} \sin\alpha) \\ Y_{P_1} &= T_Y + \mu(-X'_{P_1} \sin\alpha + Y'_{P_1} \cos\alpha) \\ X_{P_2} &= T_X + \mu(X'_{P_2} \cos\alpha + Y'_{P_2} \sin\alpha) \\ Y_{P_2} &= T_Y + \mu(-X'_{P_2} \sin\alpha + Y'_{P_2} \cos\alpha) \end{aligned} \tag{2}$$

The system of equations obtained can be easily linearised setting $a = \mu \cos \alpha$ and $b = \mu \sin \alpha$.
Once the unknown parameters are estimated, we can apply the relationship (1) to all the points known in one reference system and transform these into the other. It is usual to determine the unknown parameters using a greater number of known points than the minimally required to solve the equation system. In this case the solution of the system of equations is obtained applying the Least Squares method. The matrixes (A, x, L) used to solve the equations with the Least Squares method are the following:

$$A = \begin{bmatrix} c_{11} & c_{12} & c_{13} & c_{14} \\ \dots & \dots & \dots & \dots \\ \dots & \dots & \dots & \dots \\ c_{n1} & c_{n2} & c_{n3} & c_{n4} \end{bmatrix} \qquad x = \begin{bmatrix} a \\ b \\ T_X \\ T_Y \end{bmatrix} \qquad L = \begin{bmatrix} X_1 \\ Y_1 \\ \dots \\ \dots \\ X_n \\ Y_n \end{bmatrix} \tag{3}$$

where c_{ij} are the coefficients of the independent variables a, b, T_X, T_Y in every equation of the system. Solving the problem with the Least Squares method provides $\hat{x}$ as estimation for the parameters x and the estimated residuals $\hat{v}$ of the observations.
Based on these considerations, we tried to analyze if the Forward Search method [Atkinson et al. 2000] could be a valid approach to solve the problems related to the transformations of reference system in a precise and robust manner. In the following chapters the application of this method to planar transformations is described, including the tests performed and the results achieved.

3 Common Methods for Geodetic Data Adjustment

In geodesy most parameters are determined by statistical procedures. The Least Squares method (LS) is the most commonly used. This estimator is optimal for normal distributed observations, since it provides the best estimations for the unknown parameters (in a linear model). In matricial form it can be written as following:

$$V^T Q_{ll}^{-1} V = min \tag{4}$$

The LS method enables to consider various precisions for the observations by means of the weights matrix and allows joining different types of measure (angles, distances, etc.), thanks to the linearisation of the problem. The shortcoming of LS method resides in its strong sensibility to the presence of outliers or in general to observations which do not conform to the normal distribution. In geodesy, like in many other fields where the statistical estimation of unknown parameters is demanded, often the presence of outliers in the data set can not be avoided. In these cases two pathways leading to the solution can be followed: applying the described method purging the data set iteratively to obtain a correct solution, or using algorithms which are not sensitive to a significant deviation from the normal distribution provided that these errors do not exceed a given fraction of the entire data set. In this category we find for example the M-estimates (derived from the Maximum-Likelihood method). One of them is the Huber estimate, which is quite suitable for geodetic problems, due to its affinity to the LS method. The function to be minimized in this case is:

$$\rho(v_i) = \begin{cases} \frac{1}{2}v_i^2 & \text{for } |v_i| < c \\ c|v_i| - \frac{1}{2}c^2 & \text{for } |v_i| \geq c \end{cases} \tag{5}$$

In the interval $[-c, c]$ the function is identical to that one of the LS method, while outside it is linear. The constant c is to be set, based on the precision of the observations (e.g. $c = 3\sigma$). An improvement of this method considers a variable definition of c depending on the redundancy of the observations (proposed by Mallows and Schweppe).

An effective method developed for geodetic problems is the BIBER estimate (bounded influence by standardized residuals) [Wicki 1999] which is classified as M-estimate. The characteristic of this method is to reduce the influence of the outliers on the estimated parameters, analyzing the standardized residuals. The function to be minimized in this case is the following sum of function ρ of all the residuals:

$$\rho(\frac{v_i}{\sigma_{v_i}}) = \begin{cases} \frac{1}{2\sigma_{v_i}}v_i^2 & \text{for } |\frac{v_i}{\sigma_{v_i}}| < c \\ \frac{c}{\sigma_{v_i}}|v_i| - \frac{1}{2}c^2 & \text{for } |\frac{v_i}{\sigma_{v_i}}| \geq c \end{cases} \tag{6}$$

In the category of the estimates with a high breakdown point (BP = 0,5) we find the Least Median of Squares method (LMS). This method assumes the combination of values for which the median squared residual is the smallest as optimal estimation of the unknown parameters. Therefore the condition to be satisfied is the following:

$$med(v_i^2) \rightarrow min \tag{7}$$

This means, the estimator must yield the smallest value for the median of squared residuals computed for the entire data set. Since the possible estimates generated from the data can be very large, only a limited number of subsets of data can be analyzed using approximation methods. For the characteristics pointed out, the robust estimates are a valid instrument for the analysis of data affected by outliers.

4 The Forward Search Applied to Problems of Planar Transformations

The method of the Forward Search (FS) provides a gradual crossover from the LMS to the LS method, starting from a small robustly chosen subset of

data, incrementing it with observations that have small residuals and so are unlikely to be outliers.
The concept on which this technique is based, is to select from the data set with the LMS method an initial subset of size m free of outliers. Usually the algorithm starts with the selection of a subset of p units (number of unknown parameters), leaving n-p observations to be tested. The parameters estimated this way are applied to all the observations in order to compute the residuals. The values of the squared residuals are then ordered, selecting the $m+1$ observations with the smallest squared residual for the next step. From this subset new parameters are estimated with LS. The loop continues unit by unit, adding one observation at the time until $m = n$, that is when the end of the process is reached delivering a common LS solution [Atkinson and Riani 2000].
The innovation of the method consists, beyond its variability given by the gradual crossover from LMS to LS, in providing a continuous monitoring at every step of some diagnostic quantities (residuals, Cook's distance, estimates of the coefficients, t-statistic, etc). Controlling the variation of these indicators during the n-p steps we can identify which of the considered observations causes an abrupt alteration of them, allowing therefore to distinguish eventual "classes" inside the data. For our purpose, using this method the data can be classified in "clean" and therefore usable for the estimation of the unknown parameters, and "outliers". Supposed there are some outliers in the data, the FS will include these toward the end of the procedure. The parameters estimates and the statistic indicators will remain sensibly constant until the first outlier is included in the subset used for fitting. The final solution corresponds to the estimation of the unknown parameters obtained at the step preceding the significant variation of the indicators.
The following flowchart (Fig. 1) shows the application of the FS to geodetic adjustment:

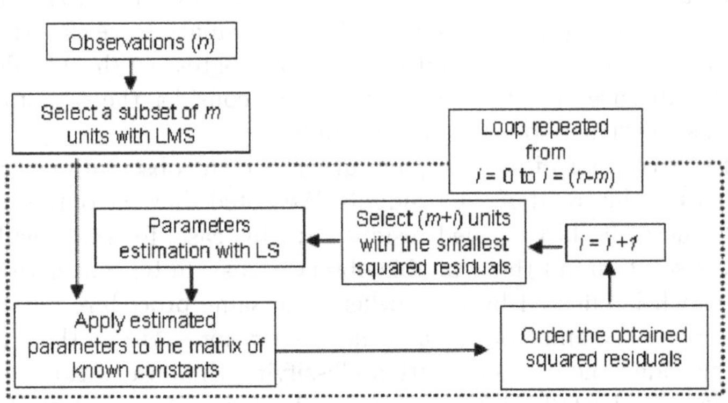

Fig. 1. Flowchart of the Forward Search

Summarising, the FS method is made up of three steps:
- *Choice of the initial subset*
- *Forward search adding an observation at every step*
- *Monitoring of the statistics during the process*

We applied the FS method to the problem concerning the transformations of reference system (*datum*) as described by equation (1). In the following paragraphs a case study is discussed.

4.1 Case Study

We consider a common geodetic problem of planar transformation between two different reference systems A and B (Fig. 2). For a given set of points the coordinates are known in both *datum*. It is hereby supposed that these points are not affected by any kind of systematic effects. The unknown parameters are, according to the linearisation of equation (1), a, b, T_X, T_Y. We calculated them based on a set of 10 points, which coordinates were known in both reference systems. In order to perform a comparison, we estimated the unknown parameters by means of the LS method, which result is comparable to that of the last step of the FS method. The comparison verifies the correct compilation of the equations and the proper progress of the FS algorithm. Due to the absence of outliers at this step we consider the so obtained parameters as the "true" solution of the problem.

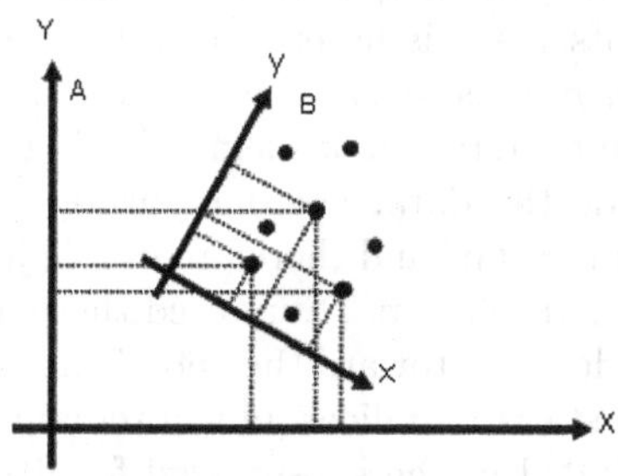

Fig. 2. Planar transformation problem

The first test consists of inserting an outlier into the observations (vector L) and to analyze the trend of the residuals. We noted that the outlier is clearly detected, due to its large scaled residual. In this case, for the calculation of the correct solution, all the available observations can be considered, except the one which is affected by the outlier. The same procedure has been followed inserting 3, 5 and 8 outliers, in the observations. Also in these cases the method provided the correct solution classifying the observations in "clean" and "outliers". The following figure (Fig. 3) shows the trend of the scaled residuals for the case with 3 outliers (observation nr. 3, 6 and 9). Remarkable is how the scaled residual of the "clean" observations remain stable on a minimum, diverging only in the last three steps, when the outliers are included in the calculation.

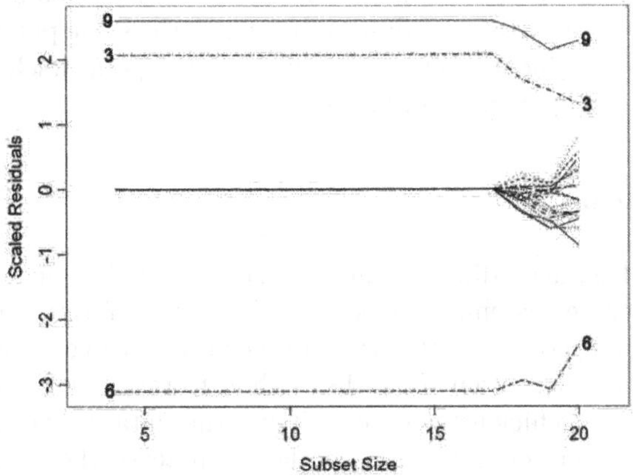

Fig. 3. Graph of scaled residuals

For each case using the estimated transformation parameters, the coordinates and the corresponding residuals of the every point were computed. In the following table (Table 1) the residuals of all observations are listed for the case with 5 outliers.

LS		FS ($m = 15$)	
$V_y[mm]$	$V_x[mm]$	$V_y[mm]$	$V_x[mm]$
2138.9	227.0	1.0	1.8
-5460.4	-1543.7	-7999.3	-3.2
2583.7	9201.7	0.1	12000.0
1293.5	-2932.8	-3.5	0.7
-9362.2	-2181.4	-10003.8	2.6
74.8	-1087.2	-0.6	1.2
-240.7	4856.0	1.5	4997.5
5964.0	-626.1	7002.7	-1.2
2361.9	-3521.0	3.2	2.4
656.6	-1938.4	-1.8	-4.4

Table 1. Comparison of the results

The left column shows the values obtained with LS, the right one those with FS. The possibility to detect the outliers in the way the FS method does, is a great advantage, because we can decide at which step the estimation of the unknown parameters has to be stopped. The FS method considers the observations (X_i, Y_i) as single independent information and not as couples of observations associated to a point. This phenomenon causes, in the considered case, the discard of only the erroneous Y component of the point, but not its X component. When working with coordinates, a certain link between X_i and Y_i components is assumed: both belong to the same point. Therefore, identifying one component as erroneous discards the point itself. This means that none of the components can be used.

5 Conclusion

The FS method is, according to our case study, a valid method to locate outliers and possible systematic effects in planar transformations problems. Considering the structure and the procedure characterizing the FS approach, which allows a crossover from the LMS method to the LS method, this technique is particularly efficient when classifying the data in two main groups: "clean data" and "outliers". So, this method can serve the computation as well as the data filtering. This method considers every observation as a single and independent input. Therefore, as in our case, the correlation between X_i and Y_i can not be taken into account during the computation. This shortcoming could be solved using a particular version of the FS, which allows the input to be groups (blocks) and not single observations. In our case a block is composed by the two coordinates (X, Y) defining a point. The future development could foresee the computation of the C_{xx} matrix of variance-covariance. Also, the estimation of the variance of the weight unit for each epoch could be implemented.

References

ATKINSON A.C., RIANI M. (2000): Robust Diagnostic Regression Analysis. *Springer, New York.*

ATKINSON A.C., RIANI M. and CERIOLI A. (2004): Exploring Multivariate Data with the Forward Search. *Springer, New York.*

HUBER, J.P. (1981): Robust statistics. *Wiley, New York.*

HUBER, J.P. et al. (1986): Robust statistics: the approach based on influence functions. *Wiley, New York.*

WICKI, F. (1999): Robuste Schätzverfahren für die Parameterschätzung in geodätischen Netzen. *ETH, Zurich.*

An R Package for the Forward Analysis of Multivariate Data

Aldo Corbellini[1] and Kjell Konis[2]

[1] Dipartimento di Studi Economici e Quantitativi,
Università Cattolica del Sacro Cuore di Piacenza, Italy
aldo@netline.it

[2] Department of Statistics
University of Oxford, United Kingdom
konis@stats.ox.ac.uk

Abstract. We describe the R package Rfwdmv (R package for the forward multivariate analysis) which implements the forward search for the analysis of multivariate data. The package provides functions useful for detecting atypical observations and/or subsets in the data and for testing in a robust way whether the data should be transformed. Additionally, the package contains functions for performing robust principal component analyses and robust discriminant analyses as well as a range of graphical tools for interactively assessing fitted forward searches on multivariate data.

1 Introduction

The forward search is a powerful general method for detecting unidentified subsets of the data and for determining their effect on fitted models. These subsets may be clusters of distinct observations or there may be one or more outliers. Alternatively, all the data may agree with the fitted model. The plots produced by the forward search make it possible to distinguish between these situations and to identify any influential observations. The Rfwdmv package implements the forward search for several methods arising in the analysis of multivariate data. This paper will show by example how the forward search can be used to test transformations of variables in multivariate data and for multivariate clustering.

2 The Forward Search in Multivariate Data

The forward search in multivariate data assumes that the data can be divided into two parts. There is a larger *clean* part that is well described by the proposed model and there may be outliers which are assumed to follow an arbitrary distribution. The goal is to estimate the parameters for the proposed model generating the *clean* data. The forward search in multivariate analysis begins by using a robust or resistant method to identify an outlier free subset of m observations where $m \ll n$. This subset is used to estimate a location vector and covariance matrix which are in turn used to compute the Mahalanobis distance for each observation in the data. A new subset of size $m+1$ is then chosen by taking the observations with the $m+1$ smallest Mahalanobis distances. This process is repeated until $m = n$. The Mahalanobis distances and other diagnostic statistics are computed and stored for each subset encountered during the forward search. The forward plots are obtained by plotting these diagnostic statistics against the subset size.

3 Obtaining the Rfwdmv Package

The Rfwdmv package is available from CRAN and can be installed straight from the R Console using the `install.packages` function. Alternatively the package can be downloaded from the following web site.

```
http://www.riani.it/arc/software.html
```

4 Example 1: Multivariate Clustering Using the Forward Search

The use of the forward search for multivariate clustering will be demonstrated by analyzing the financial data given in Zani *et al.* (1998). These data were taken from the Italian financial journal *Il Sole – 24 Ore* for May 7th, 1999 and are comprised of measurements on 103 investment funds operating in Italy since April 1996. The variables are y_1: short term performance (12 months), y_2: medium term performance (36 months), and y_3: medium term volatility (36 months). These data are included in the Rfwdmv package in the data frame `fondi.dat`.

4.1 Preliminary Analysis

We begin the analysis by looking at a pairs plot of the data. The Rfwdmv package contains the function `fwdmvPrePlot` which produces a pairs-like plot with univariate box plots along the main diagonal. The optional argument `panel = panel.bb` draws overlaid bivariate box plots on each of the off-diagonal panels.

```
> fwdmvPrePlot(fondi.dat, panel = panel.bb)
```

The plot produced is shown in figure 4.1. The right column suggests that there are two clusters in the data. We proceed by using the function `fwdmv` to fit an initial forward search to the financial data.

```
> fondi.init <- fwdmv(fondi.dat)
> plot(fondi.init)
```

The plot command prompts the user to select a plot from a menu of 11 diagnostic forward plots. Selecting `Distance Plot` generates the forward plot of Mahalanobis distances seen in figure 4.2. The majority of trajectories in the plot fall into two bands suggesting that there are two clusters in the data. Users are encouraged to read the documentation for `plot.fwdmv` and to experiment with all of the diagnostic forward plots provided in the Rfwdmv package.

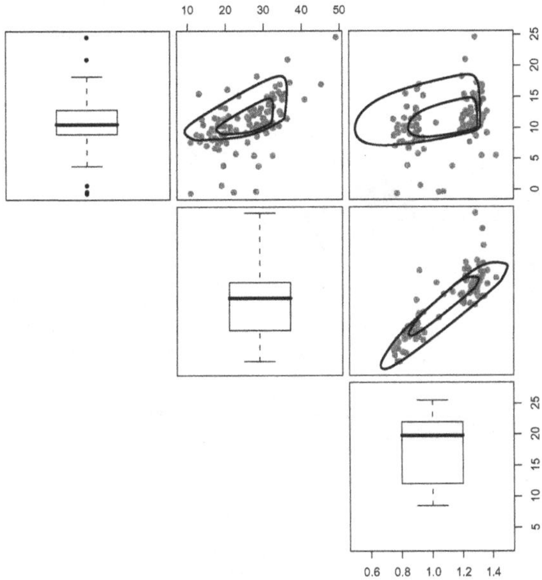

Fig. 1. The pairs plot of the financial data with overlaid bivariate boxplots produced by the function `fwdmvPrePlot` in the Rfwdmv package

4.2 Interactive Clustering

The Rfwdmv package includes the function `partition` to facilitate interactive clustering of multivariate data. The R object `fondi.init` contains a fitted initial forward search on the financial data. The partition function uses the distances in this object to allow the user to assign groups. The command

```
> p1 <- partition(fondi.init)
```

produces the forward plot of the Mahalanobis distances in figure 4.2. The user is then prompted to select a band of trajectories by drawing a line segment on the plot. All of the trajectories crossing this line segment are assigned to a tentative group. The returned object (in this example `p1`) is a fitted forward search similar to the input object except that its group element has been updated to include the newly assigned tentative group. The process can be repeated by calling the `partition` function on the object `p1` and selecting the band of trajectories lying along the bottom of the plot.

```
> p2 <- partition(p1)
```

Figure 4.2 shows a forward plot of the Mahalanobis distances after two tentative groups have been assigned.

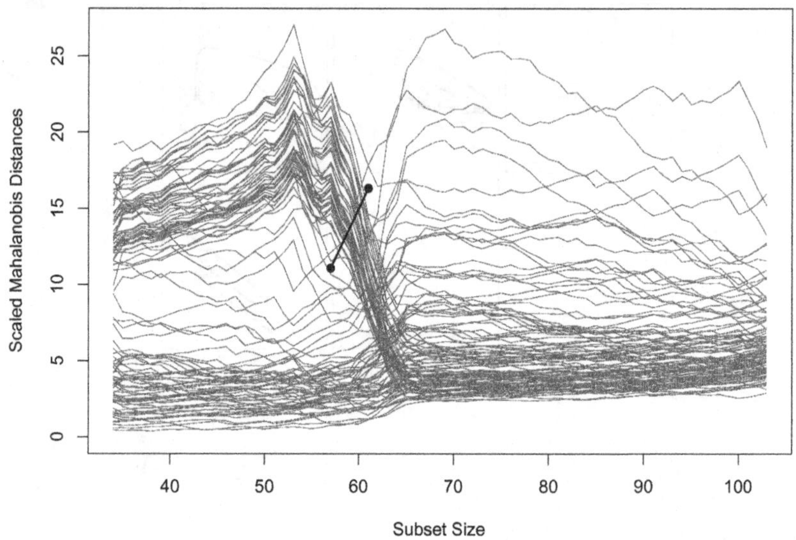

Fig. 2. A forward plot of the Mahalanobis distances computed during the initial forward search. The upward-sloping line segment represents the interactive selection of a group of trajectories

The next step is to fit a final forward search to the financial data using the user selected tentative groups. A final forward search fits a separate multivariate model to each of the tentative groups then assigns the unassigned units to the closest group. The function `fwdmv` will update an initial forward search to a final forward search if the forward search object has tentative groups assigned. The syntax is simply

```
> fondi.final <- fwdmv(p2)
```

Again, users are encouraged to use the various plot methods to assess the fit of the final forward search. Finally, the function `fwdmvConfirmPlot` should be used to examine to which group each unassigned unit should be allocated. The following command produces a so-called confirmatory plot for the last 40 steps of the forward search.

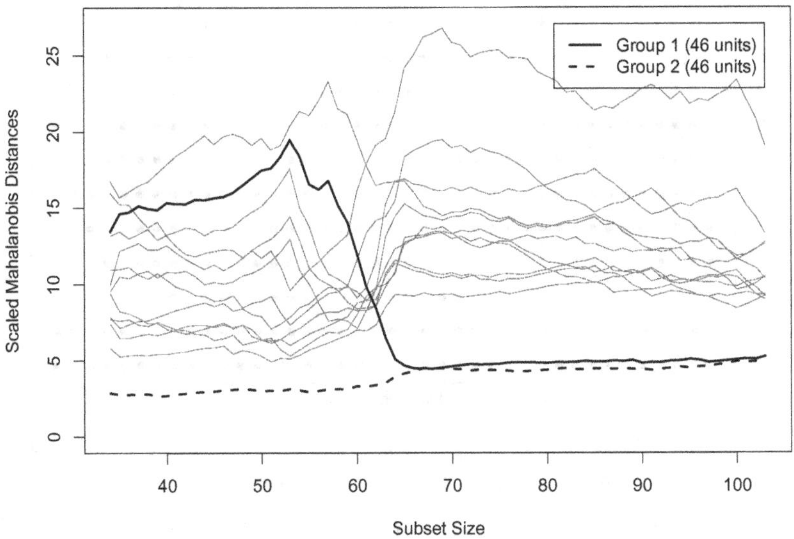

Fig. 3. A forward plot of the Mahalanobis distances with two tentative groups assigned. Each group contains 46 units leaving 12 units unassigned

```
> fwdmvConfirmPlot(fondi.final, n.steps = 40)
```

The plot is shown in figure 4.2. Units 54, 59, and 89 are closest to group 1 during the last 40 steps of the forward search. The rest are closest to group 2 except for unit 21 which switches between groups several times during the last 40 steps of the search and is thus difficult to assign.

5 Example 2: Testing Multivariate Transformations Using the Forward Search

With ungrouped data the routine to perform the forward search and test for transformation to normality is given by

```
> l.mle <- lambda.fwdmv(mussels.dat, lambda=c(1,1,1,1,1))
> plot(l.mle)
```

where `mussels.dat` is the multivariate data set chosen for this example. Default options of `lambda.fwdmv` lead to find the initial subset as the intersection of robust bivariate contours superimposed in each bivariate scatter diagram. Vector `lambda` contains the set of transformation parameters to test. The default value is 1 (i.e. $H_0 : \lambda_j = 1, j = 1, \ldots, p$) so in this case it could have been omitted. The ordering of Mahalanobis distances at each step of the forward search uses variables transformed as specified in `lambda`. The function `lambda.fwdmv` computes the p maximum likelihood estimates

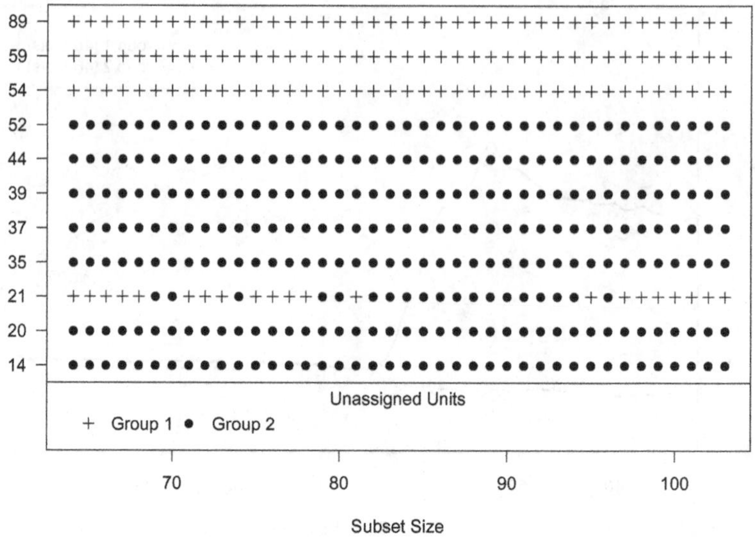

Fig. 4. A confirmatory plot of the final forward search. Notice that unit 21 switches groups several times during the last 40 steps of the forward search

$\hat{\lambda}_j$ in each step of the forward procedure. The left panel of Figure 5 shows the graphical output when the `plot` command is invoked, while the right panel shows the monitoring of the likelihood ratio test statistic, for the null hypothesis of no transformation. This plot can be obtained by

```
> l.rat <- lik.ratio.fwdmv(l.mle)
> plot(l.rat)
```

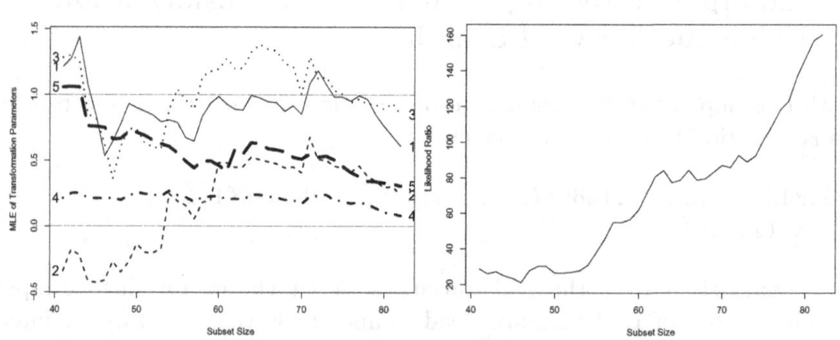

Fig. 5. Monitoring of maximum likelihood estimates (left panel) and likelihood ratio test (right panel) during the forward search of dataset "mussels"

Given that the values of the likelihood ratio test must be compared with a χ^2_5 it is clear that the data must be transformed. The monitoring of maximum likelihood estimates quickly leads to select the combination $\lambda_0 = [0.5, 0, -0.5, 0, 0]^T$ (for further details see Atkinson *et al.*(2004)). In order to confirm this combination of values of lambda we can construct a fan plot (see Atkinson and Riani (2000)) using the command:

```
> fan <- lambda.test.fwdmv(mussels.dat, c(0.5, 0, 0.5, 0, 0))
```

Basically, each transformation is first-order expanded around the most five common values of λ (that is -1, -0.5, 0, 0.5, 1). This is a way for transforming a multivariate problem into a series of p univariate ones. In each search we can test the transformation by comparing the signed square root likelihood ratio test with a standard normal distribution. By plotting `fan` we have a version of the fan plot for multivariate data, where it is also possible to have the confirmatory fan plots for each variable as panels in a single page,

```
> plot(fan)
Make a plot selection (or 0 to exit):
1:plot: width
2:plot: height
3:plot: length
4:plot: shell
5:plot: mass
```

The names of the variables are plotted as labels of the y axis in each panel. For instance, in Figure 5 we show expansion for variables `width` and `length`.

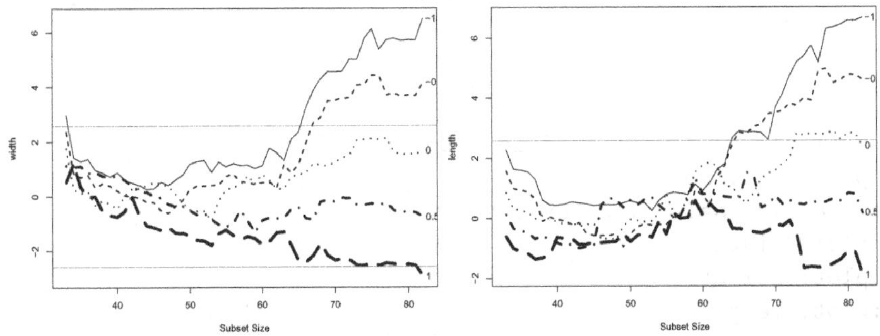

Fig. 6. Multivariate fan plot around the vector $\lambda_0 = [0.5, 0, 0.5, 0, 0]^T$ for width and length in mussels data. The horizontal lines are the 0.5 and 99.5 percentiles of the standard normal distribution

Figure 5 shows that the best value of the transformation parameter for the first variable (width) is 0.5. Log transformation is also acceptable while the hypothesis of no transformation is at the boundary of rejection region in the last 10 steps of the search.

It is also possible to build profile likelihood plots which monitor, for a given step of the forward search, the profile likelihood of a parameter λ_j, assuming fixed at their maximum likelihood estimates all parameters $\hat{\lambda}_k$, $k \neq j = 1, \ldots, p$.
For our example data we have chosen the last step of the forward search, i.e.

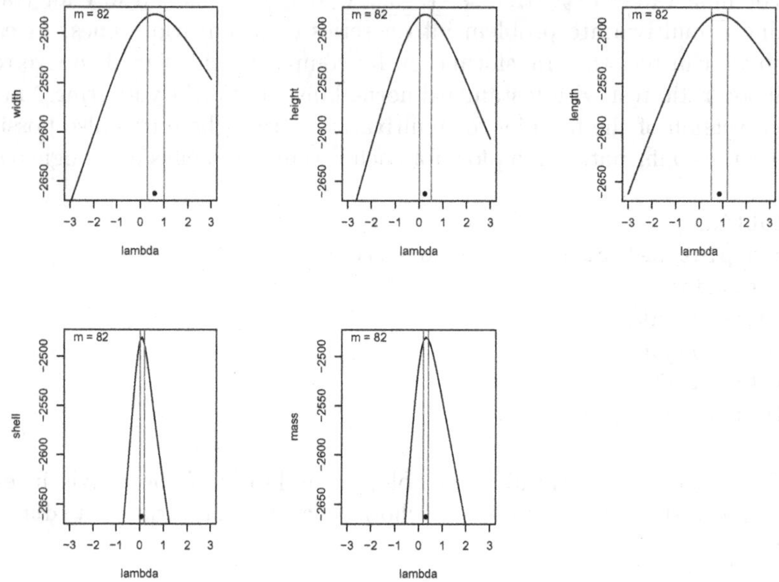

Fig. 7. Profile likelihood for `all` variables in the mussels data example. Step of the forward search $m = N = 82$

```
> prof.mle <- lambda.profile.fwdmv(l.mle)
> plot(prof.mle)
Make a plot selection (or 0 to exit):
1:plot: width
2:plot: height
3:plot: length
4:plot: shell
5:plot: mass
6:plot: all
```

The output for all variables is given in Figure 5. The dot just above the x axis corresponds to the value of maximum likelihood while the two vertical lines represent asymptotic 95% confidence intervals for each value of λ, based on χ_1^2 distribution of twice the loglikelihood ratio. This plot clearly shows what are the variables with a sharply peaked profile log-likelihood (in this example "shell" and "mass").

References

ATKINSON, A.C. and RIANI M. (2000) *Robust Diagnostic Regression Analysis*, Springer-Verlag, New York.

ATKINSON, A.C., RIANI M., and CERIOLI A. (2004) *Exploring Multivariate Data with the Forward Search*, Springer-Verlag, New York.

ZANI, S., RIANI, M. and CORBELLINI, A. (1998): Robust bivariate boxplots and multiple outlier detection. *Computational Statistics and Data Analysis, 28, 257–270.*

A Forward Search Method for Robust Generalised Procrustes Analysis

Fabio Crosilla and Alberto Beinat

University of Udine, Via Cotonificio 114, Udine (Italy)
E-mail: fabio.crosilla@uniud.it, alberto.beinat@uniud.it

Abstract. One drawback of Procrustes Analysis is the lack of robustness. To overcome this limitation a procedure that applies the Generalised Procrustes methods, by way of a progressive sequence inspired to the "forward search", was developed. Starting from an initial centroid, defined by the partial point configuration satisfying the LMS principle, this is extended by joining, at every step, a restricted subset of the remaining points. At every insertion, the updated centroid, redetermined by the new considered points, is compared with the previous by way of the common elements. If significant variations of the similarity transformation parameters occur, they reveal the presence of outliers or non stationary points among the new elements just inserted.

1 Introduction

Procrustes Analysis is a well known multivariate technique to provide L.S. matching of two or more data matrices, or for the multidimensional rotation and scaling of different matrix configurations. The matching is carried out by the direct solution of similarity transformation problems between pairs or multiple data matrices configurations. Employed at first in factor analysis (e.g. Schoenemann, 1966) today is a popular method in shape analysis (e.g. Goodall, 1991; Dryden and Mardia, 1998), in biology (e.g. Gelfand et al., 1996), in geodesy and photogrammetry (e.g. Crosilla and Beinat, 2002), in logistics (e.g. Kong and Ceglarek, 2003), in image analysis (e.g. Anderson, 1997) and in other fields. Ordinary Procrustes Analysis (OPA) directly estimates the similarity transformation parameters $\{c, \mathbf{R}, \mathbf{t}\}$, from $\mathbf{X}_1$ (*origin*) to $\mathbf{X}_2$ (*target*) data sets, which satisfy the minimum condition of the following Least Squares function:

$$\operatorname{tr}\left(c\mathbf{X}_1\mathbf{R} + \mathbf{1}\mathbf{t}^T - \mathbf{X}_2\right)^T \mathbf{W}_R \left(c\mathbf{X}_1\mathbf{R} + \mathbf{1}\mathbf{t}^T - \mathbf{X}_2\right) = \min \tag{1}$$

under the orthogonality condition $\mathbf{R}^T\mathbf{R} = \mathbf{I}$. In Formula (1) $\mathbf{X}_1$ and $\mathbf{X}_2$ are two $n \times k$ dataset matrices, containing the k-dim coordinates of the same set of n points related to two different reference frames; $\mathbf{W}_R$ is one optional $n \times n$ diagonal matrix containing the global weight assigned to every point; $\mathbf{1}$ is the $n \times 1$ auxiliary unitary vector; $\mathbf{t}$, $\mathbf{R}$ and c are the unknowns, *i.e.* the $k \times 1$ translation vector, the $k \times k$ rotation matrix and the isotropic scale factor, respectively. When $\mathbf{W}_R = \mathbf{Q}^T\mathbf{Q}$ (Cholesky decomposition), after substituting $\mathbf{X}_{1W} = \mathbf{Q}\mathbf{X}_1$, $\mathbf{X}_{2W} = \mathbf{Q}\mathbf{X}_2$ and $\mathbf{1}_W = \mathbf{Q}\mathbf{1}$, and computing once, and for all, the centering operator $\mathbf{B} = \mathbf{I} - \mathbf{1}_W\mathbf{1}_W^T / \left(\mathbf{1}_W^T\mathbf{1}_W\right)$ with $\mathbf{B}^T = \mathbf{B}$, $\mathbf{BB} = \mathbf{B}$, performing the Singular Value Decomposition of the matrix product $\mathbf{X}_{1W}^T\mathbf{B}\mathbf{X}_{2W} = \mathbf{V}\mathbf{D}_S\mathbf{W}^T$, we directly obtain the rotation matrix $\mathbf{R} = \mathbf{V}\mathbf{W}^T$, the global scale factor $c = \mathrm{tr}\left\{\mathbf{R}^T\mathbf{X}_{1W}^T\mathbf{B}\mathbf{X}_{2W}\right\} / \mathrm{tr}\left\{\mathbf{X}_{1W}^T\mathbf{B}\mathbf{X}_{1W}\right\}$ and the translation vector $\mathbf{t} = \left(\mathbf{X}_{2W} - c\mathbf{X}_{1W}\mathbf{R}\right)^T\mathbf{1}_W / \left(\mathbf{1}_W^T\mathbf{1}_W\right)$.

Generalised Procrustes Analysis (GPA) concerns instead with the multiple and simultaneous transformations of $m \geq 2$ data sets of n corresponding k-dim points, whose coordinates are referred to $m \geq 2$ different reference frames, and are characterised by measurement noise. We look for the solution that satisfies the following least squares objective function:

$$\mathrm{tr}\sum_{i<j}^{m}\left\{\left(c_i\mathbf{X}_i\mathbf{R}_i + \mathbf{1}\mathbf{t}_i^T\right) - \left(c_j\mathbf{X}_j\mathbf{R}_j + \mathbf{1}\mathbf{t}_j^T\right)\right\}^T \cdot \\ \cdot\left\{\left(c_i\mathbf{X}_i\mathbf{R}_i + \mathbf{1}\mathbf{t}_i^T\right) - \left(c_j\mathbf{X}_j\mathbf{R}_j + \mathbf{1}\mathbf{t}_j^T\right)\right\} = \min \tag{2}$$

where $\mathbf{X}_1 \ldots \mathbf{X}_m$ are $m \geq 2$ data matrices of size $n \times k$, each one containing the coordinates of the same set of n corresponding points defined in m different R_k reference frames. The solution of Equation (2) represents the GPA problem described by Kristof and Wingersky (1971), Gower (1975), ten Berge (1977) and Goodall (1991). The GPA problem has an alternative formulation. Said $\mathbf{X}_i^p = c_i\mathbf{X}_i\mathbf{R}_i + \mathbf{1}\mathbf{t}_i^T$, the following measures:

$$\sum_{i<j}^{m}\left\|\mathbf{X}_i^p - \mathbf{X}_j^p\right\|^2 = \sum_{i<j}^{m}\mathrm{tr}\left(\mathbf{X}_i^p - \mathbf{X}_j^p\right)^T\left(\mathbf{X}_i^p - \mathbf{X}_j^p\right) \tag{3}$$

$$m\sum_{i}^{m}\left\|\mathbf{X}_i^p - \mathbf{K}\right\|^2 = m\sum_{i}^{m}\mathrm{tr}\left(\mathbf{X}_i^p - \mathbf{K}\right)^T\left(\mathbf{X}_i^p - \mathbf{K}\right) \tag{4}$$

are perfectly equivalent (*e.g.* Borg and Groenen, 1997), where $\mathbf{K}$ is the unknown geometrical centroid. Therefore Eq. (4), instead of Eq. (3), can be minimised so to determine the unknowns $\{c, \mathbf{R}, \mathbf{t}\}_i$ $(i = 1 \ldots m)$ that make it possible to iteratively compute the final $\mathbf{X}_i^p$ $(i = 1 \ldots m)$. Matrix $\hat{\mathbf{K}} = \frac{1}{m}\sum_{i=1}^{m}\mathbf{X}_i^p$ represents the LS estimate of $\mathbf{K}$. Note that $\mathbf{K} + \mathbf{E}_i = \mathbf{X}_i^p$, where $\mathrm{vec}\left(\mathbf{E}_i\right) : \mathrm{N}\left\{0, \mathbf{\Sigma} = \sigma^2\left(\mathbf{Q}_n \otimes \mathbf{Q}_k\right)\right\}$, and $\mathbf{\Sigma}$ has a factored structure.

2 Procrustes Statistics

In order to propose a test to verify the significance of the transformation parameters, we introduce the so called Procrustes statistics. We consider, at first, the case of two configurations, one of which contains random errors. Let $\mathbf{X}_i$ be therefore a $n \times k$ matrix containing the k coordinates of n points defined into an arbitrary reference frame, and $\mathbf{X}_j = \mathbf{X}_i + s\mathbf{Z}$, where $\mathbf{Z}$ is a random error matrix $\mathbf{Z} : N\{0,1\}$, and s is a scale factor. The residual distance $G(\mathbf{X}_i, \mathbf{X}_j)$ between $\mathbf{X}_i$ and $\mathbf{X}_j$ (Sibson, 1979):

$$G(\mathbf{X}_i, \mathbf{X}_j) = \mathrm{tr}(\mathbf{X}_i - \mathbf{X}_j)^T (\mathbf{X}_i - \mathbf{X}_j) = s^2 \mathrm{tr}\left(\mathbf{Z}^T\mathbf{Z}\right) \tag{5}$$

follows a $s^2\chi^2$ distribution with $n \times k$ degrees of freedom. The Procrustes statistics is the residual distance $G_f(\mathbf{X}_i, \mathbf{X}_j)$ after a Procrustes transformation f has been performed on $\mathbf{X}_j$. We can define:

$$G_t(\mathbf{X}_i, \mathbf{X}_j) := \inf\left\{G\left(\mathbf{X}_i, \mathbf{X}_j + \mathbf{1t}^T\right)\right\} :\sim s^2\chi^2_{\{nk-k\}} \tag{6}$$

$$G_{tR}(\mathbf{X}_i, \mathbf{X}_j) := \inf\left\{G\left(\mathbf{X}_i, \mathbf{X}_j\mathbf{R} + \mathbf{1t}^T\right)\right\} :\sim s^2\chi^2_{\{(nk-1)/2k(k+1)\}} \tag{7}$$

$$G_{tRc}(\mathbf{X}_i, \mathbf{X}_j) := \inf\left\{G\left(\mathbf{X}_i, c\mathbf{X}_j\mathbf{R} + \mathbf{1t}^T\right)\right\} :\sim s^2\chi^2_{\{(nk-1)/2k(k+1)-1\}} \tag{8}$$

Quite similar is the case of errors in both configurations. Let $\mathbf{X}_i$ and $\mathbf{X}_j$ be two $n \times k$ matrices obtained from the same $\mathbf{X}$, to which two random error matrices $\mathbf{Z}_i : N\{0,1\}$ and $\mathbf{Z}_j : N\{0,1\}$ are summed up, respectively: $\mathbf{X}_i = \mathbf{X} + s\mathbf{Z}_i$, and $\mathbf{X}_j = \mathbf{X} + s\mathbf{Z}_j$; s is the same scale factor. The residual distance $G(\mathbf{X}_i, \mathbf{X}_j)$ between $\mathbf{X}_i$ and $\mathbf{X}_j$ (Langron and Collins, 1985) is:

$$G(\mathbf{X}_i, \mathbf{X}_j) = s^2\mathrm{tr}(\mathbf{Z}_i - \mathbf{Z}_j)^T (\mathbf{Z}_i - \mathbf{Z}_j) = s^2\mathrm{tr}(\mathbf{W}_{ij})^T (\mathbf{W}_{ij}) \tag{9}$$

Since $\mathbf{Z}_i$ and $\mathbf{Z}_j$ are independent, and $\mathbf{W}_{ij} : N\{0,2\}$, $G(\mathbf{X}_i, \mathbf{X}_j)$ follows a $2s^2\chi^2$ distribution with $n \times k$ degrees of freedom. The Procrustes statistics $G_f(X_i, X_j)$ retain the same expressions as above, except that they follow now a $2s^2\chi^2$ distribution with the same degrees of freedom of the preceding case.

3 ANOVA of Procrustes Statistics

The previous results depend on an unknown parameter s. We can get over this problem studying the statistical distribution assumed by the numerical values obtained by the ratio:

$$\frac{\text{Specific Procrustes statistics component}}{\text{Residual distance after a complete S - transformation}}$$

To investigate the significance of each geometrical transformation, i.e. translation, translation + rotation, translation + rotation + dilation, involved in the Procrustes problem we perform an analysis of variance (ANOVA) for the different Procrustes distances. In the case of two configurations $\mathbf{X}_i$ and $\mathbf{X}_j$, the Procrustes statistics after a centering translation is:

$$G_t(\mathbf{X}_i,\mathbf{X}_j) = \mathrm{tr}\left\{(\mathbf{X}_i - \mathbf{X}_j)^T (\mathbf{X}_i - \mathbf{X}_j)\right\} - n(\mathbf{x}_i - \mathbf{x}_j)^T (\mathbf{x}_i - \mathbf{x}_j) \tag{10}$$

where $\mathbf{x}_i$ and $\mathbf{x}_j$ are the positional vectors of the gravity centers of datasets $\mathbf{X}_i$ and $\mathbf{X}_j$. The effect due to the centering translation $T(\mathbf{X}_i, \mathbf{X}_j)$ is:

$$T(\mathbf{X}_i,\mathbf{X}_j) = G(\mathbf{X}_i,\mathbf{X}_j) - G_t(\mathbf{X}_i,\mathbf{X}_j) \tag{11}$$

G_t can be partitioned into the effect due to rotation $R(\mathbf{X}_i, \mathbf{X}_j)$ plus the residual distance after translation and rotation G_{tR}, that is:

$$R(\mathbf{X}_i, \mathbf{X}_j) = G_t(\mathbf{X}_i, \mathbf{X}_j) - G_{tR}(\mathbf{X}_i, \mathbf{X}_j) \tag{12}$$

Finally G_{tR} can be further partitioned into the effects due to global scaling $S(\mathbf{X}_i, \mathbf{X}_j)$ plus the residual distance after translation, rotation and dilation G_{tRc}:

$$S(\mathbf{X}_i, \mathbf{X}_j) = G_{tR}(\mathbf{X}_i, \mathbf{X}_j) - G_{tRc}(\mathbf{X}_i, \mathbf{X}_j) \tag{13}$$

The ratios S/G_{tRc}, R/G_{tRc}, and T/G_{tRc} are well approximated by the Fisher distribution (χ^2/χ^2) if the null hypothesis H_0 is satisfied. Assuming a first kind error value α and the proper degrees of freedom, the rejection of the null hypothesis H_0 shows the existence of a significant value for the transformation parameters. For instance, for the translation case:

$$H_0 = \frac{T}{G_{tRc}} \le F_{1-\alpha, k, nk-\frac{1}{2}k(k+1)-1} \Rightarrow \text{No significant translation occurs}$$

$$H_1 = \frac{T}{G_{tRc}} > F_{1-\alpha, k, nk-\frac{1}{2}k(k+1)-1} \Rightarrow \text{Significant translation exists}$$

When m configurations are involved, the Procrustes statistics become (Langron and Collins, 1985):

$$G_t(\mathbf{X}_1 \ldots \mathbf{X}_m) := \sum_{j=1}^{m} \inf\left\{G\left(\mathbf{K}, \mathbf{X}_j + \mathbf{1}\mathbf{t}_j^T\right)\right\} \tag{14}$$

$$G_{tR}(\mathbf{X}_1 \ldots \mathbf{X}_m) := \sum_{j=1}^{m} \inf\left\{G\left(\mathbf{K}, \mathbf{X}_j\mathbf{R}_j + \mathbf{1}\mathbf{t}_j^T\right)\right\} \tag{15}$$

$$G_{tRc}(\mathbf{X}_1 \ldots \mathbf{X}_m) := \sum_{j=1}^{m} \inf\left\{G\left(\mathbf{K}, c_j\mathbf{X}_j\mathbf{R}_j + \mathbf{1}\mathbf{t}_j^T\right)\right\} \tag{16}$$

where $\mathbf{K}$ is the unknown mean transformed configuration (centroid). Again, $\hat{\mathbf{K}} = \frac{1}{m}\sum_{i=1}^{m}\left(c_i\mathbf{X}_i\mathbf{R}_i + \mathbf{1}\mathbf{t}_i^T\right)$ represents the LS estimate of $\mathbf{K}$.

Table 1 summarises the degrees of freedom df_1 and df_2 for the specific F tests: m is the number of datasets, n are the points in each dataset, and k the point dimensions.

4 Robust Procrustes Analysis

The Robust solution of the Procrustes problem has always attracted many efforts. We mention, for instance, the methods by Siegel and Benson (1982), and by Rohlf and Slice (1992), based on repeated median. Recently, the use of iterative and weighted majorisation algorithms has been proposed for the

Transformation	*df*1	*df*2
Translation	$(m-1)k$	$(m-1)\{nk - 1/2k(k+1) - 1\}$
Rotation	$1/2k(m-1)(k-1)$	$(m-1)\{nk - 1/2k(k+1) - 1\}$
Dilation	$m-1$	$(m-1)\{nk - 1/2k(k+1) - 1\}$

Table 1. Degrees of freedom for the specific F tests

solution of the partial rotation problem (Kiers, 2002; Groenen et al., 2005 respectively). This paper describes a new algorithm, derived from the Robust Regression Analysis based on the Iterative Forward Search approach proposed by Atkinson and Riani (2000). In the current implementation, the procedure starts from a partial point configuration not containing outliers nor non stationary data. At each iteration, it enlarges the initial dataset by one or more observations (points), till a significant variation of the Procrustes statistics occur. At this point, if a pairwise configuration matching is executed, the process ends. Otherwise, if a generalised problem is carried out, the proposed method is able to identify the configuration containing outliers or non stationary data. In order to define the initial configuration subset not containing any outlier, it is necessary to compute the LS estimate $\hat{\mathbf{K}}^i$ of the corresponding centroid $\mathbf{K}^i$, and consequently determine the similarity transformation parameters for all the $j = 1 \ldots m$ data sub-matrices $\mathbf{X}_j^i$:

$$\hat{\mathbf{K}}^i = \frac{1}{m}\sum_{j=1}^{m}\left(c_j^i \mathbf{X}_j^i \mathbf{R}_j^i + \mathbf{1}\mathbf{t}_j^{iT}\right) \tag{17}$$

This procedure is repeated for every $i = 1 \ldots \binom{n}{s}$ possible point subset, where s is the number of points forming the generic subset. Now, the global pseudo-centroid is computed by applying the transformation parameters relative to the i-th data sub matrix $\mathbf{X}_j^i$, to the full corresponding $\mathbf{X}_j$, obtaining $\mathbf{X}_j^{P(i)}$, and averaging their values:

$$\tilde{\mathbf{K}}^i = \frac{1}{m}\sum_{j=1}^{m}\left(c_j^i \mathbf{X}_j \mathbf{R}_j^i + \mathbf{1}\mathbf{t}_j^{i^T}\right) = \frac{1}{m}\sum_{j=1}^{m}\mathbf{X}_j^{P(i)} \tag{18}$$

To define the initial point subset not containing outliers, the least median of squares (LMS) principle has been applied (Rousseeuw, 1984). As it is well known, this regression method can reach a breakdown point as high as 50%, although very recent publications (Xu, 2005) have put in evidence the fact that, under particular conditions, the method can breakdown also in the presence of one single outlier. Much attention will be devoted to this aspect in a next paper, but, as evidenced by Cerioli and Riani (2003), the forward search procedure operates properly also in the presence of outliers in the starting subset, being able to remove them at subsequent steps of the search.

Among all the possible configuration subsets, the one satisfying the following LMS condition is chosen as the initial one:

$$\text{med diag} \sum_{j=1}^{m} \left(\mathbf{X}_j^{P(i)} - \tilde{\mathbf{K}}^i\right)\left(\mathbf{X}_j^{P(i)} - \tilde{\mathbf{K}}^i\right)^T = \min \tag{19}$$

This initial subset is then enlarged joining up the point for which:

$$\text{diag} \sum_{j=1}^{m} \left(\mathbf{X}_j^{P(i)} - \tilde{\mathbf{K}}^i\right)\left(\mathbf{X}_j^{P(i)} - \tilde{\mathbf{K}}^i\right)^T = \min \tag{20}$$

selected from the remaining $(n-s)$ points of the configuration, not belonging to the initial subset. The LS estimate $\hat{\mathbf{K}}^{i(+1)}$ of the enlarged partial centroid $\mathbf{K}^{i(+1)}$, and the S-transformation parameters for the m sub-matrices $\mathbf{X}_j^{i(+1)}$, are computed again as:

$$\hat{\mathbf{K}}^{i(+1)} = \frac{1}{m}\sum_{j=1}^{m}\left(c_j^{i(+1)}\mathbf{X}_j^{i(+1)}\mathbf{R}_j^{i(+1)} + \mathbf{1t}_j^{i(+1)^T}\right) = \frac{1}{m}\sum_{j=1}^{m}\mathbf{X}_j^{i(+1),P\{i(+1)\}} \tag{21}$$

Now, to proceed to the ANOVA Procrustes statistics, it is necessary to verify whether a significant variation of the S-transformation parameters occurs by enlarging the original selected data subset. To this aim, the total distance between the partial centroid $\hat{\mathbf{K}}^{\mathbf{i}}$ and its matching elements in the m sub-matrices $\mathbf{X}_j^{i,P\{i(+1)\}}$, obtained by applying to the original $\mathbf{X}_j^i$ the S-transformation parameters relating to the $i(+1)$ dataset, is computed:

$$G = \sum_{j=1}^{m} \text{tr}\left(\mathbf{X}_j^{i,P\{i(+1)\}} - \hat{\mathbf{K}}^i\right)^T\left(\mathbf{X}_j^{i,P\{i(+1)\}} - \hat{\mathbf{K}}^i\right) \tag{22}$$

The following Procrustes distances are also computed:

$$G_t = \sum_{j=1}^{m} \text{tr}\left(\mathbf{X}_j^{i,P\{i(+1)\}} + \mathbf{1dt}_j^T - \hat{\mathbf{K}}^i\right)^T\left(\mathbf{X}_j^{i,P\{i(+1)\}} + \mathbf{1dt}_j^T - \hat{\mathbf{K}}^i\right) \tag{23}$$

$$G_{tR} = \sum_{j=1}^{m} \text{tr}\left(\mathbf{X}_j^{i,P\{i(+1)\}}d\mathbf{R}_j + \mathbf{1dt}_j^T - \hat{\mathbf{K}}^i\right)^T\left(\mathbf{X}_j^{i,P\{i(+1)\}}d\mathbf{R}_j + \mathbf{1dt}_j^T - \hat{\mathbf{K}}^i\right) \tag{24}$$

$$G_{tRc} = \sum_{j=1}^{m} \text{tr}\left(dc_j\mathbf{X}_j^{i,P\{i(+1)\}}d\mathbf{R}_j + \mathbf{1dt}_j^T - \hat{\mathbf{K}}^i\right)^T\left(dc_j\mathbf{X}_j^{i,P\{i(+1)\}}d\mathbf{R}_j + \mathbf{1dt}_j^T - \hat{\mathbf{K}}^i\right) \tag{25}$$

after having taken care of the fact that the translation components relating to the $i(+1)$ subset must be previously reduced by the difference between the gravity centers of $\hat{\mathbf{K}}^{i(+1)}$, and $\hat{\mathbf{K}}^i$. Assuming a proper first kind error α, and the proper degrees of freedom df_1 and df_2, the rejection of the null hypothesis for the following tests:

$$\left\{\frac{G - G_t}{G_{tRc}}; \frac{G_t - G_{tR}}{G_{tRc}}; \frac{G_{tR} - G_{tRc}}{G_{tRc}}\right\} > F_{1-\alpha, df_1, df_2} \tag{26}$$

indicates a significant variation of some or of all the transformation parameters $\{d\mathbf{t}_j, d\mathbf{R}_j, dc_j\}$ at this step, due to the possible entering into the $\mathbf{X}_j^{i(+1)}$ datasets of outliers or non stationary data. If instead the null hypothesis is accepted for all the tests, the iterative process continues with the insertion of a further new point ($\mathbf{X}_j^{i(+2)}$), satisfying the Equation (20) within the remaining ones of the dataset. If the outlier presence has been detected by the proposed tests, it is necessary to identify the point configuration affected by its presence. A Procrustes statistics analysis is then applied to each pair $\mathbf{X}_j^{i,P\{i(+1)\}}$, $\mathbf{X}_k^{i,P\{i(+1)\}}$ ($j = 1 \ldots k-1; k = j+1 \ldots m$) of the transformed data sub-matrices. The Procrustes statistics then become:

$$G_t\left(\mathbf{X}_j^{i,P\{i(+1)\}}, \mathbf{X}_k^{i,P\{i(+1)\}}\right) := \inf\left\{G\left(\mathbf{X}_j^{i,P\{i(+1)\}}, \mathbf{X}_k^{i,P\{i(+1)\}} + \mathbf{1}d\mathbf{t}^T\right)\right\} \quad (27)$$

$$G_{tR}\left(\mathbf{X}_j^{i,P\{i(+1)\}}, \mathbf{X}_k^{i,P\{i(+1)\}}\right) := \inf\left\{G\left(\mathbf{X}_j^{i,P\{i(+1)\}}, \mathbf{X}_k^{i,P\{i(+1)\}} d\mathbf{R} + \mathbf{1}d\mathbf{t}^T\right)\right\} \quad (28)$$

$$G_{tRc}\left(\mathbf{X}_j^{i,P\{i(+1)\}}, \mathbf{X}_k^{i,P\{i(+1)\}}\right) := \inf\left\{G\left(\mathbf{X}_j^{i,P\{i(+1)\}}, dc\mathbf{X}_k^{i,P\{i(+1)\}} d\mathbf{R} + \mathbf{1}d\mathbf{t}^T\right)\right\} \quad (29)$$

and the ratios of Equation (26) tested as usual. From these results it is possible to state that the most reliable configurations are those satisfying the null hypothesis.

5 One Numerical Example

A simple numerical example is provided to better comprehend the steps involved in the procedure. The computations have been performed by a Matlab^{TM} program, specifically implemented, based on the GPA solution described by ten Berge (1977), that solves the canonical form of the GPA problem expressed by Equation (2) in iterative way. Starting from a given configuration of $n = 5$ points, $m = 3$ different datasets $\mathbf{X}_1$, $\mathbf{X}_2$ and $\mathbf{X}_3$ have been generated, by summing up a random error component ($N\{0;1\}$) to the original template coordinates. In order to generate an outlier, a 10σ error component has been introduced in the coordinates of point 2 in dataset $\mathbf{X}_3$ (outlined in italic).

$$\mathbf{X}_1 = \begin{bmatrix} 49.3146 & 100.7678 \\ 98.8917 & 69.7942 \\ 80.1955 & 1.5008 \\ 18.3983 & -1.0386 \\ 0.0484 & 68.9035 \end{bmatrix} \mathbf{X}_2 = \begin{bmatrix} 49.6739 & 99.4620 \\ 98.9904 & 69.1479 \\ 80.8130 & 2.3155 \\ 20.9951 & -0.5222 \\ -0.2012 & 67.6479 \end{bmatrix} \mathbf{X}_3 = \begin{bmatrix} 50.9386 & 100.9085 \\ \mathit{109.9664} & \mathit{80.7631} \\ 79.2434 & 0.0795 \\ 19.4101 & 0.8358 \\ -0.0365 & 71.7294 \end{bmatrix}$$

The first step required by the method is the definition of the best initial partial centroid, satisfying Eq. (19). Assumed s=3 the size of the initial subset, the search is carried out testing all the 5!/3!(5-3)! possible point combinations, and determining each time the correspondent GPA solution (Eq. (17)). Here, from Eq. (19), subset {1,3,5} resulted the "best" initial partial configuration. Therefore the correspondent LS estimates of the centered partial centroid $\mathbf{K}^i$ (Eq. (17)), and of the global pseudo-centroid $\tilde{\mathbf{K}}^i$ (Eq. (18)) become respectively:

$$\hat{\mathbf{K}}^i = \begin{bmatrix} 6.4878 & 43.3746 \\ 36.9723 & -55.6060 \\ -43.4600 & 12.2314 \end{bmatrix} \quad \tilde{\mathbf{K}}^i = \begin{bmatrix} 6.4874 & 43.3722 \\ 59.0831 & 16.4598 \\ 36.9702 & -55.6029 \\ -23.5311 & -57.3594 \\ -43.4576 & 12.2307 \end{bmatrix}$$

The next candidate to be included in the subset is chosen by Eq. (20), looking for the minimum vector component corresponding to the points not yet included in the subset, that are {2,4}. The minimum, excluding {1,3,5}, is found for point {4}:

$$\min \begin{bmatrix} 0.6310 \; 162.7315 \; 0.1980 \; \mathit{5.5548} \; 0.6100 \end{bmatrix}$$

Following this, the new subset becomes {1,3,4,5}, to which corresponds a new LS estimate of the centered partial centroid $\mathbf{K}^{i(+1)}$ (Eq. (21)),

$$\hat{\mathbf{K}}^{i(+1)} = \begin{bmatrix} 12.2861 & 57.7303 \\ 42.9105 & -41.2030 \\ -17.5828 & -43.0473 \\ -37.6139 & 26.5199 \end{bmatrix}$$

and a new set of transformation parameters $\{c, \mathbf{R}, \mathbf{t}\}$ for each $\mathbf{X}_j^{i,P\{i(+1)\}}$. The test is performed computing $G = 2.86$ (Eq. (22)), and the proper Procrustes distances $G_t = 2.28$, $G_{tR} = 1.82$, $G_{tRc} = 1.44$ (Eq. (23,24,25)). Assuming a first kind error $\alpha = 5\%$, and the proper degrees of freedom, the Fisher tests (Eq. (26)) furnish:

Translation:	$df_1 = 4$; $df_2 = 4$	Fisher: 0.40	Passed
Rotation:	$df_1 = 2$; $df_2 = 4$	Fisher: 0.32	Passed
Dilation:	$df_1 = 2$; $df_2 = 4$	Fisher: 0.27	Passed

The procedure is repeated again, by extending the subset with the next candidate not yet included, that is {2}. The new LS estimate of the centered partial centroid $\mathbf{K}^{i(+2)}$ (Eq. (21)), is:

$$\hat{\mathbf{K}}^{i(+2)} = \begin{bmatrix} -0.6524 & 51.6346 \\ 51.9923 & 24.5358 \\ 29.9067 & -47.3867 \\ -30.6408 & -49.2093 \\ -50.6058 & 20.4256 \end{bmatrix}$$

Computing the proper $G = 42.93$, $G_t = 30.42$, $G_{tR} = 25.22$, and $G_{tRc} = 4.54$ the Fisher tests in this case result in:

Translation:	$df_1 = 4;\ df_2 = 8$	Fisher: 2.76	Passed
Rotation:	$df_1 = 2;\ df_2 = 8$	Fisher: 1.15	Passed
Dilation:	$df_1 = 2;\ df_2 = 8$	Fisher: 4.55	Failed

revealing the existence of an outlier in point {2} of one of the configurations.

6 Conclusions

The paper proposes a forward search method to perform the Robust Generalised Procrustes Analysis for multidimensional data sets. The robustness is based on the definition of an initial outlier and stationary data free subset, iteratively enlarged by the remaining points till the null hypothesis of the Procrustes statistics is rejected. The method seems suitable for many practical applications, like shape analysis, multi factorial analysis, geodetic sciences and so on. Next works will deal with an extension of the method to the cases in which outliers are present in the initial data subset, and have to be removed along the forward search.

References

ANDERSON, C.R. (1997): Object Recognition Using Statistical Shape Analysis. *PhD thesis*, University of Leeds (11, 13, 316).

ATKINSON, A.C. and RIANI M. (2000): *Robust Diagnostic Regression Analysis.* Springer, New York.

CROSILLA, F. and BEINAT, A. (2002): Use of Generalised Procrustes Analysis for Photogrammetric Block Adjustment by Independent Models. *ISPRS Journal of Photogrammetry and Remote Sensing,* Elsevier, 56(3), 195–209.

BORG, I. and GROENEN, P.J.F. (1997): *Modern Multidimensional Scaling: Theory and Applications.* Springer, New York.

CERIOLI, A. and RIANI, M. (2003): Robust Methods for the Analysis of Spatially Autocorrelated Data. *Statistical Methods & Applications,* 11, 335–358.

DRYDEN, I.L. and MARDIA, K.W. (1998): *Statistical Shape Analysis.* John Wiley & Sons, Chichester, England, 83–107.

GELFAND, M.S., MIRONOV, A.A. and PEVZNER, P.A. (1996): Gene Recognition Via Spliced Sequence Alignment. *Proc. of the National Academy of Sciences,* 93(19), 9061–9066.

GOODALL, C. (1991): Procrustes Methods in the Statistical Analysis of Shape. *Journal Royal Statistical Society, Series B-Methodological,* 53(2), 285–339.

GOWER, J.C. (1975): Generalized Procrustes Analysis. *Psychometrika,* 40(1), 33–51.

GROENEN, P.J.F., GIAQUINTO, P. and KIERS, H.A.L. (2005): An Improved Majorization Algorithm for Robust Procrustes Analysis. In: Vichi, M., Monari, P., Mignani, S. and Montanari, A. (Eds.): *New developments in classification and data analysis.* Springer, Heidelberg, 151–158.

KIERS, H.A.L. (2002): Setting up Alternating Least Squares and Iterative Majorization Algorithms for Solving Various Matrix Optimization Problems. *Computational Statistics and Data Analysis*, 41, 157–170.

KONG, Z. and CEGLAREK, D. (2003): Fixture Configuration Synthesis for Reconfigurable Assembly Using Procrustes-based Pairwise Optimization. *Transactions of NAMRI*, 31, 403–410.

KRISTOF, W. and WINGERSKY, B. (1971): Generalization of the Orthogonal Procrustes Rotation Procedure to More than Two Matrices. *Proc. of the 79-th Annual Conv. of the American Psychological Ass.*, 6, 89–90.

LANGRON, S. P. and COLLINS, A. J. (1985): Perturbation Theory for Generalized Procrustes Analysis. *Journal Royal Statistical Society*, 47(2), 277–284.

ROHLF, F.J. and SLICE, D. (1992): Methods for Comparison of Sets of Landmarks. *Syst. Zool.*, 39, 40–59.

ROUSSEEUW, P. J. (1984): Least Median of Squares Regression. *Journal of the American Statistical Association*, 79(388), 871–880.

SCHÖNEMANN, P. H. (1966): A Generalized Solution of the Orthogonal Procrustes Problem. *Psychometrika*, 31(1), 1–10.

SIBSON, R. (1979): Studies in the Robustness of Multidimensional Scaling: Perturbational Analysis of Classical Scaling. *Journ. R. Statis. Soc.*, B, 41, 217–229.

SIEGEL, A.F. and BENSON, R.H. (1982): A Robust Comparison of Biological Shapes. *Biometrics*, 38, 341–350.

TEN BERGE, J. M. F. (1977): Orthogonal Procrustes Rotation for Two or More Matrices. *Psychometrika*, 42(2), 167–276.

XU, P. (2005): Sign-constrained Robust Least Squares, Subjective Breakdown Point and the Effect of Weights of Observations on Robustness. *Journal of Geodesy*, 79, 146–159.

A Projection Method for Robust Estimation and Clustering in Large Data Sets

Daniel Peña[1] and Francisco J. Prieto[2]

[1] Departamento de Estadística,
Universidad Carlos III de Madrid, Spain
daniel.pena@uc3m.es

[2] Departamento de Estadística,
Universidad Carlos III de Madrid, Spain
franciscojavier.prieto@uc3m.es

Abstract. A projection method for robust estimation of shape and location in multivariate data and cluster analysis is presented. The key idea of the procedure is to search for heterogeneity in univariate projections on directions that are obtained both randomly, using a modification of the Stahel-Donoho procedure, and by maximizing and minimizing the kurtosis coefficient of the projected data, as proposed by Peña and Prieto (2005). We show in a Monte Carlo study that the resulting procedure works well for robust estimation. Also, it preserves the good theoretical properties of the Stahel-Donoho method.

1 Introduction

As a few outliers in multivariate data may distort arbitrarily the sample mean and the sample covariance matrix, the robust estimation of location and shape is a crucial problem in multivariate statistics. See for instance Atkinson et al. (2004) and the references therein. For high-dimensional large data sets a useful way to avoid the curse of dimensionality in data mining applications is to search for outliers in univariate projections of the data. Two procedures that use this approach are the Stahel-Donoho procedure (see Donoho, 1982), that searches for univariate outliers in projections on random directions, and the method proposed by Peña and Prieto (2001, b), that searches for outliers in projections obtained by maximizing and minimizing the kurtosis coefficient of the projected data. The first procedure has good theoretical properties, but fails for concentrated contamination and requires prohibitive computer times for large dimension problems. The second procedure works very well for concentrated contamination and it can be applied in large dimension problems, but its theoretical properties are unknown. As both procedures are based on projections, it seems sensible to explore if a combination of both could avoid their particular limitations and this has been proposed by Peña and Prieto (2005). They show that the combination of random and specific directions leads to a affine equivariant procedure which inherits the good theoretical properties of the Stahel-Donoho method and it is fast to compute so that it can be applied for large data sets.

The procedure can also be applied for cluster analysis by generalizing the approach presented in Peña and Prieto (2001,a). Then, instead of just searching for directions which are extremes of the kurtosis coefficient, we add random directions to obtain a better exploration of the space of the data.

This article summarizes the method proposed by Peña and Prieto (2005) and includes two contributions: we present new results on the relative performance of the procedure with several groups of outliers, and we discuss the application of the procedure for cluster analysis. The article is organized as follows. Section 2 summarizes the main ideas of the procedure for generating directions and presents the algorithm combining random and specific directions. Section 3 discusses the extensions of these ideas for clustering . Section 4 illustrates the performance of the proposed method as an outlier detection tool for robust estimation.

2 Finding Interesting Directions

Suppose we have a sample $(x_1, \ldots, x_n)$ of a p-dimensional vector random variable X. We are interested in searching for heterogeneity by projecting the data onto a set of directions d_j, $j = 1, \ldots, J$. The key step of the method is obtaining the directions d_j. The Stahel-Donoho procedure is based on generating these directions randomly: a random sample of size p is chosen, a hyperplane is fitted to this sample and the direction d_j orthogonal to this hyperplane is chosen. Note that if we have a set of outliers and the data is standardized to have variance equal to one, the direction orthogonal to the fitted plane is, a priori, a good one to search for outliers.

A procedure for obtaining specific directions that can reveal the presence of heterogeneity was proposed by Peña and Prieto (2001b). They showed that the projection of the data on the direction of the outliers will lead to (1) a distribution with large univariate kurtosis coefficient if the level of contamination is small and (2) a distribution with small univariate kurtosis coefficient if the level of contamination is large. In fact, if the data come from a mixture of two distributions $(1-\alpha)f_1(X) + \alpha f_2(X)$, with $.5 < \alpha < 0$ and f_i, $i = 1, 2$, is an elliptical distribution with mean μ_i and covariance matrix V_i, the directions that maximize or minimize the kurtosis coefficient of the projected data are of the form of the admissible linear classification rules. In particular, if the distributions were normal with the same covariance matrix and the proportion of contamination is not large, $0 < \alpha < 0.21$, the direction obtained by maximizing the kurtosis coefficient is the Fisher linear discriminant function whereas when the proportion of contamination is large, $0.21 < \alpha < .5$, the direction which minimizes the kurtosis coefficient is again the Fisher linear discriminant function. Thus, the extreme directions of the kurtosis coefficient seem to provide a powerful tool for searching for groups of masked outliers. Peña and Prieto (2001b) proposed an iterative procedure based on the projection on a set of $2p$ orthogonal directions obtained as extremes for the kurtosis of the projected data. Note that the first set of p directions are closely related to the independent components of the data, which are defined as a set of p variables obtained by linear transformations of the original data such that the new variables are as independent as possible. It can be shown that the independent components can be obtained by maximizing the absolute value or the square of the kurtosis coefficient and, as this coefficient cannot be smaller than one, these directions will be the same as the one obtained by maximizing the kurtosis coefficient. The performance of these directions for outlier detection was found to be very good for concentrated contamination but, as it can be expected from the previous results, it was not so good when the proportion of contamination is close to .3 and the contaminating distribution has the same variance as the original distribution. This behavior of the algorithm is explained because then the values of the kurtosis for the projected data are not expected to be either very large or very small.

Thus it seems that we may have a very powerful procedure by combining the specific directions obtained as extremes of the kurtosis with some random directions. However, as we are interested in a procedure that works in large data sets and it is well known (and it will be discussed in the next section) that the Stahel-Donoho procedure requires a huge number of directions to work as the sample size increases, the random directions are not generated by random sampling, but by using some stratified sampling scheme that is found to be more useful in large dimensions. The univariate projections onto these directions are then analyzed as previously described in a similar manner to the Stahel-Donoho algorithm. See Peña and Prieto (2005) for the justification of the method.

The algorithm that we propose is called RASP (**R**andom **A**nd **S**pecific **Pro**jetions) and works as follows. We assume that first the original data are scaled and centered. Let $\bar{x}$ denote the mean and S the covariance matrix of the original data, the points are transformed using $y_i = S^{-1/2}(x_i - \bar{x}), \quad i = 1, \ldots, n$.

Stage I: Analysis based on directions computed from finding extreme values of the kurtosis coefficient. Compute n_1 orthogonal directions and projections maximizing the kurtosis coefficient ($1 \leq n_1 \leq p$) and n_2 directions minimizing this coefficient ($1 \leq n_2 \leq p$).

1. Set $y_i^{(1)} = y_i$ and the iteration index $j = 1$.

2. The direction that maximizes the coefficient of kurtosis is obtained as the solution of the problem

$$d_j = \arg\max_d \frac{1}{n} \sum_{i=1}^{n} \left(d' y_i^{(j)} \right)^4 \qquad \text{s.t.} \quad d'd = 1. \tag{1}$$

3. The sample points are projected onto a lower dimension subspace, orthogonal to the direction d_j. Define
$$v_j = d_j - e_1, \quad Q_j = \begin{cases} I - \dfrac{v_j v_j'}{v_j' d_j} & \text{if } v_j' d_j \neq 0 \\ I & \text{otherwise,} \end{cases}$$
where e_1 denotes the first unit vector. The resulting matrix Q_j is orthogonal, and we compute the new values
$$u_i^{(j)} \equiv \begin{pmatrix} z_i^{(j)} \\ y_i^{(j+1)} \end{pmatrix} = Q_j y_i^{(j)}, \quad i = 1, \dots, n,$$
where $z_i^{(j)}$ is the first component of $u_i^{(j)}$, which satisfies $z_i^{(j)} = d_j' y_i^{(j)}$ (the univariate projection values), and $y_i^{(j+1)}$ corresponds to the remaining $p - j$ components of $u_i^{(j)}$.
We set $j = j + 1$, and if $j < n_1$ we go back to step 1(b). Otherwise, we let $z_i^{(p)} = y_i^{(p)}$.
4. The same process is applied to the computation of the directions d_j (and projections $z_i^{(j)}$), for $j = n_1 + 1, \dots, n_1 + n_2$, minimizing the kurtosis coefficient.
5. For finding outliers, as in the Stahel-Donoho approach the normalized univariate distances r_i^j are computed as
$$r_i^j = \frac{1}{\beta_p} \frac{|z_i^{(j)} - \text{median}_i(z_i^{(j)})|}{\text{MAD}_i(z_i^{(j)})}, \tag{2}$$
for each direction $j = 1, \dots, n_1 + n_2$, where β_p is a predefined reference value.

Stage II: Analysis based on directions obtained from a stratified sampling procedure as follows:

1. In iteration l, two observations are chosen randomly from the sample and the direction $\hat{d}_l$ defined by these two observations is computed. The observations are then projected onto this direction, to obtain the values $\hat{z}_i^l = \hat{d}_l^T y_i$. Then the sample is partitioned into K groups of size n/K, where K is a prespecified number, based on the ordered values of the projections $\hat{z}_i^l$, so that group k, $1 \leq k \leq K$, contains those observations i satisfying
$$\hat{z}_{(\lfloor (k-1)n/K \rfloor + 1)}^l \leq \hat{z}_i^l \leq \hat{z}_{(\lfloor kn/K \rfloor)}^l.$$
2. From each group k, $1 \leq k \leq K$, a subsample of p observations is chosen without replacement. The direction orthogonal to these observations, $\tilde{d}_{kl}$,

is computed, as well as the corresponding projections $\tilde{z}_i^{kl} = \tilde{d}_{kl}^T y_i$ for all observations i. These projections are used to obtain the corresponding normalized univariate distances r_i^j,

$$r_i^j = \frac{1}{\bar{\beta}_p} \frac{|\tilde{z}_i^{kl} - \text{median}_i(\tilde{z}_i^{kl})|}{\text{MAD}_i(\tilde{z}_i^{kl})}, \tag{3}$$

where $j = 2p + \lfloor (k-1)n/K \rfloor + l$, and $\bar{\beta}_p$ is a prespecified reference value.

3. This procedure is repeated a number of times L, until $l = L$.

Stage III: For each observation i its corresponding normalized outlyingness measure r_i is obtained from the univariate distances r_i^j defined in (2) and (3), as

$$r_i = \max_{1 \leq j \leq 2p + \lfloor Ln/K \rfloor} r_i^j.$$

Those observations having values $r_i > 1$ are labeled as outliers and removed from the sample, if their number is smaller than $n - \lfloor (n+p+1)/2 \rfloor$. Otherwise, only those $n - \lfloor (n+p+1)/2 \rfloor$ observations having the largest values of r_i are labeled as outliers.

The values of the parameters needed in the procedure are explained in Peña and Prieto (2005). We have found that $n_1 = n_2 = 1$ works very well and we call then the algorithm RASP(1). An alternative would be to use p directions, that is $n_1 = n_2 = p$, and the we cal the algorithm RASP(p). They will be compared in the next section.

3 Application to Clustering

The directions obtained in the previous section can be used for finding clusters by identifying holes in the distribution of the projected data; we use the sample spacings or first-order gaps between the ordered statistics of the projections. If the univariate observations come from a unimodal distribution, there will be large gaps near the extremes of the distribution and small gaps near the center. However, this pattern will change if there are clusters in the data. For example, with two clusters of similar size we expect a large gap separating the clusters, lying towards the center of the observations. Thus, once the univariate projections are computed for each one of the $n_1 + n_2$ projection directions, the problem is reduced to finding clusters in unidimensional samples, where these clusters are defined by regions of high probability density. We consider that a set of observations can be split into two clusters when we find a sufficiently large first-order gap in the sample. Let $z_{ki} = \boldsymbol{x}_i' \boldsymbol{d}_k$ for $k = 1, \ldots, n_1 + n_2$, and let $z_{k(i)}$ be the order statistics of this univariate sample. The first-order gaps or spacings of the sample, w_{ki}, are defined as the successive differences between two consecutive order statistics

$$w_{ki} = z_{k(i+1)} - z_{k(i)}, \qquad i = 1, ..., n-1$$

As the expected value of the gap w_i is the difference between the expected values of two consecutive order statistics, it will be in general a function of i and the distribution of the observations. For a unimodal symmetric distribution Peña and Prieto (2001a) showed that, under reasonable assumptions, the largest gaps in the sample are expected to appear at the extremes, w_1 and w_{n-1}, while the smallest ones should be those corresponding to the center of the distribution. Therefore, if the projection of the data onto $\boldsymbol{d}_k$ produces a unimodal distribution we would expect the plot of w_{ki} with respect to k to decrease until a minimum is reached (at the mode of the distribution) and then to increase again. The presence of a bimodal distribution in the projection would be shown by a new decreasing of the gaps after some point. A sufficiently large value in these gaps would provide indication of the presence of groups in the data. The cut-off for the gaps can be determined by Monte Carlo. In summary, the algorithm will be as follows:

1. For each one of the directions d_k compute the univariate projections of the original observations $u_{ki} = x_i' d_k$.
2. Standardize these observations, $z_{ki} = (u_{ki} - m_k)/s_k$, where $m_k = \sum_i u_{ki}/n$ and $s_k = \sum_i (u_{ki} - m_k)^2/(n-1)$.
3. Sort the projections z_{ki} for each value of k, to obtain the order statistics $z_{k(i)}$ and transform then using the inverse of the standard normal distribution function $\bar{z}_{ki} = \Phi^{-1}(z_{k(i)})$
4. Compute the gaps between consecutive values, $w_{ki} = \bar{z}_{k,i+1} - \bar{z}_{ki}$.
5. Search for the presence of significant gaps in w_{ki}. These large gaps will be indications of the presence of more than one cluster. In particular, we introduce a threshold $\kappa = \nu(c)$, where $\nu(c) = 1 - (1-c)^{1/n}$ denotes the c-th percentile of the distribution of the spacings, define $i_{0k} = 0$ and
$$r = \inf_j \{n > j > i_{0k} : w_{kj} > \kappa\}.$$
If $r < \infty$, the presence of several possible clusters has been detected. Otherwise, go to the next projection direction.
6. Label all observations l with $\bar{z}_{kl} \leq \bar{z}_{kr}$ as belonging to clusters different to those having $\bar{z}_{kl} > \bar{z}_{kr}$. Let $i_{0k} = r$ and repeat the procedure.

4 Simulation Results

We present in Table 4 the percentage of successes in a simulation experiment where we have compared: (1) An efficient algorithm for the implementation of the Minimum Covariance Determinant (MCD) procedure, the FASTMCD algorithm as proposed by Rousseeuw and van Driessen (1999). (2) An implementation of the Stahel-Donoho algorithm, as described in Maronna and Yohai (1995). (3) A computationally efficient algorithm recently proposed by Maronna and Zamar (2002), based on the analysis of the principal components of an adjusted covariance matrix computed from information on pairs of observations. Two iterations of the algorithm have been carried out, as suggested by the authors. (4) An algorithm based on the directions computed from the minimization and maximization of the kurtosis coefficient, as described in Peña and Prieto (2001b). (5) A stratified Stahel-Donoho sampling procedure, corresponding to the second part of the RASP algorithm described in Section 2. (6) An implementation of the RASP(1) algorithm described in Section 2. (7) An implementation of RASP(p), that is, the same algorithm as before but using now the full $2p$ directions maximizing and minimizing the kurtosis coefficient. The data for the experiment in Table 4 was generated from a standard normal multivariate distribution in dimensions 5, 10 and 20, contaminated with a proportion of outliers in a single cluster (from 10% to 40%), obtained from a second normal distribution with different covariance matrices. A total of 100 replications were carried out for each case and each algorithm.

FASTMCD	SD	MZ	kurtosis	mod-SD	RASP(1)	RASP(p)
74.9	90.1	70.2	88.0	94.9	97.5	98.0

Table 1. Overall success rates for the detection of outliers forming one cluster

In a second computational experiment, we have generated samples composed of one main cluster obtained from a standard normal distribution and two or four additional clusters. The success rates for algorithms FASTMCD, SD, MZ and RASP(p) are presented in Table 4. The number of directions generated in algorithm SD was chosen to have comparable running times for both SD and RASP(p). Note again the improvement obtained when using RASP(p) over the alternative algorithms.

FASTMCD	SD	MZ	RASP(p)
88.5	97.5	86.2	100.0

Table 2. Overall success rates for the detection of clusters

FASTMCD	SD	MZ	RASP(p)
233.8	7.0	17.8	7.9

Table 3. Average running times for the algorithms

Finally, to illustrate the computational efficiency of the different algorithms, Table 4 presents the average running times for the analysis of sets of

100 replications corresponding to the preceding algorithms, given in seconds on a Pentium M 1.6 GHz. Those for both SD and RASP(p) are significantly lower than for the other algorithms.

References

ATKINSON, A.C., RIANI, M. and CERIOLI, A. (2004): *Exploring Multivariate Data with the Forward Search.* Springer, New York.

DONOHO, D.L. (1982): Breakdown Properties of Multivariate Location Estimators.Ph.D. Qualifying paper, Harvard University. Dept. of Statistics.

MARONNA, R.A. and YOHAI, V.J. (1995): The Behavior of the Stahel-Donoho Robust Multivariate Estimator. *Journal of the American Statistical Association, 90, 330–341.*

MARONNA, R.A. and ZAMAR,R. (2002): Robust estimates of location and dispersion for high dimensional data sets. *Technometrics, 44, 307–317.*

PEÑA, D. and PRIETO, F.J. (2001a): Cluster Identification Using Projections. *Journal of the American Statistical Association, 96, 1433–1445.*

PEÑA, D. and PRIETO, F.J. (2001b): Robust Covariance Matrix Estimation and Multivariate Outlier Detection. *Technometrics, 43, 286–310.*

PEÑA, D. and PRIETO, F.J. (2005): Combining Random and Specific Directions for Outlier Detection and Robust Estimation for High-Dimensional Multivariate Data. *Manuscript.*

ROUSSEEUW, P.J. and VAN DRIESSEN, K. (1999): A Fast Algorithm for the Minimum Covariance Determinant Estimator. *Technometrics, 41, 212–223.*

Robust Multivariate Calibration

Silvia Salini

Department of Economics, Business and Statistics
University of Milan, Italy
silvia.salini@unimi.it

Abstract. Multivariate calibration uses an estimated relationship between a multivariate response Y and an explanatory vector X to predict unknown X in future from further observed responses. Up to now very little has been written about robust calibration. An approach can be based on the outliers deletion methods. An alternative is to employ robust procedures. The purpose of this paper is to present multivariate calibration methods which are able to detect and investigate those observations which differ from the bulk of the data or to identify subgroups of observations. Particular attention will be paid to the *forward search* approach.

1 Introduction

Multivariate calibration uses an estimated relationship between a multivariate response Y (of dimension q) and an explanatory vector X (of dimension p) to predict unknown X in future from further observed responses. The purpose of this paper is to present multivariate calibration methods which are able to detect and investigate those observations which differ from the bulk of the data or, more generally, to identify subgroups of observations. We are concerned not only with the identification of atypical observations, but also with the effect that they have on parameter estimates, on inferences about models, and on their suitability. In this paper particular attention will be paid to the *forward search* approach (Atkinson, Riani and Cerioli, 2004). In this method we start with a fit to very few outlier-free observations and then successively fit larger subsets. We thus order the observations by closeness to the fitted model. As a result, not only are outliers and distinct subsets of the data discovered, but the influential effect of these observations is made clear. Section 2 gives more details about multivariate calibration. Section 3 presents some possible approach on robust calibration. In section 4 the forward search procedure is applied to real data set, and some comments and remarks are given.

2 Multivariate Calibraton

Statistical calibration, potentially useful in several practical applications, deals with the inference on unknown values of explanatory variables, given a vector of response variables. Suppose for example that two different instruments for the measurement of the same phenomenon are considered. The first one $\mathbf{X}$ (*standard method*) is more difficult, accurate and expensive than the second one $\mathbf{Y}$ (*test method*). A sample of n units, in which both measures $\boldsymbol{x}$ and $\boldsymbol{y}$ are available, is considered. The set of values $(\boldsymbol{x}_i,\boldsymbol{y}_i)$ i=1,...,n is the *calibration experiment.* The statistical calibration problem arises when only the $\boldsymbol{y}_i$ obtained by the test method are known and the unknown $\boldsymbol{x}_i$ have to be estimated. The solution of this problem, *prediction experiment*, depends on the probabilistic model supposed to have generated the calibration experiment. In particular, it is assumed that the values $\boldsymbol{y}_i$ are realizations of a random variable (r.v) $\mathbf{Y}$ with known density function.

The assumptions on the values $\boldsymbol{x}_i$ may be of two types: i) the $\boldsymbol{x}_i$ are realizations of a r.v. and therefore $(\boldsymbol{x}_i,\boldsymbol{y}_i)$ are realizations of a multivariate r.v. (*random calibration*); ii) the $\boldsymbol{x}_i$ are chosen by the experimenter (*controlled calibration*).

In the classical parametric approach a linear multivariate model for both experiments is considered. Suppose that the calibration experiment is made of n observations, q response variables $Y_1, Y_2, \ldots, Y_q$ and p explanatory variables $X_1, X_2, \ldots, X_p$ with $q \geq p$, and suppose that $\mathbf{Y_1} = \mathbf{1}\alpha^T + \mathbf{XB} + \mathbf{E}_1$, where $\mathbf{Y}_1(n \times q)$, $\mathbf{X}(n \times p)$, $\mathbf{1}(n \times 1)$ are known matrices; $\mathbf{E}_1(n \times q)$ is a matrix of random errors, whose i-th row is $\mathbf{E_{1i}} \sim \mathbf{N}(\mathbf{0}, \boldsymbol{\Gamma})$; $\mathbf{B}(p \times q)$ and $\alpha(n \times 1)$ are unknown parameters. The model for the prediction experiment is given by: $\mathbf{Y}_2 = \mathbf{1}\alpha^T + \mathbf{1}\xi^T\mathbf{B} + \mathbf{E}_2$, where $\mathbf{Y}_2(m \times q)$, $\mathbf{E}_2(m \times q)$ whose j-th row is $\mathbf{E_{2j}} \sim \mathbf{N}(\mathbf{0}, \boldsymbol{\Gamma})$ and $\xi\,(q \times 1)$ is the unknown vector of calibration measures.

When $q = p$ the multivariate classical estimator for ξ is

$$\hat{\xi}_C = \left(\mathbf{\hat{B}S^{-1}\hat{B}^T}\right)^{-1} \mathbf{\hat{B}S}^{-1} (\bar{\mathbf{y}}_2 - \hat{\alpha}) \tag{1}$$

where $\hat{\mathbf{B}}$ and $\hat{\alpha}$ are least-squares estimators, $\bar{\mathbf{y}}_2$ is the mean of the observations in the predicted experiment and $\mathbf{S}$ is the pooled covariance matrix. (1) is also the maximum likelihood (ML) estimator of ξ.

When $q > p$, the ML estimator is a function of $\hat{\xi}_C$ and a quantity that depends on an inconsistency diagnostic statistic R, a measure of the consistency of $\mathbf{y}_2$ to estimate (more details in Zappa and Salini, 2004).

3 Prediction Diagnostics

Detecting outliers is an important aspect in the process of statistical modeling. Outliers, with respect to statistical models, are those observations that are inconsistent with the chosen model. Once a multivariate calibration model is built, it is used to predict a characteristic (e.g. standard measure) of new samples. Developing robust calibration procedures is important because standard regression procedures are very sensitive to the presence of atypical observations; furthermore, very little is known about robust calibration.

There are two basic approaches to robust calibration: use robust regression methods or perform classic estimators on data after rejecting outliers.

Robust estimates work well even if the data are contaminated. Several robust regression estimation methods have been proposed (Rousseeuw and Leroy, 1987): the M-estimator is the most popular; the R-estimator is based on the ranks of the residuals, the L-estimator is based on linear combination of order statistics; the Least Median of Squares (LMS) estimator minimizes the median of the squares of the residuals; the S-estimator is based on the minimization of a robust M-estimate of the residual scale; the Generalized M-estimator (GM) attempts to down-weigh the high influence points as well as large residual points; the MM-estimator is a multistage estimator which combines high breakdown with high asymptotic efficiency. In the calibration literature, generalized M-estimation techniques have been applied to the controlled calibration problem and orthogonal regression on the measurement-error model to the random calibration problem. These techniques give robust calibration estimators (Cheng and Van Ness, 1997) but extensions of these methods when $p > 1$ and $q > 1$ lead to difficulties. In addition, although robust estimators can sometimes reveal the structure of the data, they do so at the cost of down-weighting or discarding some observations. Finally, if the calibration experiment is made up of different subsets, the use of robust estimators will tend to produce a centroid which lies in between different groups. In this last case prediction will be strongly determined by the size of the subsets which make up the calibration experiment.

A second approach on robust multivariate calibration consists in performing a classical multivariate estimator on data after rejecting outliers. There are many methods to detect outliers. A single outlier can easily be detected by the methods of deletion diagnostics in which one observation at a time is deleted, followed by the calculation of new parameter estimates and residuals. With two outliers, pairs of observations can be deleted and the process can be extended to the deletion of several observations at a time. This is the basic idea of multiple deletion diagnostics. A difficulty both for computation and interpretation is the explosion of the number of combinations to be considered. A similar approach is based on the repeated application of single deletion methods (backward methods). However, such backwards procedures can fail due to masking.

The forward search appears to be more effective than the other approaches especially in the presence of multiple outliers (Atkinson and Riani, 2000). Also in the calibration context, the problem can be formulated as searching for the outlier-free data subset, the basic idea of forward search method (see next section). Genetic algorithms are proposed as a reasonable tool to select the optimum subset (Walczak, 1995). The results obtained with this genetic approach are compared with classical robust regression method of least median of squares (LMS).

Another approach, a third one, for the prediction of diagnostics in multivariate calibration problem, could be based on the inconsistency diagnostic R, mentioned in the previous section; the statistic R is central to diagnostic checking, whether or not it influences confidence intervals and point estimators (Brown and Sundberg, 1989).

4 Forward Search in Multivariate Calibration: An Example

The forward search is a general technique for robust estimation. The approach in calibration field considers the direct regression model and is based on the idea of forming a clean subset of the data, and then testing the outlyingness of the remaining points relative to the chosen clean subset. The algorithm combines robust estimation, diagnostics and computer graphics. The first step of the algorithm is based on the idea of elemental sets. The forward search starts by selecting an outlier free subset of p observations, where p is the number of parameters to be estimated in the model. To select this subset, a large number of subsets are examined, and the one with the smallest median residual is chosen - this is known as least median of squares (LMS) estimation. Having chosen this initial subset, the search moves from step p to step $p+1$ selecting $(p+1)$ units with the smallest least squares residuals. The model is re-fitted in this way until all units are included in the subset. Throughout the search, certain statistics such as the residuals, are monitored. Diagnostic plots are then constructed with the X-axis representing the subset size and the Y-axis representing the statistic of interest. In the case of calibration problem with $q > p$, q direct regression models are considered and the initial subset of dimension r, $S^{(r)}$, used to initialize the forward search, is found using the intersection of units, that have the smallest LMS residuals considering each response independently. In symbols for each response j, $S^{(p)}_{\mathbf{c}*,j}$ satisfies

$$e^2_{[\mathrm{med}],S^{(p)}_{\mathbf{c}*,j}} = \min_{\mathbf{c}}[e^2_{[\mathrm{med}],S^{(p)}_{\mathbf{c},j}}], \tag{2}$$

where $e^2_{[k],S^{(p)}_{\mathbf{c},j}}$ is the k th ordered squared residual among $e^2_{i,S^{(p)}_{\mathbf{c},j}}$, in the regression which considers the j-th variable as response, $i = 1, \ldots, n$, $\mathbf{c}$ is a collection of p units (the number of $\mathbf{c}$ collections is $\binom{n}{p}$) and med is the integer part of $(n+p+1)/2$. The initial subset is associated with the k units whose residuals at maximum have the r-th position $(r \leq n/2)$ among $e^2_{[1],S^{(p)}_{\mathbf{c}*,j}}, \ldots, e^2_{[n],S^{(p)}_{\mathbf{c}*,j}}$, $j = 1, 2, \ldots, q$. The search progresses from subset size m to $m+1$ by selecting the smallest $(m+1)$ Mahalanobis distances (MD) (Atkinson, Riani and Cerioli, 2004, p. 66) from multivariate regression $d^*_{im} = (e^T_{im}\hat{\Sigma}e^{-1}_{um}e_{im})^{1/2}$ are scaled by the square root of the estimated covariance matrix, where $\hat{\Sigma}_u = (E^T E)/(m-p)$.

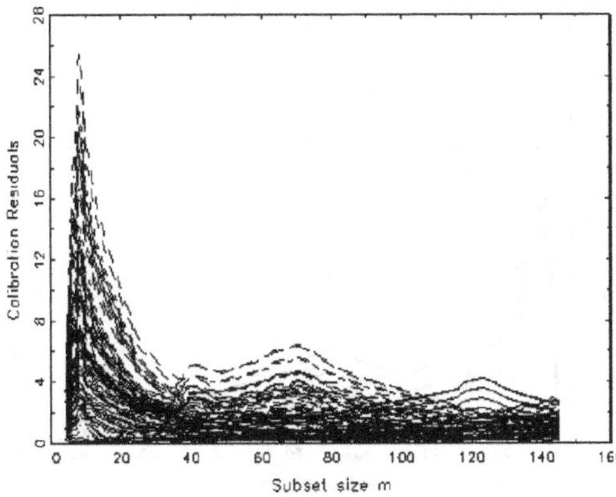

Fig. 1. Forward plot of scaled Mahalanobis distances based on residuals of calibration experiment.

This algorithm proceeds up to when all units are included in the subset $(m = k, k+1, \ldots, n)$.

In order to illustrate how the method works we can use a dataset refereed to the noise of the traffic([1]). Calibration experiment has to determine the hourly equivalent level of the noise of traffic. Time sampling techniques differ for the size of the sample: surveying every second (87000 observations), surveying every minute (1450 observations), surveying every 10 minute (145 observations), surveying every hour (24 observations). As standard measure X is considered the hourly mean obtained by surveying every second, as test measure Y is considered the hourly mean of surveying every minute and the hourly mean of surveying every 10 minute. Therefore $q = 2$ and $p = 1$ and $n = 145$. without loss of generality only 1000 subsets are considered to select the initial subset. Fig. 1 shows the typical output of forward search, it monitors the calibration residuals at each steps of the forward search, every trajectory refers to one unit. The plot evidences the potential presence of groups, corresponding to different time slots. In particular forward plot in Fig. 2 and Fig. 3 show that unit 43 and unit 115 have a different trajectory than the others. The units correspond to time 7 AM and 6.50 AM, critical time for the city traffic.

In the final part of this section we compare the forward approach with other robust estimators. It is important to notice that extensions of robust method (Cheng and Van Ness, 1997) when $p > 1$ and $q > 1$ using robust regression approach lead to difficult because the necessary robust multivariate regression theory has not been developed. In our case case $q = 2$, then two robust model are estimated. Some robust regression estimators (Huber 1981,

---

[1] I am grateful to G. Brambilla (Institute of Acoustic "O.M. Corbino" C.N.R. Roma) for providing the data.

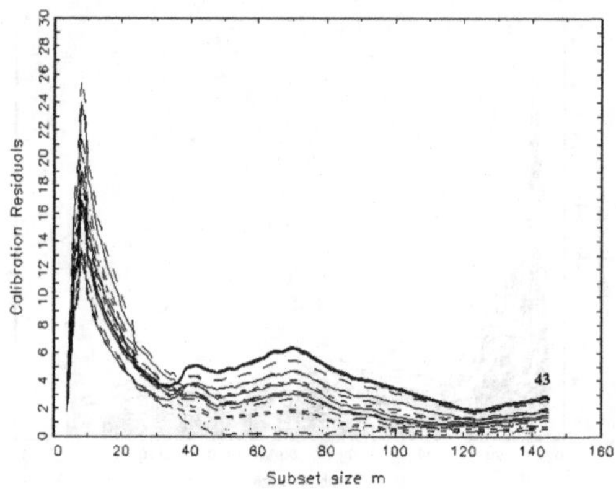

Fig. 2. Forward plot of scaled Mahalanobis distances at the step 60 of the search. Unit 43, evidenced in bold, has different trajectory than the others. The units correspond to time 7 AM, a critical time for the city traffic.

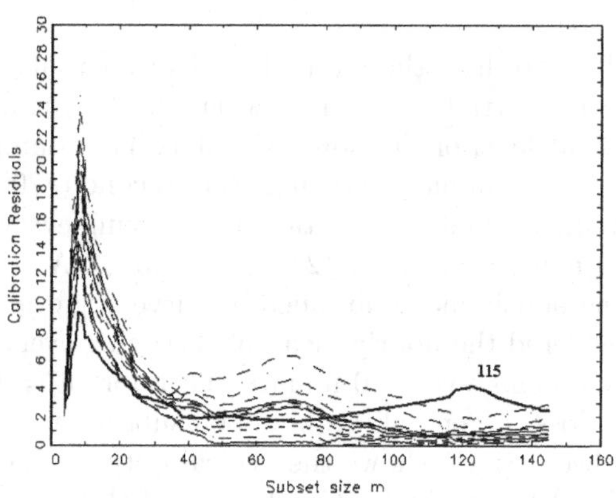

Fig. 3. Forward plot of scaled Mahalanobis distances at the step 60 of the search. Unit 115, evidenced in bold, has different trajectory than the others. The units correspond to time 6.50 PM, a critical time for the city traffic.

Hampel et al. 1986) are implemented. A combining method (Johnson and Krishnamoorthy, 1996) is applied to combine the univariate robust estimators, in fact the response variable is determined by two different measuring methods. The following equation shows the combining formula:

$$\hat{x}_c = \sum_{i=1}^{q} w_i \hat{x}_{ci} \tag{3}$$

where $w_i = (\hat{\beta}_i^2/S_i^2)/(\sum_{i=1}^{q} \hat{\beta}_i^2/S_i^2)$ in which $\hat{\beta}_i^2$ is a robust estimator and $S_i^2 = \sum_{j=1}^{n} (\hat{y}_{ij} - y_{ij})^2/(n-q)$. The estimator in (3) is a GLM estimator that minimizes $\sum_{i=1}^{q} (y_i - \hat{\alpha}_i - \hat{\beta}_i x)^2/S_i^2$ with respect to x. Table 1 shows the classical combined estimator and the robust ones in both cases of contaminated and non contaminated data.

Models	Data 1	Data 2
Classical	6.514	20.031
Huber	6.575	16.402
Tukey	6.687	15.106

Table 1. Standard Deviation of residuals for classical estimator and robust estimators for both not contaminated (Data 1) and contaminated (Data 2) data.

In the first case there are not outliers but only groups, as evidenced in forward plot in Fig. 1, robust and classical estimators perform in the same way. In presence of outliers robust estimators fit better than the classical one. Fig. 4 represents the true value of x versus the classical estimator, the Hubert and the Tukey estimator. As we expected robust estimators fit better than the classical one that it is very sensitive to the presence of atypical observations.

5 Conclusion

The problem of robust multivariate calibration is approached by the forward search method and by the classical robust regression procedures. In presence of groups the forward search performs better than classical robust procedures that are useful in presence of single outliers. It is important to notice that the combining method proposed in section 4 does not consider the robust multivariate regression theory (Rousseeuw et al., 2004) but refers to the cases in which the multivariate response variable is measured by different instruments or determined by various methods. Further, robust multivariate regression procedures can be applied on calibration problems. We want to study this extension and plan to report it elsewhere. We are currently investigating the behavior of the inconsistency diagnostic R mentioned in section 3 with forward search plots. We are interesting to create the envelopes for the R statistic, in this way we could be able to accept or reject the hypothesis that a new observation is inconsistent with the data.

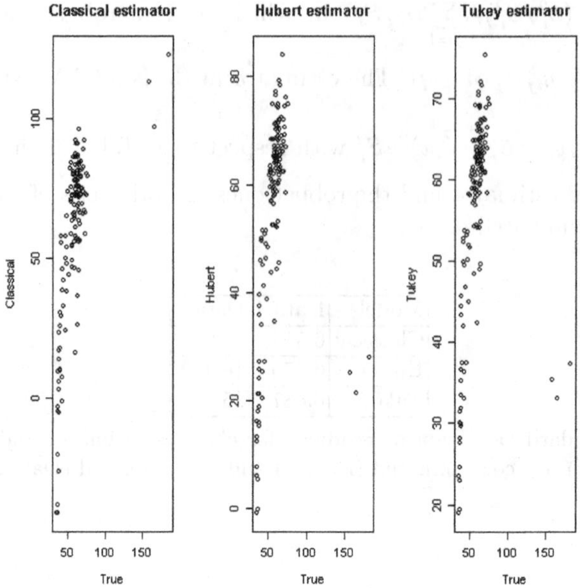

Fig. 4. True value versus classical and M-estimators in contaminated data

References

ATKINSON, A.C. and RIANI, M. (2000): *Robust Diagnostic Regression Analysis*, Springer, New York.

ATKINSON, A.C., RIANI, M. and CERIOLI, A. (2004): *Exploring Multivariate Data With the Forward Search*, Springer Verlag, New York.

BROWN, P.J. and SUNDBERG, R. (1989), Prediction Diagnostic and Updating in Multivariate Calibration, *Biometrika, 72, 349-361.*

CHENG, C.L. and VAN NESS, J.W. (1997), Robust Calibration, *Technometrics, 39, 401-411.*

HUBER P.J. (1981): *Robust Statistics.* Wiley.

HAMPEL, F.R., RONCHETTI, E.M., ROUSSEEUW, P.J. and STAHEL, W.A. (1986): *Robust Statistics: The Approach based on Influence Functions.* Wiley.

JOHNSON, D.J. and KRISHNAMOORTHY, K. (1996): Combining Independent Studies in Calibration Problem, *Journl of the American Statistical Associations, 91, 1707-1715.*

ROUSSEEUW, P. and LEROY, A. (1987): *Robust Regression and Outlier Detection*, Wiley, New York.

ROUSSEEUW, P., VAV AELST, S., VAN DRISSEN, K., and AGULLÓ, J. (2004): Robust Multivariate Regression, *Technometrics, 46, 293-305.*

WALCZAK, B. (1995): Outlier Detection in Multivariate Calibration, *Chemiometrics and Intelligent Laboratory Systems, 28, 259-272.*

ZAPPA, D. and SALINI, S. (2004): Confidence Regions for Multivariate Calibration: a proposal. In: M. Vichi et al editors: *New Developments in Classification and Data Analysis.* Springer, Bologna, 225–233.

Part IV

Data Mining Methods and Software

Procrustes Techniques for Text Mining

Simona Balbi and Michelangelo Misuraca

Dipartimento di Matematica e Statistica
Università di Napoli "Federico II", Italy
{sb, mimisura}@unina.it

Abstract. This paper aims at exploring the capability of the so called Latent Semantic Analysis applied to a multilingual context. In particular we are interested in weighing how it could be useful in solving linguistic problems, moving from a statistical point of view. Here we focus on the possibility of evaluating the goodness of a translation by comparing the latent structures of the original text and its version in another natural language. Procrustes rotations are introduced in a statistical framework as a tool for reaching this goal. An application on one year of *Le Monde Diplomatique* and the corresponding Italian edition will show the effectiveness of our proposal.

1 Introduction

Internet has deeply modified our approach to information sources compared with traditional media. In fact it is possible to consider both a wider information diffusion (from a sociodemografic, cultural and geographic viewpoint) and a wider differentiation of information contents, with respect to the subjective knowledge requirements of the users.

The increasing availability of "e-documents" makes necessary the development of tools for automatically extracting and analyzing the most informative data, in the frame of a KDD (Knwoledge Discovery in Databases) strategies.

In Text Mining procedures, and mainly in Text Retrieval, *Latent Semantic Indexing* (LSI, Deerwester *et al.* (1990)) has become a standard technique in order to reduce the high dimensionality proper of textual databases. Nowadays there is a wide literature consisting of proposals based on the primary idea of using *Singular Value Decomposition* (SVD) for analyzing a large set of documents. On the Web it is possible to find easely nice software and consulting for making one's own Latent Semantic Analysis (Landauer *et al.* (1998), e.g. `http://lsa.colorado.edu`).

From a statistical viewpoint LSI arises interesting questions (Balbi and Di Meglio (2004)), mainly derived from the use of (generalised) Singular Value Decomposition in a multivariate data analysis frame. One of the most attractive features of LSI is its producing mathematical representations of documents that does not depend directly on single terms. The main consequence is that a query can identify documents of interest, by identifying a context and not a (perfect) correspondence query/document in terms of keywords.

Frequently it is possible to obtain documents translated into different languages, known as *multilingual corpora*. In particular it is possible to consider *parallel corpora*, when the translation mate is an exact translation (e.g. United Nations or EU multilingual *corpora*), and *comparable corpora*, when the translation mate is an approximate translation (e.g. e-documents published on the Web). Parallel *corpora* are objects of interest at present because of the opportunity offered to "align" original and translation and gain insights into the nature of translation.

In this direction, a promising research path concerns the so-called *Cross Language - LSI* (CL-LSI, Littman *et al.* (1998)), developed for retrieving documents written in different languages. Procrustes CL-LSI has been envisaged as a further development.

The comparison of different languages by (generalised) Procrustes rotations was proposed by Balbi and Esposito (1998) in a classic Textual Data Analysis frame, dealing with Italian advertisement campaigns of the same product in three different periods, for exploring the language evolution.

In this paper we aim at reviewing the problem of applying Procrustes rotations in the frame of Textual Data Analysis, when we deal with different terms as in the case of different languages. The substantive problem we deal with is the comparison between two *corpora* in different languages, when one is the translation of the other. The effectiveness of the proposal will be tested, by comparing one year of *Le Monde Diplomatique* and its Italian translation.

2 Methodological Background

The analysis of multilingual *corpora* has been discussed both in Textual Statistics and in Text Mining with different goals:

- in Textual Data Analysis dealing with the open-ended questions in multinational surveys, by visualizing in a same referential space the different association structures (Akuto and Lebart (1992), Lebart (1998));
- in Information Retrieval for developing language-indipendent representations of terms, in the frame of natural language and machine translation applications (Grefenstette (1998)).

In order to visualize the relationships between document and between terms a factorial approach is commonly performed. A $p \times n$ "lexical table", cross-tabulating terms and documents, is built by juxtaposing the n document/vectors obtained after encoding the *corpus* through the Bag-of-Words scheme:

$$D_j = (w_{1j}, w_{2j}, \ldots, w_{ij}, \ldots, w_{pj}) \qquad (1)$$

where D_j is the j-th document of the *corpus* ($k = 1, \ldots, n$) and $w_{i,j}$ is the importance of the i-th term in the document ($i = 1, \ldots, p$).

The axes of factorial representations are calculated by applying to the obtained matrix the SVD, used in several Text Retrieval techniques like LSI as well as in Data Analysis techniques like *Correspondence Analysis* (CA).

2.1 The Use of SVD

Let $\mathbf{A}$ be a matrix with p rows and n columns, where $p > n$. For sake of simplicity we consider that $rank(\mathbf{A}) = n$.

By SVD $\mathbf{A}$ can be decomposed in the product:

$$\mathbf{A} = \mathbf{U}\mathbf{\Lambda}\mathbf{V}' \quad \text{with the constraints} \quad \mathbf{U}'\mathbf{U} = \mathbf{V}'\mathbf{V} = \mathbf{I} \tag{2}$$

where $\mathbf{\Lambda}$ is a $n \times n$ diagonal matrix of λ_α positive numbers ($\alpha = 1, \ldots, n$), called *singular values* and arranged in decreasing order, while $\mathbf{U}$ and $\mathbf{V}$ are ortho-normalized matrices having in columns the left and right *singular vectors*, respectively.

The left singular vectors define an ortho-normal basis for the columns of $\mathbf{A}$ in a p-dimensional space. In a similar fashion the right singular vectors define an ortho-normal basis for the rows of $\mathbf{A}$ in a n-dimensional space. Multiplying only the first $q << p$ components of $\mathbf{U}$, $\mathbf{\Lambda}$ and $\mathbf{V}$, SVD allows the best (least squares) lower rank approximation of $\mathbf{A}$.

2.2 Latent Semantic Indexing vs Correspondence Analysis

In a LSI framework we compute the SVD of the *term by document* matrix $\mathbf{F}$, where the general element f_{ij} is the frequency of the i-th term ($i = 1, \ldots, p$) in the j-th document ($j = 1, \ldots, n$), according to the Bag-of-Words encoding.

Frequencies are often normalized in a *tf/idf* (term frequency/inverse document frequency) scheme:

$$f_{ij}^{\star} = \frac{f_{ij}}{\max f_j} \cdot \log \frac{n}{n_i} \tag{3}$$

where $\max f_j$ is the higher frequency in the j-th document and n_i is the number of documents in which term i appears (Salton and Buckley (1988)).

Balbi and Misuraca (2005) have shown how normalization can be embedded in a generalised SVD frame, by considering different ortho-normalized constraints to singular vectors.

LSI success is due to its capability of finding similarities between documents even if they have no terms in common. A consequent step has been the proposal of its application to cross-language text retrieval. CL-LSI automatically finds a language-independent representation for documents.

CA aims at analyzing the association between qualitative variables, by identifying a latent structure. The core analysis consists of factorial displays in a peculiar Euclidean distance, based on χ^2, with some nice properties as the distributional equivalence.

From an algebraic viewpoint CA is a *generalised* SVD (Greenacre (1984)), where the ortho-normalized constraints state the χ^2-metric. Correspondence Analysis on lexical tables is nowadays a standard technique, mainly applied for analyzing open-ended questions in social surveys.

In studying some linguistic phenomena it seems more suitable to consider a non symmetric scheme, in a antecedence/consequence framework, by introducing a different metric in the two spaces in which the terms and the documents are spanned (Balbi (1995)).

In both cases, in a LSI and in a CA scheme, one of the most powerful features is given by the graphical representation of the terms and the documents: LSI is useful in searching a specific knowledge in a textual database, CA is more indicated to search a general knowledge, according with the exploratory nature of the multidimensional techniques.

3 Procrustes Analysis

In statistical literature, when the aim is to investigate the agreement between totally or partially paired tables, different versions of *Procrustes Analysis* are carried out (Gower (1975)). Procrustes Analysis has been often utilised in the geometric frame of multidimensional data analysis, for comparing factorial configuration obtained analysing different data sets. Specifically, aim of this technique is to compare two sets of coordinates by optimising a *goodness-of-fit* criterion, i.e. by translating, rotating (and in case reflecting), and dilating one configuration, in order to minimize its distance from the other one.

Given two configurations of n points $\mathbf{Y_1}$ and $\mathbf{Y_2}$ in a p-dimensional space, shifted to the origin, the best rotation (in a least square sense) of $\mathbf{Y_2}$ to $\mathbf{Y_1}$ is obtained by considering the *Polar Decomposition* of $\mathbf{Z} = \mathbf{Y_2}'\mathbf{Y_1}$:

$$\mathbf{Z} = \mathbf{U}\Gamma\mathbf{V}' \quad \text{with the constraints} \quad \mathbf{U}'\mathbf{U} = \mathbf{V}'\mathbf{V} = \mathbf{I} \tag{4}$$

in which $\mathbf{U}$ and $\mathbf{V}$ are square matrices of order p, having in columns the left and right singular vectors of $\mathbf{Z}$, respectively, normalized to 1.

From the (4) it is possible to derive the so called *rotation factor* $\mathbf{R} = \mathbf{U}\mathbf{V}'$. It is proved that:

$$\mathbf{R} = \mathbf{U}\mathbf{V}' = \mathbf{Z}(\mathbf{Z}'\mathbf{Z})^{-1/2} \tag{5}$$

A *dilation factor* is sometimes necessary to "re-scale" the coordinates of $\mathbf{Y_2}$, by considering:

$$\delta = \frac{\text{trace}(\mathbf{\Gamma})}{\text{trace}(\mathbf{Y_2}'\mathbf{Y_2})} \tag{6}$$

By applying the rotation and the dilation factor to $\mathbf{Y_2}$ it is possible to obtain the new coordinates $\mathbf{Y_2}^\star = \delta\mathbf{Y_2}\mathbf{R}$.

4 A Textual Statistics Approach: CL - CA

The CL-LSI main aim is to reduce high-dimensional configurations of multilingual textual data and it is very suitable in finding document similarities, even if they have no terms in common. In a similar fashion CA aims to represent the meaningful information in reduced spaces, but it considers a weighted Euclidean metric based on χ^2, and so preserves the importance of infrequent terms.

In comparing the association structures of a *corpus* and its translation in another languages CL-LSI seems to be not suitable, as term frequencies are deeply influenced by grammatical and linguistic features. Here we propose a *Cross Language - CA* (CL-CA), aiming at comparing the latent structure of a monolingual *corpus* and its translation in a different language. The motivation is connected with the peculiar metric adopted by CA, and the basic idea that each textual problem should be faced by a careful choice of the way of measuring distances between elements.

4.1 The Strategy in Practice

Let us consider a collection of documents and its translation in another language. A classic CA is performed on each corpus for calculating the principal coordinates and so obtaining $\mathbf{X}$ and $\mathbf{Y}$, coordinate matrices of each language.

The coordinates matrices are standardized for ensuring that the centroids of the two configuration coincide and lie on the origin of the coordinates. In particular, we use the standard deviation of $\mathbf{X}$ and $\mathbf{Y}$ for normalizing each matrix, respectively, thus we consider a symmetrical role for the two languages (Mardia *et al.* (1995)), and a re-scaling factor is not necessary.

Assuming that $\mathbf{X}$ contains the CA coordinates of the "original" documents and $\mathbf{Y}$ the CA coordinates of their translation, we want to evaluate the $\mathbf{Y}$ goodness of fit on $\mathbf{X}$, by performing a Polar Decomposition of $\mathbf{Z} = \mathbf{Y}'\mathbf{X}$.

The goodness-of-fit measure Δ^2 between the two configuration, projected on the same space, is given by:

$$\Delta^2(\mathbf{X}, \mathbf{YR}) = \text{trace}(\mathbf{X}'\mathbf{X}) + \text{trace}(\mathbf{Y}'\mathbf{Y}) - 2\text{trace}(\mathbf{Z}'\mathbf{Z})^{1/2} \tag{7}$$

The Δ^2 assumes null value if $\mathbf{X} = \mathbf{Y}^\star$, therefore the smaller is the distance the more similar are the two configurations. In this way we can evaluate how the $\mathbf{Y}^\star$ plot fits the $\mathbf{X}$ plot, and consequently how good is the translation of the documents.

5 A Worldwide Review: Le Monde Diplomatique

Le Monde Diplomatique (LMD) is a monthly review edited in France since 1954, characterized by a critical viewpoint dealing with worldwide economical, political, social and cultural issues. At present, LMD is published in 20 different languages and distributed in about 30 countries, on paper or electronic support.

Regarding the Italian edition, some book reviews are drawn up by the Italian editorial staff, together with the translation from the French edition.

The language is quite homogeneous because the revue is translated from French by the same few persons. Since 1998, an electronic edition is available on the website `http://www.ilmanifesto.it/MondeDiplo/`, together with a paper edition sold as monthly supplement of the newspaper *Il Manifesto*.

The two *corpora* we deal with are a sub-set of 240 articles published in 2003 on the French edition and the corresponding articles in the Italian edition. The whole collection has been downloaded from the newspaper website and it has been automatically converted in text format with a script written in Java language. For each article, only the body has been considered (i.e. titles and subheadings have been ignored).

The articles have been normalised in order to reduce the possibility of data splitting, for example by converting all the capital letters to the lower case or conforming the transliteration of words coming from other alphabets, mainly proper nouns, or using the same notations for acronyms or dates. After the pre-treatment step two vocabularies of about 2400 terms for each language have been obtained.

The two separate Correspondence Analysis, performed on the French and Italian lexical tables, show very similar structures in terms of *explained inertia* (Figure 1). As usual in LSI we retain only the first 100 dimensions.

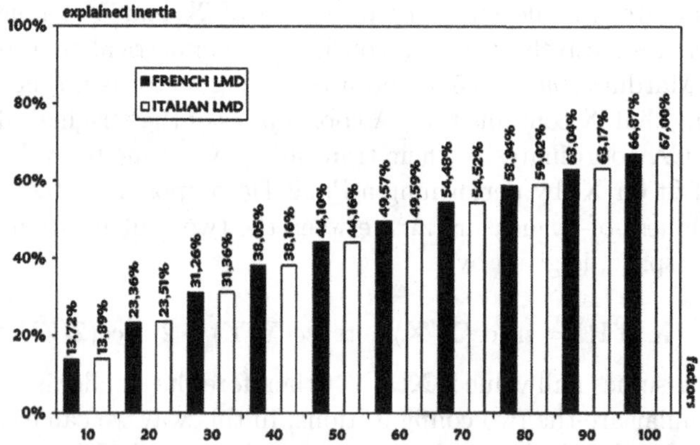

Fig. 1. Explained inertia distribution in the French and Italian LMD.

The explained inertia of the first k factors can be expressed in percentage by the ratio τ_k:

$$\tau_k = \frac{\sum_{\alpha=1}^{k} \sqrt{\lambda_\alpha}}{\sum_{\alpha=1}^{n} \sqrt{\lambda_\alpha}} \tag{8}$$

in which λ_α is the α-th singular value obtained by the SVD. The first 100 factors explain about 2/3 of total inertia, both for French and Italian.

After performing the Procrustes rotation it is possible to jointly represent the French and Italian documents on the same space. Even if the original distances between the articles represented in each monolingual document space are different, we can approximately assume a same common metric in the joint plot, because of the similar lexical and grammatical structure of the two languages.

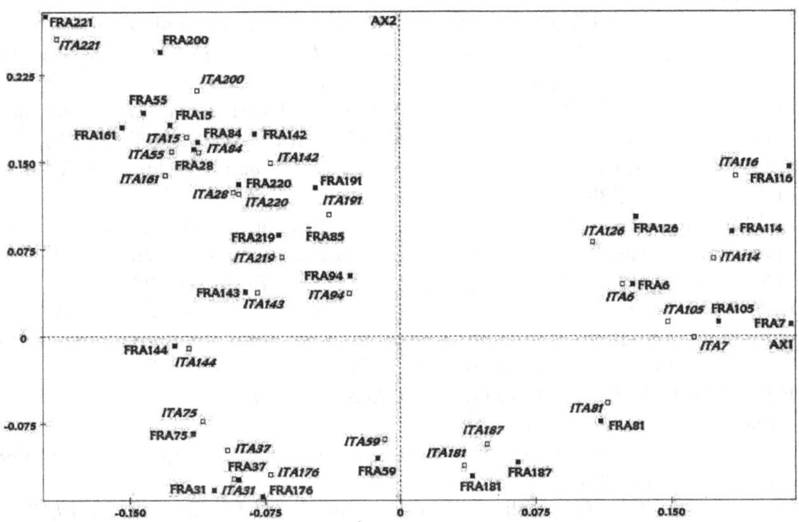

Fig. 2. Joint documents space after Procrustes Analysis.

In Figure 2 it can be seen that the original French articles are very close to the corresponding Italian ones, gathering a good quality of the translation, with a goodness-of-fit $\Delta^2 = 40.20$.

6 Final Remarks

In this paper we faced the problem of comparing two *corpora* when one is the translation of the other. The Procrustes Analysis on the results of CA seems to be more suitable for evaluating the goodness of the translation with respect to the LSI.

The use of *ad-hoc* metrics for taking into account the differences in the language structures would be a substantive improvement, specially when different linguistic groups are compared. Furthermore, the analysis can be improved by introducing external information, in terms of meta-data collectable on the documents (*extra-textual information*) or in terms of weights based on the data belonging to the analyzed *corpora* (*intra-textual information*).

References

AKUTO, H. and LEBART, L. (1992): Le repas idéal. Analyse de réponses libres en anglais, fran ccais, japonais. *Les Cahiers de l'Analyse des Données, 17, 327–352.*

BALBI, S. (1995): Non symmetrical correspondence analysis of textual data and confidence regions for graphical forms. In: S. Bolasco et al. (Eds.), *Actes des 3es Journées internationales d'Analyse statistique des Données Textuelles.* CISU, Roma, 2, 5–12.

BALBI, S. and DI MEGLIO, E. (2004): Contributions of Textual Data Analysis to Text Retrieval. In: D. Banks et al. (Eds.), *Classification, Clustering and Data Mining Applications.* Springer-Verlag, Berlin, 511–520.

BALBI, S. and ESPOSITO, V. (1998): Comparing advertising campaigns by means of textual data analysis with external information. In: S. Mellet (Ed.), *Actes des 4es Journées internationales d'Analyse statistique des Données Textuelles.* UPRESA, Nice, 39–47.

BALBI, S. and MISURACA, M. (2005): Visualization Techniques in Non Symmetrical Relationships. In: S. Sirmakessis (Ed.), *Knowledge Mining (Studies in Fuzziness and Soft Computing).* Springer-Verlag, Berlin, 23–29.

DEERWESTER, S., DUMAIS, S.T., FURNAS, G.W., LANDAUER, T.K., HARSHMAN, R. (1990): Indexing by latent semantic analysis. *Journal of the American Society for Information Science, 41, 391–407.*

GREFENSTETTE, G. (Ed.) (1998): *Cross Language Information Retrieval.* Kluwer Academic Publishers, London.

GREENACRE, M. (1984): *Theory and Application of Correspondence Analysis.* Academic Press, London.

GOWER, J.C. (1975): Generalised Procrustes analysis. *Psychometrica, 40, 33–51.*

LANDAUER, T.K., FOLTZ, P.W., LAHAM, D. (1998): Introduction to Latent Semantic Analysis. *Discourse Processes, 25, 259–284.*

LEBART, L. (1998): Text mining in different languages. *Applied Stochastic Models and Data Analysis, 14, 323–334.*

LITTMAN, M.L., DUMAIS, S.T., LANDAUER, T.K., (1998): Automatic cross-language information retrieval using latent semantic indexing. In: G. Grefenstette (Ed.): *Cross Language Information Retrieval.* Kluwer Academic Publishers, London, 51–62.

MARDIA, K.W., KENT, J.T., BIBBY, J.M. (1995): *Multivariate Analysis.* Academic Press, London.

SALTON, G. and BUCKLEY, C. (1988): Term-weighting approaches in automatic text retrieval. *Information Processing & Management, 24, 513–523.*

Building Recommendations from Random Walks on Library OPAC Usage Data

Markus Franke, Andreas Geyer-Schulz, and Andreas Neumann

Department of Information Services and Electronic Markets
Universität Karlsruhe (TH), Germany
{Markus.Franke|Andreas.Geyer-Schulz|Andreas.Neumann@em.uni-karlsruhe.de}

Abstract. In this contribution we describe a new way of building a recommender service based on OPAC web-usage histories. The service is based on a clustering approach with restricted random walks. This algorithm has some properties of single linkage clustering and suffers from the same deficiency, namely bridging. By introducing the idea of a walk context (see Franke and Thede (2005) and Franke and Geyer-Schulz (2004)) the bridging effect can be considerably reduced and small clusters suitable as recommendations are produced. The resulting clustering algorithm scales well for the large data sets in library networks. It complements behavior-based recommender services by supporting the exploration of the revealed semantic net of a library network's documents and it offers the user the choice of the trade-off between precision and recall. The architecture of the behavior-based system is described in Geyer-Schulz et al. (2003).

1 Introduction

Recommender services for which amazon.com's "Customers who bought this book also bought ..." is a prominent example ideally create a win-win situation for all involved parties. The customer is assisted in navigating through a huge range of books offered by the shop, the bookseller automatically gets the benefit of cross-selling by complementary recommendations to its customers (Schafer et al. (2001)).

In this article we present an innovative recommender system based on a fast clustering algorithm for large object sets (Franke and Thede (2005)) that is based on product co-occurrences in purchase histories. These histories stem from the users of the Online Public Access Catalog (OPAC) of the university's library at Karlsruhe. A purchase is defined as viewing a document's detail page in the WWW interface of the OPAC. The co-occurrence between two documents is defined as viewing their detail pages together in one user session.

In behavior-based recommender systems, we tacitly assume that a higher than random number of co-occurrences implies complementary documents and that the number of co-occurrences is a kind of similarity measure in the context of behavior-based recommender systems. The recommender service described in this article itself is a complementary service: it supports the exploration of the revealed semantic net of documents and it gives the user control of choosing his own trade-off between precision and recall.

The outline of the paper is as follows. We survey existing approaches in section 2. In section 3 we introduce restricted random walk clustering before addressing the generation of recommendations from clusters in section 4. A few preliminary results on cluster quality will be discussed in section 5 and, finally, we summarize our findings and give an outlook in section 6.

2 Recommender Systems and Cluster Algorithms for Library OPACs

Recommender systems and web personalization are still active research areas with an increasing potential in applications, for a survey of recent activities, see Uchyigit (2005). According to the classification of Resnick and Varian (1997) we concentrate on implicit recommender systems based on user-behavior without user cooperation (e.g. purchase histories at the (online) store, Usenet postings or bookmarks). The reason for this is that this class of recommender systems has considerably fewer incentive-related problems (e.g. bias, lies, free riding) than explicit or mixed recommender systems (see e.g. Geyer-Schulz et al. (2001) or Nichols (1997)).

Query, content, and metadata analysis based on relevance feedback, ontology and semantic web analysis, natural language processing, and linguistic analysis for instance by Semeraro et al. (2001), Yang (1999) and others suffer in the context of hybrid library systems as e.g. the library of the Universität Karlsruhe (TH) from two major disadvantages. Less than one percent of the content is in digital form and less than thirty per cent of the corpus have metadata like thesaurus categories or keywords. Therefore, recommender systems of this type currently cannot be applied successfully in libraries.

Today, two behavior-based recommender systems for books are in the field: the famous system of amazon.com and the one used e.g. by the university library of Karlsruhe (Geyer-Schulz et al. (2003)).

Amazon.com's approach is to recommend the books that have been bought (or viewed) most often together with the book the customer is currently considering. Details of the method, e.g. how the number of books to recommend is determined or whether product management may change the ranking remain undisclosed. The implementation challenge is to achieve scalability for huge data sets, even if the co-occurrence matrix is quite sparse.

The method implemented at the university library of Karlsruhe is statistically more sophisticated, it is based on Ehrenberg's repeat buying theory (Ehrenberg (1988); Geyer-Schulz et al. (2003)). Its advantage lies in a noticeably better quality of the recommendations, because the underlying assumption of independent Poisson processes with Γ-distributed means leads to a logarithmic series distribution (LSD) which allows to efficiently distinguish between random and meaningful co-occurrences in a more robust way.

However, both of these recommender systems exploit only local neighborhoods in the similarity graph derived from the purchase histories. Extensions that aim at a more global picture of the similarity graph must include the neighbors of the neighbors. For example, cluster-based recommender systems give a more global view of the similarity graph. Cluster recommendations do not only contain directly related documents, but also indirect relations. Unfortunately, algorithms for such recommendations quickly become computationally intractable.

For a survey of clustering and classification algorithms, we refer the reader to Duda et al. (2001) or Bock (1974) and to the books shown in figure 3. Examples for recommender systems or collaborative filtering algorithms based on cluster algorithms can be found e.g. in Sarwar et al. (2002) and Kohrs and Merialdo (1999).

In the context of libraries Viegener (1997) extensively studied the use of cluster algorithms for the construction of thesauri. Viegener's results are encouraging, because he found semantically meaningful patterns in library data. But all standard cluster algorithms proved to be computationally expensive – Viegener's results were computed on a supercomputer at the Universität Karlsruhe (TH) which is unavailable for the library's day to day operations. In addition, the quality of the clusters generated by the algorithms tested may not be sufficient for recommendations. Single linkage clustering for example tends to bridging, i.e. to connecting independent clusters via an object located between clusters, a bridge element.

A new idea for a fast clustering algorithm based on sampling of the similarity graph is restricted random walk clustering as proposed by Schöll and Schöll-Paschinger (2003). We selected this algorithm for two reasons. Its ability to cope with large data sets that will be discussed in section 3.4 and the quality of its clusters with respect to library purchase histories.

The bridging effect is much weaker with restricted random walk clustering as shown in Schöll and Schöll-Paschinger (2003) and it is even smaller with the modifications proposed in Franke and Thede (2005). In addition, the cluster size can be adapted for generating recommendations with an appropriate trade-off between precision and recall as shown in section 3.3. A comprehensive comparison of the performance of restricted random walk clustering with other cluster algorithms can be found in the appendix of Schöll (2002).

3 Restricted Random Walks

Clustering with restricted random walks on a similarity graph as described by Schöll and Paschinger (2002) is a two step process: Start at a randomly chosen node, and advance through the graph by iteratively selecting a neighbor of the current node at random as successor. When walking over the document set, the neighborhood (as defined by the similarity measure) is shrinking in each step, since the similarity must be higher than the edge taken in the last step. This is repeated until the neighborhood is empty. Then another walk is started. The second phase is the cluster construction from the walk histories. The basic assumption is that the nearer the position of an edge in a walk is to the end of the walk, the higher is the probability that the two documents connected by the edge are in the same cluster.

3.1 Input Data: Purchase Histories

Input data are the purchase histories generated by users of the OPAC of the library of the Universität Karlsruhe (TH). Users browsing through the catalogues contribute to constructing raw baskets: Each session with the OPAC contains a number of documents whose detail pages the user has inspected. This data is aggregated and stored in raw baskets such that the raw basket of a document contains a list of all other documents that occurred in one or more sessions together with it. Furthermore, the co-occurrence frequency of the two documents, i.e. the number of sessions that contain both documents, is included in the raw basket. Revealed preference theory suggests that choice data of this kind reveals the preferences of users and allows a complete reconstruction of the users' utility function.

Co-occurrence frequencies measure the similarity of two documents and a similarity graph $G = (V, E, \omega)$ may be constructed as follows: The set of vertices V denotes the set of documents in the OPAC with a purchase history. If two documents have been viewed together in a session, $E \subseteq V\text{x}V$ contains an edge between these documents, the weight ω_{ij} on the edge between documents i and j is the number of co-occurrences of i and j. ω_{ii} is set to zero in order to prevent the walk from visiting the same document in two consecutive steps. The neighborhood of a document consists of all documents that share an edge with it.

3.2 Restricted Random Walks

Formally, a restricted random walk is a finite series of nodes $R = (i_0, \ldots, i_r) \in V^r$ of length r. i_0 denotes the start node.

$$T_{i_{m-1}i_m} = \{(i_m, j)|\omega_{i_m j} > \omega_{i_{m-1}i_m}\} \tag{1}$$

is the set of all possible successor edges whose weight is higher than (i_{m-1}, i_m) and thus can be chosen in the $m + 1$st step.

For the start of the walk, one of the start node's neighbors is selected as i_1 with equal probability. The set of possible successor edges is the set of all incident edges of i_1 with a higher weight than $\omega_{i_0i_1}$: $T_{i_0i_1} = \{(i_1j)|\omega_{i_1j} > \omega_{i_0i_1}\}$. From this set, i_2 is picked at random using a uniform distribution and $T_{i_1i_2}$ is constructed accordingly. This is repeated until $T_{i_{r-1}i_r}$ is empty. For an example of such a walk, consider Fig. 1.

Suppose a walk starting from node A, the first successor is either B or C with equal probability. If C is chosen, the only successor edge is CB and then BA. As we see in figure 1, at this point no edge with a higher weight than 8

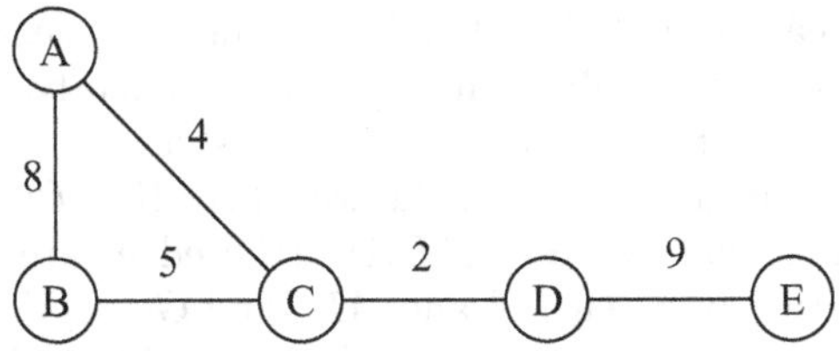

Fig. 1. An similarity graph

exists and the walk ends here. Other walks are e.g. BC, CDE, $DCAB$, and ED.

A sufficient covering of the complete document set requires starting several walks from each node. In this article five walks per node are used. More sophisticated methods are being developed using random graph theory (Erdös and Renyi (1957)).

The formulation of the walk as a stochastic process on the edges of the graph and the introduction of an "empty" transition state shown in Franke and Thede (2005) lead to an irreducible and infinite Markov chain. Restricted random walks are based solely on local information, namely the neighborhood of the current node. This greatly facilitates the implementation and reduces the time and space requirements of the algorithm, when compared with other cluster algorithms.

3.3 The Construction of Clusters

For cluster construction, two main variants have been suggested in the literature: The approach by Schöll and Paschinger or the walk context introduced in Franke and Thede (2005).

It is important to note that clustering with restricted random walks generates a hierarchy of clusters. This implies to choose a threshold (level) l, i.e. a height at which a cut is made through the treelike structure (dendrogram) in order to determine the cluster for a given node. Cluster hierarchies allow the user to interactively explore clusters by adapting the threshold. This property is exploited in in the user interface for zooming recommendations in section 4. Sorting cluster members by the minimum level at which they become members of the cluster allows the construction of recommendation lists as the m top members.

The original idea by Schöll and Paschinger is to generate, for a given node, component clusters as follows. A series of graphs $G_k = (V, E_k)$ is constructed from the data generated by all walks. V is the set of objects visited by at least one walk. An edge (i, j) is present in E_k if the transition (i, j) has been made in the k-th step of any walk.

Then, the union

$$H_l = \cup_{k=l}^{\infty} G_k \tag{2}$$

is constructed for each level l. Schöll and Paschinger define a cluster at level l as a component (connected subgraph) of H_l. Consequently, if a path between two nodes exists in H_l, they are in the same cluster.

In the example given above with the walks $ACBA$, BC, CDE, $DCAB$, and ED this means that $G_3 = (V, \{BA\})$ (the edges are undirected, thus there is no distinction between BA and AB) and $G_2 = (V, \{CB, DE, CA\})$. As a consequence, the only cluster at level 4 is $\{A, B\}$, at level 3 we get the clusters $\{A, B, C\}$ and $\{D, E\}$, reflecting nicely the structure of the original graph.

The main drawback of this clustering approach is that very large clusters were generated with purchase histories from the library's OPAC, sometimes containing several hundred documents even at the highest step level available. We conjecture that the reason is a bridging effect to documents covering more than one subject or documents read in connection with documents from different domains or random search behavior in a single session.

Furthermore, the step number as level measure has two major disadvantages. First, it mixes final steps from short walks that have a relatively high significance with steps from the middle of long walks where the random factor is still strong. This is evident for the clusters at $l = 3$. Although C and D have a high similarity, they do not appear in the top-level cluster because the walks containing them are too short. Second, the maximum step level is dependent on the course of the walks as well as the underlying data set and cannot be fixed a priori.

Walk context clusters solve the problem of large clusters. Instead of including all documents indirectly connected to the one in question, we only consider those nodes that have been visited in the same walk as the node whose cluster is to be generated (the central node), respecting the condition that both nodes have a higher step level than the given threshold in the corresponding walk. This has the advantage of reducing the cluster size on the one hand and the bridging effect on the other since it is less probable that some bridge between different clusters has been crossed in the course of one of the walks containing the document in question. Even if a bridge element is included in the walk, the number of documents from another cluster that are falsely included in the currently constructed cluster is limited since only members of the walk are considered that are located relatively near the bridge element. Obviously, walk context clusters are non-disjunctive.

For a discussion of different measures for the cluster level for walk context clusters we refer the reader to Franke and Thede (2005) and Franke and Geyer-Schulz (2005). For the results in section 5 the following measure which defines the level as a relative position of the step in a walk was used:

$$l^+ = \frac{\text{step number}}{\text{total steps in this walk} + 1} \tag{3}$$

The adjusted level l^+ converges asymptotically to one for the last step in a walk.

In our example, the clusters at $l = 1$ are as follows: $\{A, B\}$, $\{B, A, C\}$, $\{C, B\}$, $\{D, E\}$, $\{E, D\}$ where the first node is the central node for the respective cluster. As can be seen, a cluster-based recommendation for B includes both A and C whereas C's recommendation does not contain B.

3.4 Complexity

n denotes the number of documents or, more generally, nodes. According to Schöll and Paschinger (2002) the time complexity per walk is $O(\log n)$ with the base of the logarithm undefined; $\log_2$ seems to be a good estimate on our data. Executing c walks per document the total complexity of the walk phase is $O(cn \log n) = O(n \log n)$. Several experiments indicate that setting $c = 5$ is sufficient and that increasing c does not improve cluster quality.

For the usage data of the last two years, an analysis of the data showed that in practice the size of the neighborhood – and thus the degree of the nodes – is bounded by a constant and independent of n. Although the number of documents has grown over this period, the important factor for the complexity, namely the maximum size of the neighborhood of a node, remained constant. Since in practice the walk complexity is thus decoupled from the total size of the graph, even a linear complexity is possible if further theoretical developments can confirm this conjecture (Franke and Thede (2005)).

For the cluster construction phase the data structures for storing the walk data determine the complexity of this phase. This implies that an efficient implementation of this data structure is required. With an ideal hash table (with $O(1)$), the construction of a cluster for a given document is $O($ number of walks visiting the document $*1)$. A constant neighborhood size implies a constant walk length with growing n and that the number of walks that have visited a certain document is also constant. Otherwise, it is $O(\log n)$. Assuming that the number of walks visiting a node more than once is negligible, a total of $O(n \log n)$ nodes is visited during n walks of length $O(\log n)$, leading to an average of $O(\log n)$ walks visiting a node (Franke and Geyer-Schulz (2005)).

Per 31/10/2005 the input data consists of 1,087,427 library purchase histories for documents in the university library of the Universität Karlsruhe (TH). 614,943 of these histories contain sufficient data for clustering. The documents are connected by 44,708,674 edges, the average degree of a node is about 82. On an Intel dual Xeon machine with 2.4 GHz, the computation of 5 walks per document, that is more than 3 million walks in total, takes about 1 day.

4 Zooming Recommendations

The basis for zooming recommendations are clusters generated by restricted random walks. Figures 2 and 3 show the prototype of the user interface for

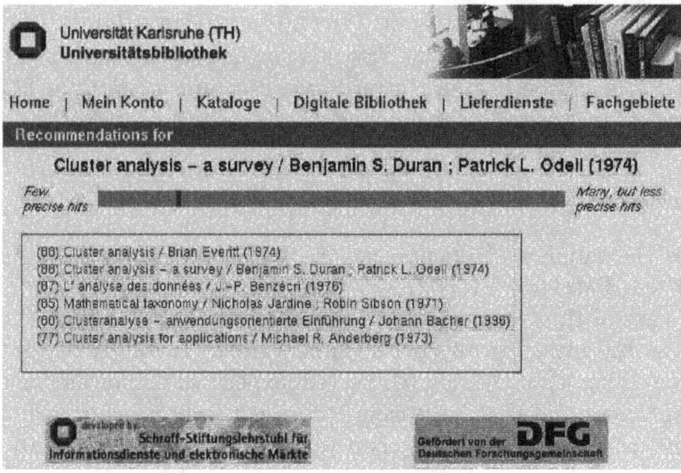

Fig. 2. Recommendations for Duran and Odell, *Cluster Analysis - A Survey* (High Precision)

recommendations with zoom. By positioning his pointing device (mouse) on the scale ranging from *few, precise hits* to *many, but less precise hits*, the user may perform zooming operations on the current cluster. Zooming means that by choosing an appropriate threshold level the cluster size is adapted. Since a cluster exactly corresponds to the basin of attraction of a random walk on the similarity graph, a reduction in the threshold increases the size of the cluster (and thus recall) and it reduces precision because of the random noise in the walks near the start of the walks. This effect is not yet visible in figure 3. However, for the example shown, precision drops when moving the slider all the way to the right hand side.

Clusters resulting from the walk context method are not disjunctive (allowing overlapping clusters) and generate a hierarchy of clusters which is exploited for the zoom effect. If documents A and B are both in the cluster for document C, A need not be in the cluster generated for B and vice versa. For book recommendations this is desirable because of the resulting clusters for bridge documents (documents belonging to more than one cluster). Recommendations for bridge documents (e.g. document C) should contain books from all domains that are concerned (e.g. A and B), while document A normally is not connected with B and thus should not be in the recommendation for B, if A and B are not in the same domain.

5 Evaluation

Since the zooming recommender service is still in a prototype stage, we have performed three evaluations, namely a very small user evaluation, and two keyword based evaluations of the cluster quality.

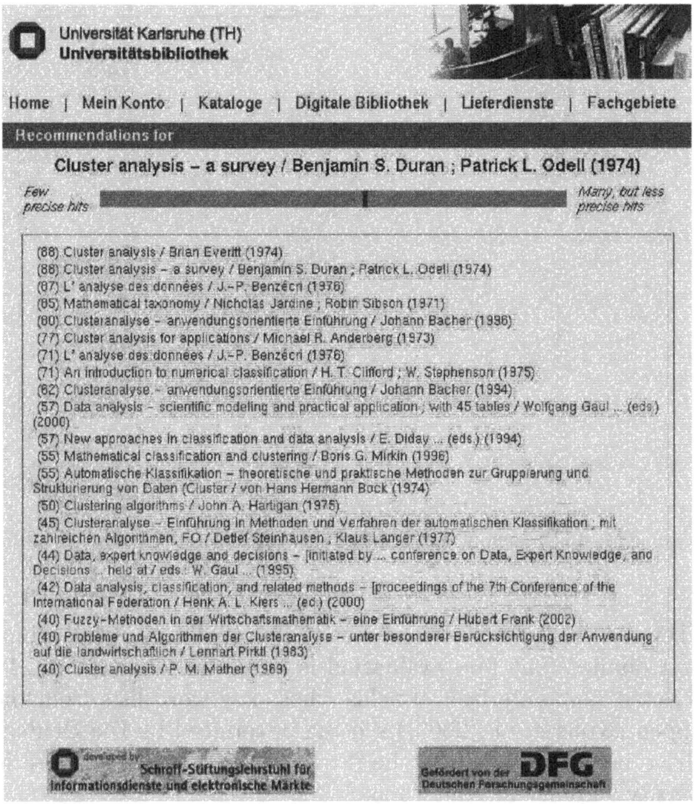

Fig. 3. Recommendations for Duran and Odell, *Cluster Analysis - A Survey* (Medium Precision)

Good Recommendations	5 of 5	4 of 5	3 of 5	2 of 5	1 of 5	0 of 5
Number of lists	17	5	3	1	2	2
Percent	56.6	16.6	10.0	3.3	6.6	6.6

Table 1. Top five evaluation 30 recommendations lists

The user evaluation is based on a random sample of 30 recommendation lists from the field of business administration and economics. The top five recommendations of each list have been judged by the authors as "good recommendations" or as "should not have been recommended". Precision has been defined as precision $= \frac{\text{number of correctly recommended documents}}{\text{total number of recommended documents}}$.

Overall precision on the top five recommendations was 78.66 %. More than 90% of the recommendation list contain at least one useful recommendation in the top five. The distribution is shown in table 1.

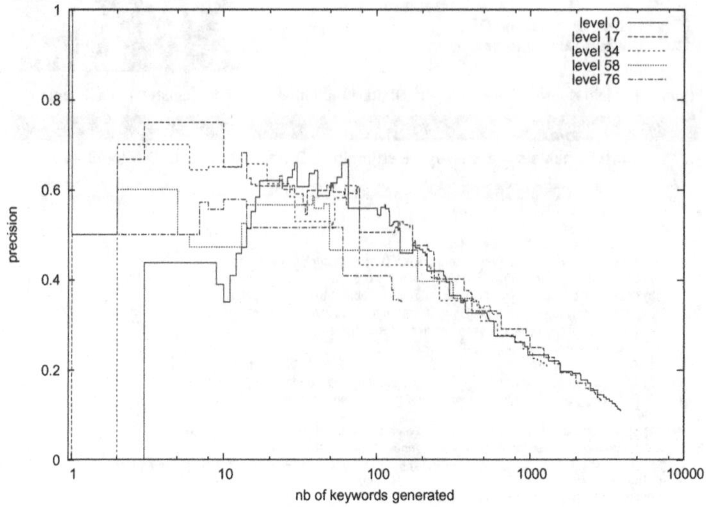

Fig. 4. Precision versus number of generated keywords

The keyword based evaluations serve as an additional approximation of the cluster quality. The first evaluation is based on a sample of 212 documents selected randomly from 9 fields whose key word lists from the SWD Sachgruppen (Kunz et al. (2003)) schema introduced by Die Deutsche Bibliothek have been controlled by the 9 reference librarians of these fields at the university's library at Karlsruhe.

Figure 4 plots the precision versus the number of generated keywords for a number of cut-off levels of the cluster. The trade-off between the number of generated keywords and the precision as well as the reduction of precision with growing cluster size (decreasing level) clearly shows in figure 4.

A second key-word based evaluation can be found in Franke and Geyer-Schulz (2005). For each document in the training sample of 40,000 documents, we counted the documents in the cluster that share at least one category in the manual classification. This is the number of correctly recommended documents. Furthermore, the manual classification only covers about 55% of the documents in the university's catalog so that the number of documents that "should" be recommended could not be determined without a considerable error. Due to this fact, the precision as described tends to be a rather conservative lower bound, especially if we consider the fact that the quality of the manual classification system at Karlsruhe differs strongly between topics.

The maximum precision that was reached by using l^+ was 0.95 at level 0.95, but then, keyword recommendations could only be generated for 11 documents out of nearly 40,000. On the other hand, in order to have keyword recommendations for more than 50% (26,067 in this case) of the documents, a precision of 67% is feasible.

Note, that while the keyword-based evaluations give a lower bound, the user evaluation performed by the authors indicates an upper bound of the quality of the recommender system. Nevertheless, a more complete follow-up study of the user evaluation by independent experts is planned.

6 Summary

In this article we have presented a new method for generating recommendations based on restricted random walks on large data sets in an efficient way. The precision and performance we were able to achieve are promising. However, open questions for further research are:

In order to increase the operational efficiency of the system intelligent updating of clusters when new usages histories arrive is currently under investigation. The idea is to reuse as much as possible from the existing walks. Furthermore, coverage of the graph should be improved by more intelligent decisions on the number of walks that are started from a node and a better understanding of the asymptotic behavior of the algorithm as the number of walks approaches infinity. In addition, filtering the random co-occurrences of documents from the similarity graph with stochastic process models Geyer-Schulz et al. (2003) should be investigated.

Acknowledgment

We gratefully acknowledge the funding of the project "RecKVK" by the Deutsche Forschungsgemeinschaft (DFG) and of the project "SESAM" by the BMBF.

References

BOCK, H. (1974): *Automatische Klassifikation*, Vandenhoeck&Ruprecht, Göttingen.

DUDA, R.O., HART, P.E. and STORK, D.G. (2001): *Pattern Classification*, Wiley-Interscience, New York, 2 edition.

EHRENBERG, A.S. (1988): *Repeat-Buying: Facts, Theory and Applications*, Charles Griffin & Company Ltd, London, 2 edition.

ERDÖS, P. and RENYI, A. (1957): On random graphs I, *Publ. Math., 6, 290–297.*

FRANKE, M. and GEYER-SCHULZ, A. (2004): Automated Indexing with Restricted Random Walks on Large Document Sets. In: Heery R. and Lyon L. (Eds) *Research and Advanced Technology for Digital Libraries – 8th European Conference, ECDL 2004*, Springer, Berlin, volume 3232 of *LNCS*, 232–243.

FRANKE, M. and GEYER-SCHULZ, A. (2005): Using restricted random walks for library recommendations, in: Uchyigit 2005, 107–115.

FRANKE, M. and THEDE, A. (2005): Clustering of Large Document Sets with Restricted Random Walks on Usage Histories. In: Weihs C. and Gaul W. (Eds): *Classification – the Ubiquitous Challenge*, Springer, Heidelberg, 402–409.

GEYER-SCHULZ, A., HAHSLER, M. and JAHN, M. (2001): Educational and scientific recommender systems: Designing the information channels of the virtual university, *International Journal of Engineering Education, 17, 2, 153 - 163.*

GEYER-SCHULZ, A., NEUMANN, A. and THEDE, A. (2003): An architecture for behavior-based library recommender systems – integration and first experiences, *Information Technology and Libraries, 22, 4, 165 - 174.*

KOHRS, A. and MERIALDO, B. (1999): Clustering for collaborative filtering applications, *Computational Intelligence for Modelling, Control & Automation 1999, IOS Press, Amsterdam, 199–204.*

KUNZ, M., HEINER-FREILING, M. and BEE, G. (2003): SWD Sachgruppen, Technical report, Die Deutsche Bibliothek.

NICHOLS, D.M. (1997): Implicit rating and filtering, in: *Fifth DELOS Workshop: Filtering and Collaborative Filtering, ERCIM, 28–33.*

RESNICK, P. and VARIAN, H.R. (1997): Recommender Systems, *CACM, 40, 3, 56 - 58.*

SARWAR, B.M., KARYPIS, G., KONSTAN, J. and RIEDL, J. (2002): Recommender systems for large-scale e-commerce: Scalable neighborhood formation using clustering. *Proceedings of the Fifth International Conference on Computer and Information Technology*, Bangladesh.

SCHAFER, J.B., KONSTAN, J.A. and RIEDL, J. (2001): E-commerce recommendation applications, *Data Mining and Knowledge Discovery, 5, 1/2, 115–153.*

SCHÖLL J. (2002): *Clusteranalyse mit Zufallswegen*, Ph.D. thesis, TU Wien, Wien.

SCHÖLL J. and PASCHINGER E. (2002): Cluster Analysis with Restricted Random Walks. In: Jajuga K., Sokolowski A. and Bock H.H., (Eds) *Classification, Clustering, and Data Analysis*. Springer-Verlag, Heidelberg, 113–120.

SCHÖLL J. and SCHÖLL-PASCHINGER E. (2003): Classification by restricted random walks, *Pattern Recognition, 36, 6, 1279–1290.*

SEMERARO G., FERILLI S., FANIZZI N. and ESPOSITO F. (2001): Document classification and interpretation through the inference of logic-based models. In: P. Constantopoulos I.S. (Ed.): *Proceedings of the ECDL 2001*. Springer Verlag, Berlin, volume 2163 of *LNCS*, 59–70.

UCHYIGIT G. (Ed.) (2005): *Web Personalization, Recommender Systems and Intelligent User Interfaces, Proc. WPRSIUI 2005*, INSTICC Press, Portugal.

VIEGENER J. (1997): *Inkrementelle, domänenunabhängige Thesauruserstellung in dokumentbasierten Informationssystemen durch Kombination von Konstruktionsverfahren*, infix, Sankt Augustin.

YANG Y. (1999): An evaluation of statistical approaches to text categorization, *Information Retrieval, 1, 1, 69–90.*

A Software Tool via Web for the Statistical Data Analysis: R-php

Angelo M. Mineo and Alfredo Pontillo

Dipartimento di Scienze Statistiche e Matematiche "S. Vianelli",
Università di Palermo, Italy
{elio.mineo; alf}@dssm.unipa.it

Abstract. The spread of Internet and the growing demand of services from the web users have changed and are still changing the way to organize the work or the study. Nowadays, the main part of information and many services are on the web and the software is going toward the same direction: in fact, the use of software implemented via web is ever-increasing, with a client-server logic that enables the "centralized" use of software installed on a server. In this paper we describe the structure and the running of `R-php`, an environment for statistical analysis, freely accessible and attainable through the World Wide Web, based on the statistical environment `R`. `R-php` is based on two modules: a base module and a point-and-click module. By using the point-and-click module, the so far implemented statistical analyses include also ANOVA, linear regression and some data analysis methods such as cluster analysis and PCA (Principal Component Analysis).

1 Introduction

`R-php` is a project developed into the Department of Statistical and Mathematical Sciences of Palermo and has the aim to realise a statistical web-oriented software, i.e. a software that a user reaches through Internet and uses by means of a browser. `R-php` is an open-source project; the code is released by the authors and could be freely installed. The internet site where can be found information on this project and where is possible to download the source code is *http://dssm.unipa.it/R-php/*.

The idea to design a statistical software that can be used through Internet comes out from the following considerations: it is a fact that the growing propagation of Internet and the demand of new services from its users have changed and still are thoroughly changing the way to access the structures of daily use on work, on study and so on; nowadays, the most of information and services goes through the web and in the same direction the software philosophy is moving: let's think, for example, about the `PhpMyAdmin` tool, that is a graphical interface via web of the well known database management system `MySQL`; in the business field, another example is given by the development of home-banking systems on the web.

A basic feature of `R-php` is that all the statistical computations are done by exploiting as "engine" the open-source statistical programming environment `R` (R Development Core Team, 2005), that is used more and more from Universities, but also from consulting firms (Scwartz, 2004), for example. Then R-php is classifiable as a web project based on R. Besides this project, there are others, such as Rweb, R_PHP_ONLINE and so on (see the web page *http://franklin.imgen.bcm.tmc.edu/R.web.servers/*). Shortly, the main difference between R-php and the other similar projects is that R-php has an interactive module (R-php point-and-click) that allows to make some of the main statistical analyses without the user has necessarily to know R.

The potential users of R-php are several: for example, let's think about the use from students either inside didactic facilities, such as informatics laboratories, or from home by means of a simple connection to Internet, or from an user that not knowing and wanting not to know a programming environment, such as R, wants to do simple statistical analyses without having to use expensive statistical software.

In this paper we shall describe the needed software to implement a R-php server with its two modules: R-php base and R-php point-and-click. Then, we shall present some other projects related to the installation of web server using R by pointing out the main differences with R-php. At the end, we shall show an example of use of R-php.

2 Needed Software to Install R-php

The needed software to implement a R-php server is the following: R, PHP, MySQL, Apache, ImageMagick and htmldoc. All this software is open-source. A short description of each software follows.

- **R**: R is a well known language and environment for statistical computing and graphics; R can be considered as a "dialect" of S. The term "environment" is intended to characterize a fully planned and coherent system, rather than a system with an incremental accretion of very specific and inflexible tools, as it is frequently the case with other statistical software. R provides a wide variety of statistical and graphical techniques and is highly extensible. R compiles and runs on a wide variety of UNIX platforms and similar systems (including FreeBSD and Linux), Windows and MacOS.
- **PHP**: PHP, recursive acronym for PHP Hypertext Preprocessor, is a server-side scripting language to create web pages with a dynamic content. It is widely used to develop software via web and it is usually integrated with dedicated languages, such as HTML and JavaScript. Moreover, it gives a set of functions to interact with many relational database management systems, such as MySQL, PostgreeSQL, Oracle, and so on.
- **MySQL**: MySQL, acronym for My Standard Query Language, is a multiuser client/server database. MySQL is, maybe, the most popular open source database management system. One of the more appreciated features of MySQL is the speed to access data: this is the reason why it is the more used database system for web-oriented software.

- **Apache**: `Apache` is a powerful web server for the `http` protocol; it is flexible and highly configurable to meet any requirements. At September 2005, about 70% of the internet web servers uses it (source: Netcraft Web Server Survey, *http://news.netcraft.com/archives/web_server_survey.html*).
- **ImageMagick**: `ImageMagick` is a good software to manipulate images; it has been created to run on `UNIX` alike platforms, but nowadays is distributed also for `Windows` and `MacOS` systems. It gives a set of commands that can be used also in batch mode and that allows, among other things, to convert images in many formats, to modify the image size, to cut some parts of an image, to manage colors and transparency, to realize thumbnails and editing of images. A great feature of this software is exactly the opportunity to use it by prompt.
- **htmldoc**: `htmldoc` is a package to create `.pdf` files from `.html` files.

3 Description of R-php

In this paragraph we shortly show the way `R-php` works.

3.1 R-php Base

When a user goes to the home page of this module, automatically a temporary directory is created; the name of this directory is generated in the following way: first, it is generated a pseudo-random alpha-numeric character string, which is associated to the `UNIX` time stamp. Inside this temporary directory there are all the files of the current session, generated by the connected user; this fact guarantees the multi use, i.e. the contemporary connection of several users with no confusion among the files generated in these different concurrent sessions. This feature of `R-php` does not seem implemented in other similar systems. The maximum number of contemporary users does not depend on `R-php`, but from the setting of the `Apache` server that contains it. Besides, the `UNIX` time stamp is useful to delete automatically the temporary directories from the server after six hours from the beginning of the working session: this fact avoids to have an excessive number of directories on the server and then allows to have a "clean" server.

This `R-php` module allows to introduce `R` code in a text area; this code is reported in a text file. As it is known, `R` has commands that allow the interaction with the operating system of the used computer; this could be dangerous in an environment like this one, because an user could input (voluntarily or not) harmful commands. To avoid this drawback, we have decided to implement a control structure that does not allow the use of a set of commands that we think are dangerous for the server safety. These commands are contained in a `MySQL` database, that has also a short description what the banned command does of. It is clear that a user interested to install a `R-php` server can modify this list of commands in every moment. For a greater safety, this control is made, first, client side in `JavaScript` and then (only if the first control is evaded) also server side in `PHP`. About the data input, besides the possibility to input data by using `R` commands, there is the possibility to read data from an `ASCII` file of the user's computer. The `R` commands, contained on the previously created text file, are processed by `R` that in the meanwhile has been invocated in batch mode by `PHP`. Then `R` gives the output in two formats; a textual one with the requested analysis and a `.ps` format that contains possible graphs. At this point, the two files are treated in two different ways:

- the text file is simply formatted by using style sheet to have a more readable web page;
- the `.ps` file is divided to obtain a file for each image; then each image is converted by `ImageMagick` in a more suitable format (`.png` format) for the web visualization.

At this point, the output with possible graphs is visualized in a new window. In this window there is the chance to save the output in `.pdf` format: this operation is made by using the `htmldoc` software; the user can also decide to save the graphs one by one.

3.2 R-php Point-and-Click

The `R-php` module that we describe in this paragraph is not a simple `R` web interface, but has in every way the features of a Graphical User Interface (GUI); this fact makes `R-php` different from the other web-projects based on R. In this module, the organization of the concurrent sessions is the same as that one described for `R-php base`. The data input is made by loading an `ASCII` file from the user's computer: this operation is made by means of `JavaScript`. Afterwards, the file contents are visualized in a new page as a spreadsheet; this data set is managed by `MySQL`: this allows to make some interactive operations directly on the data; for example, it is possible to modify the variable names or the value of every single cell of the spreadsheet. After we have loaded the data set, it is possible to choose what kind of analysis we have to do among those proposed. Each of these analyses can be made by means of a GUI, where the user can choose the options he wants to use; such options are translated in `R` code and processed. With more details, this phase provides the following steps:

1. the code is transcribed in an `ASCII` file that will be the `R` input;
2. the database containing the data is exported to an `ASCII` file;
3. `R`, called in batch mode, processes the code contained in the previously created text file;
4. `R` gives the output in two files; the first one is an `ASCII` file that presents the results in a text mode, the second one is a `.ps` file that contains the possible graphs;

5. the `ASCII` file is formatted to be compatible with the web standards of visualization;
6. the `.ps` file is divided in more files containing each one a graph and then each graph is converted in `.png` format (this operation is the same of that one of the `R-php base` module);

At this point, it is generated a web page that has the text and the graphs of the analysis. Moreover, the generated page containing the output allows other interesting operations, such as the output saving, with the graphs too, in `.pdf` format and the saving of the single graphs by means of a simple click.

We do not go longer on the description of each GUI for the several analyses that we can make with `R-php point-and-click`, because the design of the GUI's is formally the same for each analysis. The GUI implemented so far are the following:

- descriptive statistics;
- linear regression;
- analysis of variance;
- cluster analysis;
- principal component analysis;
- metric multidimensional scaling;
- factor analysis.

4 Some Web Server Projects Using R

In this section we describe some interesting web server projects using `R` and we shortly show the main differences of these projects in respect of `R-php`.

- **RWeb**: `RWeb` (Banfield, 1999) is probably the oldest project of this kind and it has been the start up of many web-oriented applications using `R` as "engine". `RWeb` is composed by two parts that require a different knowledge of the `R` environment. The first part is a simple `R` interface that works properly on many browsers, such as `Internet Explorer`, `Mozilla Firefox`, `Opera`, and so on. The use of this module requires the knowledge of the `R` environment. The main page has a text area where the user inputs `R` code and by clicking on the submit button the software gives the output page, included possible graphs. The second part, called `RWeb modules`, is designed as a point-and-click interface and the user could not know the `R` programming language. To use this module it is required to choose the data set, the type of analysis and the options for that analysis. The currently implemented analyses include regression, summary statistics and graphs, ANOVA, two–way tables and a probability calculator. Among the considered projects, `RWeb`, seems for some aspects the most similar to `R-php`. The aspects that distinguish `R-php` from `Rweb` are, for example, the following: `R-php` gives more analyses that a user can perform by means of the point-and-click module and presents more tools to manage the output, such as the possibility to save it in a `.pdf` format.

- **CGIwithR**: `CGIwithR` (Firth, 2003) is a package which allows `R` to be used as a CGI scripting language (CGI stands for Common Gateway Interface). In substance, it gives a set of `R` functions that allows any `R` output to be formatted and then used with a web browser. Although `CGIwithR` is oriented to the realization of web pages containing `R` output, this project is different from `R-php` that, on the contrary, is a `R` front end exploiting its features by calling it in batch mode by means of a GUI.
- **Rpad**: `Rpad` (Short and Grosjean, 2005) is a web–based application allowing the `R` code insert and the output visualization by means of a web browser. `Rpad` installs as an `R` package and can be run locally or can be used remotely as a web server. If an user wants to use `Rpad` as a service, he has to open `R` and then he needs to load the `Rpad` package and to run the `Rpad()` function that activates a mini web server; moreover, a new web session is started up by using the default browser and by showing the `Rpad` interactive pages. In brief, `Rpad` allows the use of `R` by means of a browser (both locally and remotely), but anyway requires the knowledge of the `R` language; in this way, it can be compared only with the `R-php base` module.
- **R_PHP_ONLINE**: `R_PHP_ONLINE` is a `PHP` web interface to run `R` code on line, including graphic output. In brief, it allows to insert code and to visualize the output by means of a web browser. It also allows the graph visualization, but this operation is not so easy; on the project web site (*http://steve-chen.net/R_PHP/*) the author suggests different ways to do this; surely, the most instinctive method seems to be this one: add the functions `bitmap()` and `dev.off()` before and after the real function call which generate graphics; in this way, only one graph at once can be generated. `R_PHP_ONLINE` does not have a point-and-click module and then can be compared only with the `R-php base` module. The latter is anyway more "user friendly" when a user has to produce some graphs; indeed, in `R-php` it is possible to use directly the `R` graph functions even when there are several graphs to produce.

Anyway, there are other similar projects worthy of mention, both regarding GUI and web server; for more details the reader can visit the web page: *http://franklin.imgen.bcm.tmc.edu/R.web.servers/*. In this page the reader can find information on the projects cited on this paper, too.

5 An Example of Use of R-php

In this section, we show a simple example of use of `R-php point-and-click`, by using the cluster analysis GUI. When the user chooses this GUI, he can choose either a hierarchical method or the k-means algorithm. Let's consider

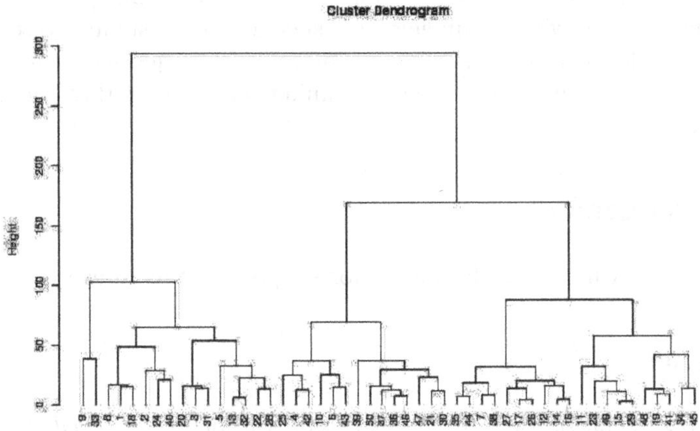

Fig. 1. Dendrogram obtained by using the module point-and-click of R-php.

the data set `USArrests` contained in `R`. This data set contains statistics, in arrests per 100.000 residents for assault, murder, and rape in each of the 50 US states in 1973 (McNeil, 1977). By loading this data set to `R-php` and by clicking on the cluster analysis GUI and on `Hierarchical`, `R-php` shows a windows where the user can choose the type of distance matrix (the options are: `euclidian`, `maximum`, `manhattan`, `canberra`, `binary` and `minkowski`), the agglomeration method (the options are: `ward`, `single`, `complete`, `average`, `mcquitty`, `median` and `centroid`), the numbers of groups and the height of the dendrogram (the user can not choose both these two options, of course). It is possible to say if we want to visualize the distance matrix or not. By clicking on the send button we have the output containing: the names of the involved variables, the distance matrix, the indication of the number of cluster whom the unit belongs to, the number of units for each cluster, the dendrogram and a graph showing the number of groups according to the height of the dendrogram. In Figure 1 it is shown the resulting dendrogram. If in the input window we have chosen a value for the field `height of dendrogram`, in the output there is the dendrogram with a horizontal red line in correspondence of the input height value.

6 Conclusion

In this paper we have shortly described the open-source web oriented statistical software `R-php`. In our opinion, the main feature of this software is the possibility to use a statistical software via web with a well developed GUI, i.e. the possibility for a user to make statistical analysis having only a connection via Internet and no statistical software installed on his own computer. It is intention of the authors to develop further this tool by correcting eventual bugs and by developing new GUI for other statistical analyses. Moreover, new functionalities will be implemented, such as the possibility to see and to save the R code used by the software to perform the requested analysis, or to use case names in the spreadsheet containing the data set that the user has to analyse.

Aknowledgement

Thanks to the University of Palermo for supporting this research.

References

BANFIELD, J. (1999): Rweb: Web–based Statistical Analysis. *Journal of Statistical Software, 4(1).*

FIRTH, D. (2003): CGIwithR: Facilities for processing web forms using R. *Journal of Statistical Software, 8(10).*

McNEIL, D.R. (1977): *Interactive Data Analysis.* Wiley & Sons, New York.

R DEVELOPMENT CORE TEAM (2005): *R: A Language and Environment for Statistical Computing.* R Foundation for Statistical Computing, Vienna.

SCWARTZ, M. (2004): The Decision to use R: a Consoulting Business Perspective. *R–News, 4(1), 2–5.*

SHORT, T. and GROSJEAN, P. (2005): *The Rpad package.* Comprehensive R Archive Network, Vienna.

Evolutionary Algorithms for Classification and Regression Trees

Francesco Mola[1] and Raffaele Miele[2]

[1] Dipartimento di Economia,
Università di Cagliari, Italy
mola@unica.it

[2] Dipartimento di Matematica e Statistica,
Università degli Studi di Napoli Federico II, Italy
rafmiele@unina.it

Abstract. Optimization Problems represent a topic whose importance is getting higher and higher for many statistical methodologies. This is particularly true for Data Mining. It is a fact that, for a particular class of problems, it is not feasible to exhaustively examine all possible solutions. This has led researchers' attention towards a particular class of algorithms called Heuristics. Some of these Heuristics (in particular Genetic Algorithms and Ant Colony Optimization Algorithms), which are inspired to natural phenomena, have captured the attention of the scientific community in many fields. In this paper Evolutionary Algorithms are presented, in order to face two well-known problems that affect Classification and Regression Trees.

1 Introduction

It is often needed, in order to apply many statistical techniques, to make use of a discrete optimization algorithm in order to find solutions to a problem whose solution space cannot be completely explored. Binary Segmentation[1] techniques allow to recursively partition a data set on the basis of some criteria in order to get more homogeneous subsets in the sense of a particular variable called *response variable.* The output of the analysis can be easily represented in a graphical tree structure. The way in which the data set is progressively partitioned is based on a (splitting) rule that divides the units in two offspring groups. Splitting rules are generated on the basis of the explanatory variables, also called *predictors.* In CART methodology, in order to choose the best split, all possible dichotomizations for each predictor must be evaluated at each step. This may lead to a combinatorial problem because, when examining an unordered nominal predictor, it comes that the number of all possible splits[2] is $2^{k-1} - 1$. It follows that evaluating all possible splits for a predictor whose number of modalities is high is computationally infeasible. This is the reason why many statistical packages don't allow to process

[1] Classification and Regression Trees, Breiman et al. (1984); C4.5, Quinlan (1993); Two-Stage Binary Segmentation, Mola and Siciliano (1992).

[2] Being k the number of modalities of the predictor.

nominal unordered predictors whose number of modalities is higher than a certain threshold. Even if we consider the FAST (Mola and Siciliano (1997)) algorithm (that provides a smaller computational effort to grow the binary tree), the combinatorial problem is still relevant, especially when complex data sets have to be analyzed. Another problem lies in the fact that the choice of any split, for each node of the tree, conditions the choice of all the splits below that one, thus reducing the solutions space that is being examined. This happens because, due to another combinatorial optimization problem, it is not possible to evaluate all possible trees that could be built from a data set. Many segmentation algorithms (CART, C4.5, etc.) are forced to make use of "greedy" procedures that allow to get results in a reasonable time by renouncing to the global optimality of the obtained solution. This paper proposes two algorithms that can deal with the forementioned problems. The following sections (2 and 3 respectively) focus on Genetic Algorithms and Ant Colony Optimization techniques, that are two heuristics that are particularly useful for "attacking" combinatorial optimization problems. In section 4 these heuristics are used to look for the best split from a high level nominal predictor and for extracting exploratory trees.

2 Genetic Algorithms

Genetic Algorithms (GA, Holland (1975)) are powerful and broadly applicable stochastic search and optimization techniques based on principles of evolution theory. GAs allow to start from an initial set of potential solutions to a problem and progressively improve them on the basis of an objective function, which is called *fitness* function. The solutions' quality is improved by recombining the "genetic code" of the best ones by making use of some genetic operators until some stopping criterion is met. A GA starts with an initial *population*[3] in which each individual is called *chromosome*, and represents a solution to the problem at hand. A chromosome is a string of (usually) binary symbols. Such a population of chromosomes evolves through successive iterations, called *generations*. During each generation the chromosomes are evaluated, using some measure of fitness (the objective function). In order to create a new generation the new chromosomes are formed by either merging two chromosomes' genetic code from current generation using a crossover operator, or by modifying some chromosomes using a mutation operator. A new generation is formed by selecting (using the fitness values) some of the parents and offspring and, therefore, rejecting the others. Chromosomes whose fitness value is greater have higher probability of being selected. After several generations the algorithms converge to the best chromosome which, hopefully, represents the optimal solution to the problem. The computational advantage that is obtained by using a GA is due to the fact that the search is oriented by the fitness function. This orientation is not based on the structure

[3] Usually chosen randomly.

of the whole chromosome, but just on some parts of it (called building blocks) which are highly correlated with high values of the fitness function. It has been demonstrated (Holland (1975)) that a GA evaluates, at each iteration, a number of building blocks whose order is the cube of the number of individuals of the population. If we let $P(t)$ and $C(t)$ be parents and offspring in current generation t, a genetic algorithm can be described as follows:

- $t = 0$;
- initialize and evaluate P(t);
- while (not termination condition) do
  - recombine P(t) to yield C(t) and evaluate C(t);
  - select P(t + 1) from P(t) and C(t);
  - $t = t + 1$;
- end while

3 Ant Colony Optimization

An artificial Ant Colony System (ACS) is an agent-based system which simulates the natural behavior of ants, which develop mechanisms of cooperation and learning. ACS was proposed by Dorigo et al. (1991a) as a new heuristic to solve combinatorial optimization problems. This heuristic, called Ant Colony Optimization (ACO), has been shown to be both robust and versatile in the sense that it can be applied to a wide range of combinatorial optimization problems. One of the most important advantages of ACO is that it is a population-based heuristic. This allows the system to use a mechanism of positive feedback between agents as a search mechanism. ACS makes use of artificial ants which are agents characterized by their imitation of the behavior of real ants with some exceptions (see Dorigo et al. (1996)). First they have, while searching for the path to follow, probabilistic preference for paths with a larger amount of pheromone. Artificial ants also tend to use an indirect communication system based on the amount of pheromone deposited in each path. The whole ant colony will tend to leave more pheromone on shorter paths. Those characteristics, from one side, allow real ants to find the shortest path between food and the nest and, on the other side, make artificial ants particularly useful for finding solutions to the Travelling Salesman Problem (TSP), which was the first application for these systems. An ACO algorithm can be applied if some conditions are verified[4] but one of the most problematic is that the solution space must be modeled as a graph. In order to give an idea about the behavior of an ACO algorithm we consider, as an example, the TSP for which we describe the corresponding ACO algorithm:

- Put ants on the different nodes of the graphs and initialize pheromone trails.

[4] see Dorigo et al. (1996) for details.

- while (not termination condition)
  - Make all ants move from a node to another one. The direction is stochastically chosen on the basis of a greedy rule (based on a visibility parameter η, which is an inverse measure of the distance) and the amount of pheromone (usually called τ) left on any possible direction.
  - Update pheromone trails.
  - Keep in memory the shortest path ever found.
- end while

4 A Genetic Algorithm for High Level Unordered Nominal Predictors

A Segmentation procedure has to find, for each node, the best split between all possible ones. The CART methodology reaches this target by making use of a brute-force procedure in which all possible splits from all possible variables are generated evaluated. This procedure can lead to computational problems, because the number of possible splits that can be generated by a nominal unordered predictor grows exponentially with its number of modalities. In order to design a Genetic Algorithm to solve this combinatorial problem it is necessary to identify:

- a meaningful representation (coding) for the candidate solutions (possible splits)
- a way to generate the initial population
- a fitness function to evaluate any candidate solution
- a set of useful genetic operators that can efficiently recombine and mutate the candidate solutions
- values for the parameters used by the GA (population size, genetic operators parameters values, selective pressure, etc.)
- a stopping rule for the algorithm.

These points have been tackled as follows. For the *coding* it has been chosen the following representation: a solution is coded in a string of bits (chromosome) called x where each bit (gene) is associated to a modality of the predictor as follows:

$$x_i = \begin{cases} 0 \text{ if the i-th modality goes to left} \\ 1 \text{ if the i-th modality goes to right} \end{cases}$$

The choice of the fitness function is straightforward: the split evaluation function of the segmentation procedure will be used (i.e. the impurity decrease proposed by Breiman et al. (1984)). Since the canonical (binary) coding has been chosen, it has also been decided to use the corresponding two parents single-point crossover and mutation operators. As stopping rule, a maximum number of iterations has been chosen on the basis of empirical studies. The rest of the GA specifications are the classic ones: elitism is used (at each iteration the best solution is kept in memory) and the initial population is chosen randomly. The algorithm can be summarized as follows:

- t = 0;
- Generate initial population P(t) by randomly assigning the bits;
- Evaluate population P(t) by calculating the impurity reduction;
- ElitistSol = the best generated solution;
- while not (max number of iterations) do
  - $t = t + 1$;
  - Select P(t) from P(t-1) by using the roulette wheel;
  - Do crossover and mutation on P(t);
  - Evaluate population P(t) by calculating the impurity reduction;
  - If (fitness of ElitistSol) > (fitness of best solution in the population) then put ElitistSol in the population.
- end while
- return to best solution

The proposed algorithm has been initially applied to many simulated and real datasets with a reasonably high number of modalities (i.e. 16) in order to extract, by exhaustive enumeration and evaluation of the solutions, the globally optimal solution. In all experiments the GA was able to find the optimum in a small number of generations (between 10 and 30). When the complexity of the problem grows, many iterations seem to be required. The GA has been tested also on the *adult* dataset from the UCI Machine Learning website. This dataset has 32561[5] units and some nominal unordered predictors with many modalities. In particular the "native-country" predictor has 42 modalities. The GA has been run to try to find a good split by making use of this predictor that R and SPSS, for instance, refused to process. As said before, 30 iterations seemed to be not enough because, in many runs of the algorithm, the "probably best" solution appeared after iteration 80.

5 Ant Colony-Based Tree Growing

The phase of Tree construction, in order to evaluate all possible trees generable from a given data set, would require an unavailable amount of computation. This is the reason why many Classification and Regression Trees algorithms use a local (greedy) search strategy that leads to finding local optima. In this section an Ant Colony based tree growing procedure is proposed. In order to attack a problem with ACO the following design task must be performed:

1. Represent the problem in the form of a weighted graph, on which ants build solutions.

[5] Just the training set has been used.

2. Define the heuristic reference (η) for each decision an ant has to take while constructing a solution.
3. Choose a specific ACO algorithm.
4. Tune the parameters of the ACO algorithm.

One of the most complex task is probably the first one, in which a way to represent the problem in the form of a weighted graph must be found. The representation used in this case is based on the following idea: let us imagine to have two nominal predictors $P_1 = \{a_1, b_1, c_1\}$ and $P_2 = \{a_2, b_2\}$ having two and three modalities, respectively[6]. In this case the set of all possible splits, at root node, is the following:

- $S_1 = [a_1] - [b_1, c_1]$
- $S_2 = [a_1, b_1] - [c_1]$
- $S_3 = [a_1, c_1] - [b_1]$
- $S_4 = [a_2] - [b_2]$

Any time a split is chosen it generates two child nodes. For these nodes the set of possible splits is, in the worst case[7], the same as the parent node except the one that was chosen for splitting. This consideration leads to the representation shown in Figure 1 in which, for simplicity, only the first two levels of the possible trees are considered. The figure also gives an idea of the problem complexity when dealing with predictors that generate hundreds or even thousands of splits (which is a common case). In this way the space of all possible trees is represented by a connected graph where moving from a level to another one means splitting a variable. The arcs of this graph have the same meaning of the arcs of the TSP graph (transition from a state to another one or, better, adding a component to a partial solution), so it is correct to deposit pheromone on them. The pheromone trails meaning, in this case, would be the desirability to choose the corresponding split from a certain node. As for the heuristic information η, it can be used the impurity decrease obtained by adding the corresponding split to the tree. This measure has a meaning which is similar, in some way, to the one that visibility has in the TSP. An arc is much more desirable as higher the impurity decrease is. In order to make analogy with the TSP, the impurity decrease can be seen as an inverse measure of the distance between two nodes. Once the graph has been constructed, pheromone trails meaning and heuristic function have been defined it is possible to attack the problem using an ACO algorithm. The proposed algorithm is shown as follows:

- Put all ants on the root node of all generable (and not fully explorable) trees;

[6] Such simple predictors are being considered, just to explain the idea, because of the combinatorial explosion of the phenomenon.

[7] When a node is split in two child nodes, it may happen that one or more of the modalities of a predictor are not present at all in the corresponding child subset. This would make the number of possible splits in the child node to decrease.

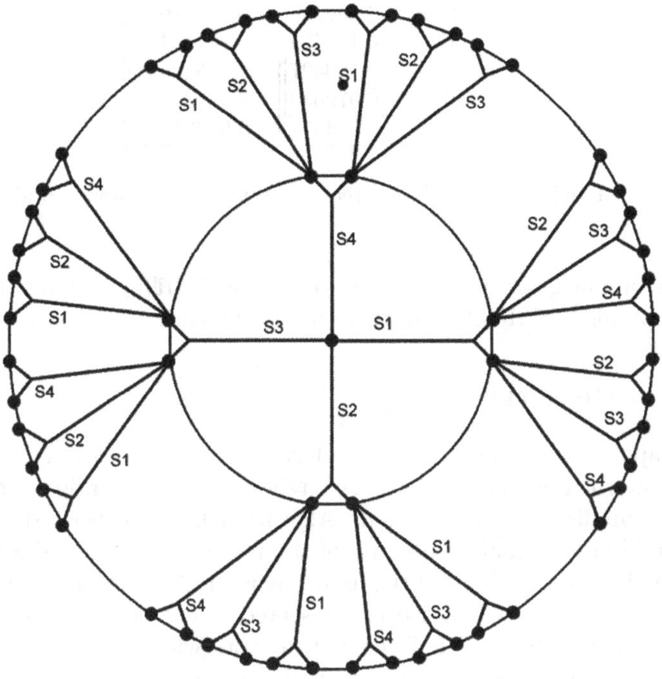

Fig. 1. The problem graph representation for building a tree

- f = 0;
- while (not termination condition)
  - let the f-th ant move from the actual node by stochastically choosing the direction (split) on the basis of the Two-stage splitting criterion (greedy rule) and on the amount of pheromone found on any possible direction. When the ant moves it generates two nodes, so another ant is necessary.
  - Repeat recursively the previous step until the ants reach some stopping condition (es. minimum number of units in a node);
  - update Pheromone trail generated by the ants;
  - f = f+1;
- end while;
- Extract the best tree from all the explored solutions;

The ACO algorithm which has been chosen is the Elitist Ant System (see Dorigo et al. (1991b)) and has been tested on simulated datasets. Table 1 shows the obtained results by applying the algorithm on a generated dataset of 500 observations with 11 nominal unordered predictors (with a number of modalities that ranges between 2 and 9) and 2 continuous predictors. It can be seen that, when the required tree depth increases, the differences between

Tree Depth	CART	Ant System
4	0.158119	0.153846
5	0.147435	0.121794
6	0.100427	0.085477
7	0.079059	0.059829
8	0.044871	0.029911

Table 1. Global impurity of the Trees extracted by the proposed algorithm

the global impurity of the tree obtained by the CART greedy heuristic and the one obtained by the Ant System tend to increase.

6 Final Remarks

In this paper evolutionary algorithms have been presented to solve some of the well known combinatorial problems related to Classification and Regression Trees application. The Genetic Algorithm has also been shown to be robust and scalable, mainly because of the fact that it is not affected at all by the number of modalities of the used predictor. An Ant Colony Optimization based algorithm has also been presented to extract better exploratory Classification and Regression Trees. Future work will focus on extending the ACO algorithm to the case of ternary and multiple segmentation and to the pruning phase. An integration with the Genetic Algorithm could also bring benefits.

References

BREIMAN, L. and FRIEDMAN, J. and OLSHEN, R. and STONE, C. (1984): *Classification and Regression Trees.* Belmont C.A. Wadsworth.

DORIGO, M. and MANIEZZO, V. and COLORNI, A. (1991a): The Ant System: An autocatalytic optimizing process. Technical Report 91-016 Revised, Politecnico di Milano, Italy.

DORIGO, M. and MANIEZZO, V. and COLORNI, A. (1991b): Positive feedback as a search strategy. Technical Report 91-016 Revised, Politecnico di Milano, Italy.

DORIGO, M. and MANIEZZO, V. and COLORNI, A. (1996): Ant system: Optimization by a colony of cooperating agents. *IEEE Transactions on Systems, Man, and Cybernetics-Part B*, 26(1),29-41.

HOLLAND, J. (1975): *Adaptation in Natural and Artificial Systems.* Ann Arbor, University of Michigan Press.

MOLA, F. and SICILIANO, R. (1992): A two stage predictive algorithm in Binary Segmentation. *Computational Statistics*, 1, 179-184.

MOLA, F. and SICILIANO, R. (1997) A Fast Splitting Procedure for Classification and Regression Trees, *Statistics and Computing*, 7, 208-216.

QUINLAN, J. (1993): *C4.5: Programs for Machine Learning.* Morgan Kaufmann.

Variable Selection Using Random Forests

Marco Sandri and Paola Zuccolotto[1]

Dipartimento Metodi Quantitativi,
Università di Brescia
c.da S. Chiara, 50 - 25122 Brescia, Italy
zuk@eco.unibs.it

Abstract. One of the main topic in the development of predictive models is the identification of variables which are predictors of a given outcome. Automated model selection methods, such as backward or forward stepwise regression, are classical solutions to this problem, but are generally based on strong assumptions about the functional form of the model or the distribution of residuals. In this paper an alternative selection method, based on the technique of Random Forests, is proposed in the context of classification, with an application to a real dataset.

1 Introduction

In many empirical analyses a crucial problem is the presence in the data of a set of variables not significatively contributing to explain the analyzed phenomenon, but capable to create a random noise which prevents from distinguishing the main effects and the relevant predictors. In this context proper methods are necessary in order to identify variables that are predictors of a given outcome. Many automatic variable selection techniques have been proposed in the literature, for example the backward or forward stepwise regression (see Miller (1984) and Hocking (1976)) or the recent stepwise bootstrap method of Austin and Tu (2004). These methods are for the most part based on assumptions about the functional form of the models or on the distribution of residuals. These hypothesis can be dangerously strong in presence of one or more of the following situations: (*i*) a large number of observed variables is available, (*ii*) collinearity is present, (*iii*) the data generating process is complex, (*iv*) the sample size is small with reference to all these conditions. Data analysis can be basically approached by two points of view: data modeling and algorithmic modeling (Breiman (2001b)). The former assumes that data are generated by a given stochastic model, while the latter treats data mechanism as unknown, a *black box* whose insides are complex and often partly unknowable. The aim of the present paper is to propose a variable selection method based on the algorithmic approach and to examine its performance on a particular dataset. In the mid-1980s two powerful new algorithms for fitting data were developed: neural nets and decision trees, and were applied in a wide range of fields, from physics, to medicine, to economics, even if in some applications (see e.g. Ennis et al. (1998)) their performance was poorer than that of simpler models like linear logistic regression. The main shortcomings of these two methods were overfitting and instability, the latter with particular reference to decision trees.

While overfitting has been long discussed and many techniques are available to overcome the problem (stopping rules, cross-validation, pruning, ...), few has been made to handle instability, a problem occurring when there are many different models with similar predictive accuracy and a slight perturbation in the data or in the model construction can cause a skip from one model to another, close in terms of error, but distant in terms of the meaning (Breiman (1996a)). The proposal of Random Forests (Breiman (2001a)), a method for classification or regression based on the repeated growing of trees through the introduction of a random perturbation, tries to manage these situations averaging the outcome of a great number of models fitted to the same dataset. As a subproduct of this technique, the identification of variables which are important in a great number of models provides suggestions in terms of variable selection. The proposal of this paper is to use the technique of Random Forests (RF) as a tool for variable selection, and a procedure is introduced and evaluated on a real dataset. The paper is organized as follows: in section 2 the technique of RF is briefly recalled, confining the attention to the case of classification, in section 3 a variable selection method based on RF is proposed, the application to a real dataset is reported in section 4, conclusive remarks follow in section 5.

2 Random Forests

A population is partitioned into two or more groups, according to some qualitative feature. It follows that each individual in the population belongs to (only) one group. The information about the group is contained in the categorical variable Y, while relevant further information is collected in a set of exogenous variables $\mathbf{X}$, always known, which is assumed to somewhat affect Y. Given a random sample $S = \{(y_1, \mathbf{x}_1); \cdots ; (y_n, \mathbf{x}_n)\}$, several statistical techniques are available in order to determine an operative rule $h(\mathbf{x})$ called *classifier*, used to assign to one group an individual of the population, not contained in the sample, for which only the exogenous variables $\mathbf{x}_{n+1}$ are known. A *random classifier* $h(\mathbf{x}, \boldsymbol{\theta})$ is a classifier whose prediction about y depends, besides on the input vector $\mathbf{x}$, on a random vector $\boldsymbol{\theta}$ from a known distribution $\boldsymbol{\Theta}$. Given a i.i.d. sequence $\{\boldsymbol{\theta}_k\} = \{\boldsymbol{\theta}_1, \boldsymbol{\theta}_2, \cdots, \boldsymbol{\theta}_k, \cdots\}$ of random vectors from a known distribution $\boldsymbol{\Theta}$, a Random Forest $\mathrm{RF}(\mathbf{x}, \{\boldsymbol{\theta}_k\})$ is itself a random classifier, consisting of a sequence of random classifiers $\{h(\mathbf{x}, \boldsymbol{\theta}_1), h(\mathbf{x}, \boldsymbol{\theta}_2), \cdots, h(\mathbf{x}, \boldsymbol{\theta}_k), \cdots\}$ each predicting a value for y at input $\mathbf{x}$. The RF prediction for y is expressed in terms of probability of Y assuming the value y, $Pr\{Y = y\}$. By definition a RF is composed by an infinite number of classifiers, but from an operational point of view the term is used to indicate a finite set of classifiers $\{h(\mathbf{x}, \boldsymbol{\theta}_1), h(\mathbf{x}, \boldsymbol{\theta}_2), \cdots, h(\mathbf{x}, \boldsymbol{\theta}_k)\}$. The k-set's prediction for y corresponds to the prediction whose frequency exceeds a given threshold[1]. Asymptotic results have been derived in order to know the behavior of the set as the number of classifiers increases. Limiting laws and statistical features of RF have been developed by Breiman (2001a) and a detailed explanation can be found in Sandri and Zuccolotto (2004). The theory of RF is quite general and can be applied to several kinds of classifiers and randomizations: examples are already present in literature, for instance the bagging technique of Breiman (1996b) or the random split selection of Dietterich (2000). Moreover other well-known techniques, like bootstrap itself, although introduced in different contexts, can be led back to the RF framework. Nevertheless by now the methodology called Random Forests is used uniquely with reference to its original formulation, due to Breiman (2001a), which uses CART-structured (Classification And Regression Trees, Breiman et al. (19984)) classifiers. RF with randomly selected inputs are sequences of trees grown by selecting at random *at each node* a small group of F input variables to split on. This procedure is often used in tandem with *bagging* (Breiman (1996b)), that is with a random selection of a subsample of the original training set at each tree. The trees obtained in this way are a RF, that is a k-set of random classifiers $\{h(\mathbf{x}, \boldsymbol{\theta}_1), h(\mathbf{x}, \boldsymbol{\theta}_2), \cdots, h(\mathbf{x}, \boldsymbol{\theta}_k)\}$ where the vectors $\boldsymbol{\theta}_i$ denote the randomization injected by the subsample drawing and by the selection of the F variables at each node.

2.1 Variable Importance Measures

The main drawback of using a set of random classifiers lies in its explanatory power: predictions are the outcome of a *black box* where it is impossible to distinguish the contribution of the single predictors. With RF this problem is even more crucial, because the method performs very well especially in presence of a small number of informative predictors hidden among a great number of noise variables. To overcome this weakness the following four measures of variable importance are available in order to identify the informative predictors and exclude the others (Breiman (2002)):

- **Measure 1**: at each tree of the RF all the values of the h-th variable are randomly permuted and new classifications are obtained with this new dataset, over only those individuals who have not contributed to the growing of the tree. At the end a new misclassification error rate $\hat{e}_h$ is

[1] In the standard case, the k-set's prediction for y corresponds to the most voted prediction, but a generalization is needed, as sometimes real datasets are characterized by extremely unbalanced class frequencies, so that the prediction rule of the RF has to be changed to other than majority votes. The optimal cutoff value can be determined for example with the usual method based on the joint maximization of sensitivity and specificity.

then computed and compared with $\hat{e}$. The $M1$ measure for h-th variable is given by

$$M1_h = \max\{0; \hat{e}_h - \hat{e}\}.$$

- **Measure 2**: for an individual $(y, \mathbf{x})$ the margin function $mg(y, \mathbf{x})$ is defined as a measure of the extent to which the proportion of correct classifications exceeds the proportion of the most voted incorrect classifications. If at each tree all the values of the h-th variable are randomly permuted, new margins $mg_h(y, \mathbf{x})$ can be calculated over only those trees which have not been grown with that subject. The $M2$ measure of importance is given by the average lowering of the margin across all cases:

$$M2_h = \max\{0; av_S[mg(y, \mathbf{x}) - mg_h(y, \mathbf{x})]\}.$$

- **Measure 3**: in the framework just described for $M2$, the $M3$ measure is given by the difference between the number of lowered and raised margins:

$$M3_h = \max\{0; \#[mg(y, \mathbf{x}) < mg_h(y, \mathbf{x})] - \#[mg(y, \mathbf{x}) \geq mg_h(y, \mathbf{x})]\}.$$

- **Measure 4**: at each node z in every tree only a small number of variables is randomly chosen to split on, relying on some splitting criterion given by a heterogeneity index such as the Gini index or the Shannon entropy. Let $d(h, z)$ be the decrease in the heterogeneity index allowed by variable $\mathbf{X}_h$ at node z, then $\mathbf{X}_h$ is used to split at node z if $d(h, z) > d(w, z)$ for all variables $\mathbf{X}_w$ randomly chosen at node z. The $M4$ measure is calculated as the sum of all decreases in the RF due to h-th variable, divided by the number of trees:

$$M4_h = \frac{1}{k} \sum_z [d(h, z) I(h, z)]$$

where $I(h, z)$ is the indicator function that is equal to 1 if h-th variable is used to split at node z and 0 otherwise.

3 Variable Selection Using Random Forests

In this paper the possible use of RF as a method for variable selection is emphasized, relying on the above mentioned four importance measures. A selection procedure can be defined, observing that the exogenous variables described by the four measures can be considered as points in a four-dimensional space, with the following steps: (1) calculate a four-dimensional centroid with coordinates given by an average (or a median) of the four measures; (2) calculate the distance of each point-variable from the centroid and arrange the calculated distances in non-increasing order; (3) select the variable whose distance from the centroid exceeds a given threshold, for example the average distance. This simple method is often quite effective, because the noise variables represented in the four-dimensional space tend to cluster together in a unique group and the predictors appear like outliers. A refinement of this proposal, which provides a useful graphical representation, can be proposed observing that the four measures are often correlated and this allows a dimensional reduction of the space where the variables are defined. With a simple Principal Component Analysis (PCA) the first two factors can be selected and a scatterplot of the variables can be represented in the two-dimensional factorial space, where the cluster of noise variables and the "outliers" can be recognized. The above described procedure based on the calculation of the distances from an average centroid can be applied also in this context and helps deciding which points have to be effectively considered outliers[2]. Simulation studies show that these methods very favorably compare with a forward stepwise logistic regression, even when the real data generating mechanism is a logistic one. Their major advantage lies in a sensibly smaller number of wrongly identified predictors. The main problem of these methods consists in the definition of the threshold between predictive and not predictive variables. To help deciding if this threshold exists and where it could be placed, a useful graphical representation could be a sort of scree-plot of the distances from the centroid, where the actual existence of two groups of variables, and the positioning of the threshold between them, can be easily recognized.

4 Case Study

A prospective study was conducted from January 1995 to December 1998 by the First Department of General Surgery (Ospedale Maggiore di Borgo Trento, Verona, Italy) in patients affected by acute peptic ulcer who underwent endoscopic examination and were treated with a particular injection therapy. The aims of the study were to identify risk factors for recurrence of hemorrhage, as early prediction and treatment of rebleeding would improve the overall outcome of the therapy. The dataset consists of 499 cases, observed according to 32 exogenous variables related to patient history (gender, age, bleeding at home or during hospitalization, previous peptic ulcer disease, previous gastrointestinal hemorrhage, intake of nonsteoridal anti-inflammatory drugs, intake of anticoagulant drugs, associated diseases, recent - within 30 days - or past - more than 30 days - surgical operations), to the magnitude of bleeding (symptoms: haematemesis, coffee-ground vomit, melena, anemia; systolic blood pressure, heart rate, hypovolemic shock, hematocrit

[2] In this case the distance function can take into consideration the importance of the two factors and a weight can be introduced given, for example, by the fraction of total variance accounted for by each factor or by the correspondent eigenvalue. Actually this procedure could be redundant, as the space rotation implied by the PCA, already involves a overdispersion of points along the more informative dimensions.

and hemoglobin level, units of blood transfused), to endoscopic state (number, size, location of peptic ulcers, Forrest classification, presence of gastritis or duodenitis). The values of all the variables are classified into categories according to medical suggestions. We think that the use of the raw data could allow a more detailed analysis. The results were presented in a paper (Guglielmi et al. (2002)) where a logistic regression with variables selected relying both on statistical evidences and on medical experience was able to provide a (in-sample) 24% misclassification error with sensitivity and specificity equal to 76%[3]. In this paper two logistic regressions are fitted to the same data, with variables selected respectively by a AIC stepwise procedure[4] (Model A) and by our RF-based method (Model B).

The AIC stepwise variable selection method identifies nine relevant predictors[5], while using the RF procedure, eight predictors are selected[6]. In both cases the resulting predictors has been judged reasonable on the basis of medical experience. In the left part of Figure 1 the scree plot of variable distances from the centroid are represented for three approaches (the basic method in the four-dimensional space, the refinement based on the PCA of the four measures with Euclidean distance or weighted Euclidean distance), while the right part of Figure 1 shows the two-dimensional scatterplot of the variables in the first two principal components space, with a virtual line separating the outlier variables selected as predictors.

In order to evaluate the performance of the two models, a cross-validation study has been carried out with validation sets of size 125 (25% of the sample) and $r = 1000$ repeated data splittings. The estimated probabilities of the two models are used to classify a patient being or not at risk of rebleeding, according to a cutoff point determined by minimizing the absolute difference between sensitivity and specificity in each validation set. Results are reported in Table 4, where also the corresponding in-sample statistics are shown.

The two models exhibit a substantially equal goodness-of-fit and also have a high agreement rate (in the in-sample analysis 91.58% of the individuals is classified in the same class by the two models). However it has to be noticed that Model B, built with the RF variable selection, has a reduced number of

---

[3] The predictors included in the model were: associated diseases/liver cirrhosis (`livcir`), recent surgical operations (`recsurg`), systolic blood pressure (`sbp`), symptoms/haematemesis (`hematem`), ulcer size (`size`), ulcer location (`location(2)`), Forrest class (`Forrest`).

[4] Coherently with our previous simulative studies, a forward selection is used. Anyway, the backward option was experimented: it leads to a less parsimonious model with substantially the same predictive performance.

[5] Forrest class, systolic blood pressure, ulcer size, recent surgical operations, ulcer location, units of blood transfused (`uobt`), age (`age`), symptoms/haematemesis, intake of anticoagulant drugs (`anticoag`).

[6] Systolic blood pressure, Forrest class, hypovolemic shock (`shock`), recent surgical operations, age, ulcer size, symptoms/haematemesis and-or melena (`symptoms`), ulcer location.

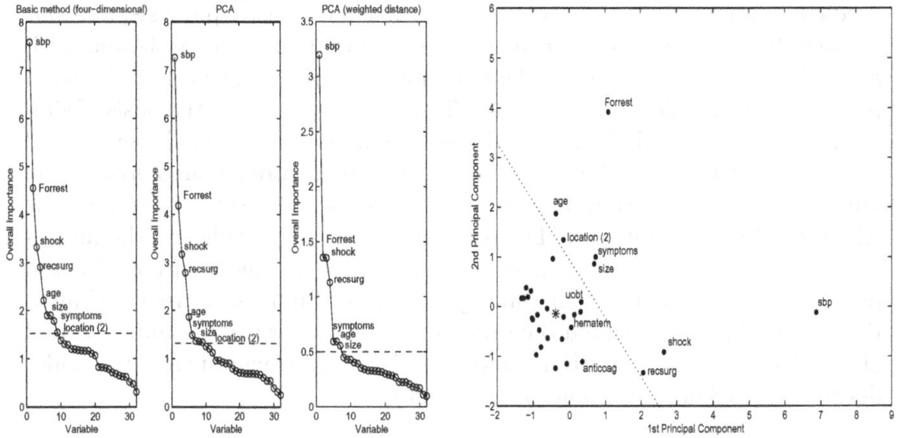

Fig. 1. Left: Scree plots of variable distances from centroid (thresholds: average distances); Right: Scatterplot of the variables in the first two principal components space.

	Misclassification error	Cohen k	Sensitivity	Specificity	Cutoff
	Model A/Model B - In-sample analysis				
Model A	23.05%	0.3482	77.03%	76.94%	0.1783
Model B	22.65%	0.3556	77.03%	77.41%	0.1642
	Model A - Out-of-sample analysis				
25th percentile	21.60%	0.2224	55.55%	73.15%	0.1629
Median	24.00%	0.2751	63.63%	76.19%	0.1742
75th percentile	26.40%	0.3320	70.65%	79.25%	0.1841
	Model B - Out-of-sample analysis				
25th percentile	22.40%	0.2138	55.55%	71.96%	0.1558
Median	24.80%	0.2686	63.63%	75.23%	0.1658
75th percentile	27.20%	0.3196	72.22%	78.57%	0.1771

Table 1. Misclassification error, Cohen k, sensitivity, specificity, cutoff value of the two logistic models (In- and out-of-sample analyses)

predictors, coherently with the simulation results assessing a good capability of the method in the false variables selection rate.

5 Concluding Remarks

In this paper a variable selection method based on Breiman's Random Forests is proposed and applied to a real dataset of patients affected by acute peptic ulcers, in order to identify risk factors for recurrence of hemorrhage. The main advantage of selecting relevant variables through an algorithmic modeling technique is the independence from any assumptions on the relationships among variables and on the distribution of errors. After having selected the predictors, a model could be developed with some given hypothesis, and this outlines Random Forests as a technique for preliminary analysis and variable selection and not only for classification or regression, which are its main purposes. The results on real data confirm what expected on the basis of simulation studies: the RF-based variable selection identifies a smaller number of relevant predictors and allows the construction of a more parsimonious model but with predictive performance similar to the logistic model selected by the AIC stepwise procedure. Further research is currently exploring the advantages deriving from the combination of measures coming from model-based prediction methods and algorithmic modeling techniques. Moreover simulation studies have highlighted the presence of a bias effect in a commonly used algorithmic variable importance measure. An adjustment strategy is under development (Sandri and Zuccolotto (2006)).

References

AUSTIN, P. and TU, J. (2004): Bootstrap methods for developing predictive models. *The American Statistician, 58, 131–137.*

BREIMAN, L., FRIEDMAN, J.H., OLSHEN, R.A. and STONE, C.J. (1984): *Classification and Regression Trees.* Chapman & Hall, London.

BREIMAN, L. (1996a): The heuristic of instability in model selection. *Annals of Statistics, 24, 2350–2383.*

BREIMAN, L. (1996b): Bagging predictions. *Machine Learning, 24, 123–140.*

BREIMAN, L. (2001a): Random Forests. *Machine Learning, 45, 5–32.*

BREIMAN, L. (2001b): Statistical modeling: the two cultures. *Statistical Science, 16, 199–231.*

BREIMAN, L. (2002): Manual on setting up, using, and understanding Random Forests v3.1. *Technical Report,* `http://oz.berkeley.edu/users/breiman`.

DIETTERICH, T. (2000): An experimental comparison of three methods for construction ensembles of decision trees: bagging, boosting and randomization. *Machine Learning, 40, 139–157.*

ENNIS, M., HINTON, G., NAYLOR, D., REVOW, M. and TIBSHIRANI, R. (1998): A comparison of statistical learning methods on the gusto database. *Statistics in Medicine, 17, 2501–2508.*

GUGLIELMI, A., RUZZENENTE, A., SANDRI, M., KIND, R., LOMBARDO, F., RODELLA, L., CATALANO, F., DE MANZONI, G. and CORDIANO, C. (2002): Risk assessment and prediction of rebleeding in bleeding gastroduodenal ulcer. *Endoscopy, 34, 771–779.*

HOCKING, R.R. (1976): The analysis and selection of variables in linear regression. *Biometrics, 42, 1–49.*

MILLER, A.J. (1984): Selection of subsets of regression variables. *Journal of the Royal Statistical Society, Series A, 147, 389–425.*

SANDRI, M. and ZUCCOLOTTO, P. (2004): Classification with Random Forests: the theoretical framework. *Rapporto di Ricerca del Dipartimento Metodi Quantitativi, Università degli Studi di Brescia, 235.*

SANDRI, M. and ZUCCOLOTTO, P. (2006): Analysis of a bias effect on a tree-based variable importance measure. Evaluation of an empirical adjustment strategy. *Manuscript.*

Boosted Incremental Tree-based Imputation of Missing Data

Roberta Siciliano, Massimo Aria, Antonio D'Ambrosio

Dipartimento di Matematica e Statistica,
Università degli Studi di Napoli Federico II, Italy
{roberta,aria,antdambr}@unina.it

Abstract. Tree-based procedures have been recently considered as non parametric tools for missing data imputation when dealing with large data structures and no probability assumption. A previous work used an incremental algorithm based on cross-validated decision trees and a lexicographic ordering of the single data to be imputed. This paper considers an ensemble method where tree-based model is used as learner. Furthermore, the incremental imputation concerns missing data of each variable at turn. As a result, the proposed method allows more accurate imputations through a more efficient algorithm. A simulation case study shows the overall good performance of the proposed method against some competitors. A MatLab implementation enriches Tree Harvest Software for non-standard classification and regression trees.

1 Introduction

In the framework of missing data imputation, it is worthwhile to distinguish between missing data completely at random and missing data at random. More precisely, when we say that data are missing completely at random (MCAR) we mean that the probability an observation is missing is unrelated to the value of the variable or to the value of any other variables. Instead, data can be considered as missing at random (MAR) if the data meet the requirement that missingness does not depend on the value of the variable after controlling for another variable. Last condition requires a model-based imputation for that the missing value can be understood as the sum of the model function and the error term. Classical approaches are linear regression (Little, 1992), logistic regression (Vach, 1994), generalized linear models (Ibrahim et al., 1999), whereas more recent approaches are nonparametric regression (Chu and Cheng, 1995) and tree-based models (Siciliano and Conversano, 2002).

Parametric and semi-parametric approaches can be unsatisfactory for nonlinear data yielding to biased estimates if model misspecification occurs. As an alternative, tree-based models do not require the specification of a model structure, deal with numerical and categorical inputs, consider conditional interactions among variables so that they can be used to derive simple imputation rules. Automatic tree-based procedures have been already considered for data validation (Petrakos et al., 2004) as well as for data imputation (Siciliano and Conversano, 2002). In this paper, we provide a general tree-based methodology for missing data imputation as well as specific algorithms to obtain the final estimates. The proposed algorithms have been implemented in the software Tree Harvest (TH) (Aria and Siciliano, 2004, Siciliano et al., 2004) characterized by an integrated graphical user interface for exploring data and results interactively. Special issue of TH is to provide nonstandard methods suited for analyzing specific structures of data (i.e., multi-responses, multi-block predictors, missing data, etc.).

2 Key Concepts

Previous work (Siciliano and Conversano, 2002) shows that Incremental Non Parametric Imputation (INPI) using trees is preferred to a single tree-based model imputation. Main idea of INPI is to rearrange columns and rows of the data matrix according to a lexicographic ordering of the data (with respect to both rows and columns), that matches the order by value, corresponding to the number of missing data occurring in each record. Any missing input is handled using the tree-based model fitted to the current data which is iteratively updated by the imputed data. The imputation is incremental because, as it goes on, more and more information is added to the data matrix. As a result, cross-validated trees are used to impute data and the algorithm performs an incremental imputation of each single data at time.
The above-mentioned approach is revised in this paper by considering two new concepts: first, the use of ensemble methods (in place of cross-validation) should provide more robust estimates; second, the incremental imputation of each variable at time (instead of each single data at time) allows for a more efficient algorithm, thus reducing the computational cost of the overall procedure.

3 The Methodology

For a $n \times k$ data matrix $\mathbf{Y}$ we define a lexicographic ordering of the variables as the k-dimensional vector $\ell = [l_{(1)}, \ldots, l_{(j)}, \ldots, l_{(k)}]$ such that $l_{(j)}$ points the column of the variable that is at the j-th position in the increasing order of all variables in terms of the number of missing values. It is assumed that at least the first ordered variable presents no missing values. Main issue of the incremental approach is that following the positions defined by ℓ each column presenting missing values at turn plays the role of dependent variable to be imputed by the complete set of variables with no missing values and once that this variable is imputed it concurs to form the complete set of predictors used for the subsequent imputation. For the imputation algorithm we can consider cross-validated trees providing an Incremental Non Parametric

Imputation of variable ($INPI_{var}$) rather than of single data ($INPI$). Alternatively, we can consider an imputation algorithm by an ensemble of trees, using stump (i.e., a tree with only one split) for imputation of a qualitative variable and fast trees (Mola and Siciliano, 1997) for the other cases. Ensemble methods such as Boosting (i.e., an ensemble method to define an accurate learner) (Freund and Schapire, 1997) and Bagging (Breiman, 1996) allows to define a robust imputation procedure. Whereas bagging is convenient when the learner is unstable in terms of accuracy variation, boosting is convenient when the learner has an high bias (Freund and Schapire, 1997). Since the Stump can be noveled as an accurate but biased learner (Hastie et al., 2002), boosting has been preferred to bagging in the proposed algorithm when the imputation procedure use stump as weak learner. The proposed Boosted Incremental Non Parametric Imputation is named $BINPI$ when trees are used and $BINPI_{stump}$ when stump is considered.

4 The Algorithm

Figure (1) describes an example of the main steps of the basic imputation algorithm. It is worth noting that numbers associated to the columns correspond to the values of the lexicographic order vector ℓ. Main steps of the algorithm can be outlined as follows. Let Y a $N \times K$ matrix bearing missing data where y_k is the $k-$th variable of Y.

0. Set $r = 1$.
1. Find $y_{k^*}^{(r)}$ as the variable (the column) with the smallest number of missing data, where $k^* : \#mis_{k^*} \leq mis_k$, for $k = 1, 2, ..., K$ and $\#mis_k > 0$;
2. Sort columns such that the first p variables are complete and the $p+1$-th is $y_{k^*}^{(r)}$;
3. Sort rows such that the first l rows are complete and the remaining $N - l$ are missing in the $p+1$-th column;
4. Use Stump or classical tree as *weak learner* for v-fold AdaBoost iterations to impute the $N - l$ missing data in variable $y_{k^*}^{(r)}$ on the basis of the learning sample $\mathcal{L}^{(r)} = \{y_{nk^*}^{(r)}, \mathbf{x}_n^{(r)} = (x_{n1}, \ldots, x_{np})'\}$ for $n = 1, \ldots, l$.
5. Set $r = r + 1$. Go to step 1 until all missing data are imputed.

5 Simulation Study

The proposed algorithms, namely Boosted Incremental Non Parametric Imputation using either trees ($BINPI$) or stump ($BINPI_{stump}$), have been evaluated with respect to standard methods such as Unconditional Mean Imputation (UMI) and Parametric Imputation (PI). A further comparison takes account of Incremental Non Parametric Imputation of single data ($INPI$) as well as of single variable ($INPI_{var}$). A simulation study has been

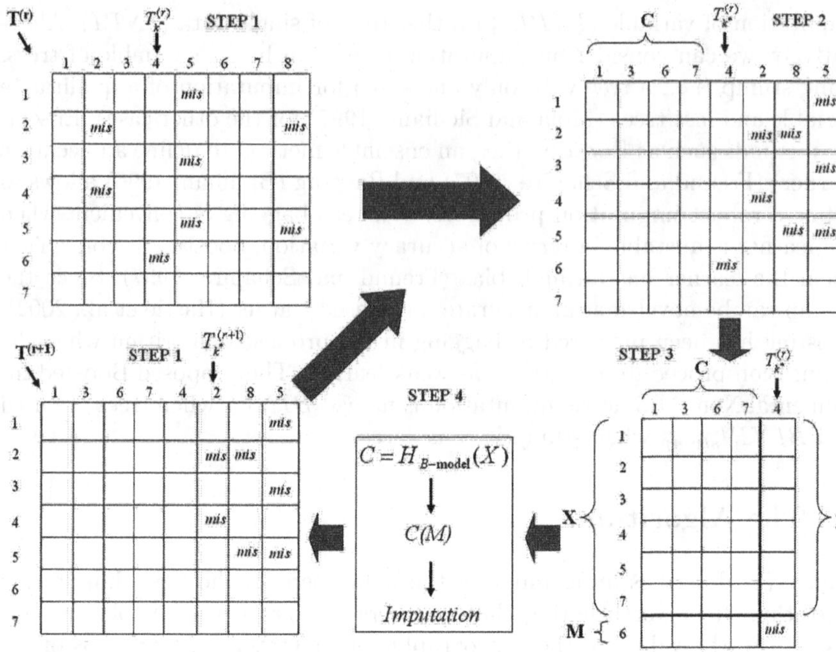

Fig. 1. Boosted Incremental Non Parametric Imputation Algorithm

designed. The basic assumption is that missing data are generated according to a missing at random schema so that a dependence relationship structure among variables is defined. The simulation setting consists of varying both the number of missing values in the uncompleted variables and the type of relationships between the uncompleted variables and the complete ones. In each setting, covariates are uniformly distributed in [0, 10] and values are missing with conditional probability:

$$\Psi = [1 + exp\,(\alpha + \beta)]^{-1}.$$

We distinguish the case of missing values generation in nominal covariates (i.e., nominal response case) from the case of missing values generation in numerical covariates (i.e., numerical missing case).

5.1 Binary Missing Case

In each simulation setting, we consider two different cases, linear as well as non-linear relationships. We generated five data structures for the nominal response case. The variables under imputation are simulated according to the binomial distribution. In the *simulation 1*, *simulation 2* and *simulation 3* the parameters characterizing the distribution of the missing values are expressed as a linear combination of some of the covariates.

Simulation 1:

$$Y_1 \sim Bin\left(n, \tfrac{2+0.35(X_1+X_2)}{10}\right),\ m_1 = \{1 + exp\left[-3 + 0.5\left(X_1 + X_2\right)\right]\}^{-1};$$
$$Y_2 \sim Bin\left(n, \tfrac{4+0.35(X_2-X_3)}{10}\right),\ m_2 = \{1 + exp\left[-3 + 0.5\left(X_2 - X_3\right)\right]\}^{-1}.$$

Simulation 2 (to simulation 1 the following variable is added up):

$$Y_3 \sim Bin\left(n, \tfrac{3+0.35(X_3+X_4)}{10}\right),\ m_3 = \{1 + exp\left[-3 + 0.5\left(X_3 + X_4\right)\right]\}^{-1}.$$

Simulation 3 (with respect to simulation 1 the second variable is replaced by):

$$Y_2 \sim Bin\left(n, \tfrac{5+0.35(X_2-X_3)}{10}\right),\ m_3 = \{1 + exp\left[-3 + 0.5\left(X_3 + X_4\right)\right]\}^{-1}.$$

In the *simulation 4* and *simulation 5* the parameters characterizing the distribution of the missing values depend on some covariates in a non linear way.

Simulation 4:

$$Y_1 \sim Bin\left(n, \left|\sin\left(0.3X_1 + 0.9X_2\right)\right|\right)$$
$$m_1 = \{1 + exp\left[1.5 + 0.5\left(X_1 + X_2\right)\right]\}^{-1};$$
$$Y_2 \sim Bin\left(n, \left|\sin\left(0.9X_2 + 0.3X_3\right)\right|\right)$$
$$m_2 = \{1 + exp\left[1.5 + 5\left(0.3X_2 + 0.9X_3\right)\right]\}^{-1}.$$

Simulation 5 (To simulation 4 the following variable is added up):

$$Y_3 \sim Bin\left(n, \left|\sin\left(0.5X_3 + 0.5X_4\right)\right|\right)$$
$$m_3 = \{1 + exp\left[1.5 + 0.5\left(0.5X_3 + 0.5X_4\right)\right]\}^{-1}.$$

Performance of $BINPI$ is compared with $INPI$ as well as with the trivial approach UMI. Figure (2) shows the results of the five simulations concerning the case of missing data presented in dummy variables. Each simulation was performed with two goals: estimating the expected value parameter of each binomial distribution (to be compared with the true value) as well as calculating the number of uncorrect imputations in each variables. It is worth noting that an estimation of the probability of success near to the true value does not imply necessarily a correct imputation. The empirical evidence demonstrates the overall good performance of $BINPI$ over $INPI$ in terms of accuracy. This can be justified with two properties of $BINPI$: first, by definition a larger sample is used to build up the classifier; second, a more accurate learner is considered. Finally, $BINPI$ provides a variable imputation (instead of a single data imputation) yielding to a computationally more efficient procedure that can be recommended in the analysis of large data sets such as in statistical offices surveys.

Sim 1				
# errors	Y_1	Y_2	π_1	π_2
UMI	0	81	0,2130	0,1620
INPI	2	0	0,2150	0,2430
BINPI	**0**	0	**0,2130**	**0,2430**
TRUE			**0,2130**	**0,2430**
# missings	203	81		

Sim 4				
# errors	Y_1	Y_2	π_1	π_2
UMI	25	**17**	0,1630	**0,1710**
INPI	45	95	0,1870	0,2630
BINPI	**24**	94	**0,1640**	0,2620
TRUE			**0,1760**	**0,1880**
# missings	151	138		

Sim 2						
# errors	Y_1	Y_2	Y_3	π_1	π_2	π_3
UMI	**0**	80	804	0,2120	0,1980	0,0150
INPI	1	**0**	432	0,2130	**0,2780**	0,3730
BINPI	**0**	**0**	84	**0,2120**	**0,2780**	**0,7350**
TRUE				**0,2120**	**0,2780**	**0,8190**
# missings	169	80	808			

Sim 5						
# errors	Y_1	Y_2	Y_3	π_1	π_2	π_3
UMI	76	86	77	0,6070	0,6320	0,6190
INPI	70	86	84	0,4510	0,6320	0,5216
BINPI	**62**	**72**	**66**	**0,4670**	**0,6200**	**0,5600**
TRUE				**0,5310**	**0,5460**	**0,5420**
# missings	180	170	180			

Sim 3						
# errors	Y_1	Y_2	Y_3	π_1	π_2	π_3
UMI	164	78	191	0,6260	0,4350	0,6470
INPI	2	1	1	0,4640	0,5120	0,4570
BINPI	**0**	**0**	0	**0,4620**	**0,5130**	**0,4560**
TRUE				**0,4620**	**0,5130**	**0,4560**
# missings	164	78	191			

Fig. 2. Main results of simulations: binary missing case

5.2 Numerical Missing Case

We generated two differents data structures for the numerical response case. The variables under imputation are obtained according to the normal distribution. *Simulation 1* presented missing values in two variables, whereas in *simulation 2* missing data occur in three covariates.

Simulation 6:

$$Y_1 \sim N\left(X_1 - X_2^2, exp\left(0.3X_1 + 0.1X_2\right)\right)$$
$$m_1 = \left\{1 + exp\left[-1 + 0.5\left(X_1 + X_2\right)\right]\right\}^{-1}.$$
$$Y_2 \sim N\left(X_3 - X_4^2, exp\left(-1 + 0.5\left(0.3X_3 + 0.1X_4\right)\right)\right)$$
$$m_2 = \left\{1 + exp\left[-1 + 0.5\left(X_3 + X_4\right)\right]\right\}^{-1}.$$

Simulation 7 (to simulation 6 the following variable is added up):

$$Y_3 \sim N\left(X_5 - X_6^2, exp\left(-1 + 0.5\left(0.2X_5 + 0.1X_6\right)\right)\right)$$
$$m_3 = \left\{1 + exp\left[-1 + 0.5\left(X_5 + X_6\right)\right]\right\}^{-1}.$$

Figure (3) shows that $BINPI$ algorithm works better than $BINPI_{stump}$ and $INPI_{var}$. Stump, even if it performs better than UMI and PI, is not suitable for numerical case. Use of boosting improve imputation performance.

Sim 6

	μ_1	μ_2	σ_1	σ_2
True	**-28,5621**	**-27,6313**	**30,4572**	**29,8676**
UMI	-37,3740	-36,2881	23,5282	24,0916
PI	-24,1569	-23,4112	36,0030	35,1718
INPI_var	-30,6277	-29,6004	30,4762	30,2005
BINPI	**-29,2998**	**-27,9536**	29,1428	29,0064
BINPI Stump	-30,1244	-28,6836	28,8604	28,7842

Sim 7

	μ_1	μ_2	μ_3	σ_1	σ_2	σ_3
True	**-28,7185**	**-28,8756**	**-28,8833**	**30,2523**	**30,1449**	**30,0950**
UMI	-37,3008	-38,0560	-37,9888	21,3271	22,8859	24,8487
PI	-24,6298	-24,5213	-24,5822	35,3920	35,6813	35,5709
INPI_var	-30,2286	-29,7767	-29,8391	30,6879	30,1415	30,3985
BINPI	**-29,0670**	**-29,8364**	**-28,6490**	29,8219	30,2431	30,4518
BINPI Stump	-29,7880	-30,0284	-29,0653	29,4203	30,0786	29,3984

Fig. 3. Main results of simulations: numerical missing values

6 Concluding Remarks

Incremental Non Parametric Imputation provides a more accurate imputation compared to other standard methods. Main results of the simulation study can be outlined as follows:

- Imputation of a variable at turn is preferred to the imputation of a single data at turn ($INPI_{var}$ is preferred to $INPI$);
- Boosting algorithm allows for a more accurate imputation ($BINPI$ is preferred to $INPI$ and $INPI_{var}$);
- Stump is ideal for a two-class problem in terms of computation efficiency ($BINPI_{stump}$ is preferred to $BINPI$ for missing data in two-class variables);
- Fast tree is preferred to Stump for imputation of numerical missing values in terms of accuracy ($BINPI$ is preferred to $BINPI_{stump}$ for missing data in numerical variables).

Acknowledgements

Authors wish to thank anonymous referees for their helpful comments. Research supported by MIUR Funds COFIN05.

References

ARIA, M. and SICILIANO, R. (2003): Learning from Trees: Two-Stage Enhancements. *CLADAG 2003, Book of Short Papers, CLUEB, Bologna, 21-24.*

BREIMAN, L. (1996). Bagging Predictors. *Machine Learning, 36.*

CHU, C.K. and CHENG, P.E. (1995): Nonparametric regression estimation with missing data. *Journal of Statistical Planning and Inference, 48, 85-99.*

EIBL, G. and PFEIFFER, K.P. (2002): How to make AdaBoost.M1 work for weak base classifiers by changing only one line of the code. *Machine Learning: ECML 2002, Lecture Notes in Artificial Intelligence.*

FREUND, Y. and SCHAPIRE R.E. (1997): A decision-theoretic generalization of on-line learning and an application to boosting. *Journal of Computer and System Sciences, 55.*

FRIEDMAN, J.H. and POPESCU, B.E. (2005): *Predictive Learning via Rule Ensembles*, Technical Report of Stanford University.

HASTIE, T., TIBSHIRANI, R. and FRIEDMAN J. (2002): *The Elements of Statistical Learning*, Springer Verlag, New York.

IBRAHIM, J.G., LIPSITZ, S.R. and CHEN, M.H. (1999): Missing Covariates in Generalized Linear Models when the missing data mechanism is non- ignorable. *Journal of the Royal Statistical Society, B, 61, 173-190.*

LITTLE, J.R.A and RUBIN, D.B. (1987): *Statistical Analysis with Missing Data.* John Wiley and Sons, New York.

LITTLE, J.R.A. (1992): Regression with Missing X's: a Review. *Journal of the American Statistical Association, 87, 1227-1237.*

MOLA, F. and SICILIANO, R. (1992): A two-stage predictive splitting algorithm in binary segmentation. In Dodge, Y. and Whittaker, J. (Eds.): *Computational Statistics.* Physica Verlag, Heidelberg, 179-184.

MOLA, F. and SICILIANO, R. (1997): A Fast Splitting Procedure for Classification and Regression Trees. *Statistics and Computing, 7, 208-216.*

PETRAKOS, G., CONVERSANO, C., FARMAKIS, G., MOLA, F., SICILIANO, R. and STAVROPOULOS, P. (2004): New ways to specify data edits, *Journal of Royal Statistical Society, Series A, 167, 249-274.*

SICILIANO, R., ARIA, M. and CONVERSANO, C. (2004): Tree Harvest: Methods, Software and Some Applications. In Antoch J. (Ed.): *Proceedings in Computational Statistics.* Physica-Verlag, 1807-1814.

SICILIANO, R. and CONVERSANO, C. (2002): Tree-based Classifiers for Conditional Missing Data Incremental Imputation, *Proceedings of the International Conference on Data Clean, University of Jyvaskyla.*

SICILIANO, R. and MOLA, F. (2000): Multivariate Data Analysis through Classification and Regression Trees. *Computational Statistics and Data Analysis, 32, 285-301.*

VACH, W. (1994): Logistic Regression with Missing Values and covariates. *Lecture notes in statistics*, vol. 86, Springer Verlag, Berlin.

Sensitivity of Attributes on the Performance of Attribute-Aware Collaborative Filtering

Karen H. L. Tso and Lars Schmidt-Thieme

Computer-based New Media Group (CGNM),
Institute of Computer Science,
University of Freiburg, 79110 Freiburg, Germany
{tso,lst}@informatik.uni-freiburg.de

Abstract. Collaborative Filtering (CF), the most commonly-used technique for recommender systems, does not make use of object attributes. Several hybrid recommender systems have been proposed, that aim at improving the recommendation quality by incorporating attributes in a CF model.

In this paper, we conduct an empirical study of the sensitivity of attributes for several existing hybrid techniques using a movie dataset with an augmented movie attribute set. In addition, we propose two attribute selection measures to select informative attributes for attribute-aware CF filtering algorithms.

1 Introduction

For recommender systems, nearest-neighbor methods, called CF (Goldberg et al. (1992)), is the prevalent method in practice. On the other hand, methods that regard only attributes and disregard the rating information of other users, are commonly called the Content-Based Filtering (CBF). They have shown to perform very poorly. Yet, attributes usually contain valuable information that could improve the performance of recommender systems; hence it makes it desirable to include attribute information in CF models – so called hybrid collaborative/content-based filtering methods.

Although there are several hybrid methods that consider attribute information in CF for predicting ratings — how much a given user will like a particular item; to our best knowledge there is no prior approach for predicting items — which N items a user will be interested in. Please note that predicting good items, i.e. items that have been rated with 4 or 5 on a scale from 1 to 5, by its nature is a rating prediction problem (on a more coarse scale bad/good).

In addition, the behavior of hybrid algorithms is to be investigated as the number of informative attributes increases. Thus, quantitative measures for attribute selection are needed to eliminate irrelevant ones.

In this paper, we will make the following contributions: (i) propose two methods for attribute selection and (ii) evaluate the impact of attributes on existing hybrid algorithms that predict items and CF methods that do not consider attributes.

2 Related Work

Before we discuss the related works, we introduce notations being used in this paper. Let

- U be a set of **users**,
- I be a set of **items**,
- B be a set of (binary) **item attributes**,
- $D_{i,b} \in \{0, 1\}$ specify whether item $i \in I$ has attribute $b \in B$,
- $O_{u,i} \in \{0, 1\}$ specify whether item $i \in I$ occurred with user $u \in U$

There are many proposals on how to integrate attributes in collaborative filtering for ratings. They can be roughly categorized into four groups:
(i) Methods that add a pseudo-item i_b for each item attribute $b \in B$ that for each user $u \in U$ gets a pseudo-rating

$$R_{u,i_b} := f(\{(i, R_{u,i}) \in I \mid O_{u,i} = 1 \text{ and } D_{i,b} = 1\})$$

where f is some function on the user's ratings of items having attribute b. Ziegler et al. (2004) presented a more complex function that considers a taxonomic relation between original items.
(ii) Methods that add a pseudo-user u_b (often called agent) for each item attribute $b \in B$ with a pseudo-rating for each item $i \in I$

$$R_{u_b,i} := D_{i,b}$$

e.g., Good et al. (1999). These methods perform standard user- or item-based CF on top of the rating matrix enriched by pseudo-items or -users.
(iii)a) Methods that combine linearly the predictions of a pure CBF model and a pure CF model (Claypool et al. (1999), Pazzani (1999), Good et al. (1999), Li and Kim (2003))

$$\hat{R}^{\text{combined}} := \lambda \hat{R}^{\text{cbf}} + (1 - \lambda)\hat{R}^{\text{cf}}$$

where the weight coefficient $\lambda \in [0, 1]$ is learned either by regression, simple iterative update schemes or grid search. Some other existing methods also use a user-specific λ.
b)Apply the nearest neighbor models to both models and combining the attribute-depended with the rating-depended similarity and use CF with the combined similarity (Delgado et al. (1998)).
(iv) Methods that apply a CBF and a CF model sequentially, i.e. predict ratings by means of CBF and then re-estimate them from the completed rating matrix by means of CF (Melville et al. (2002)).

There are also further proposals on how to integrate attributes when the problem is viewed as a classification problem (Basilico and Hofmann (2004), Basu et al. (1998)). As we lose the simplicity of CF, we do not consider those more complex methods here.

Many methods appear to mix simple ideas with more complex components as clustering, rule-based learners etc., often without investigating whether the additional effort pays off in the quality at the end. Therefore, we have selected three basic methods that try to keep the simplicity of CF, but still should improve prediction results: a sequential CBF-CF method (iv), a linear-combination CBF-CF method (iii a) and a combination of similarities methods (iii b). The first approach is an adapted form of Content-Boosted CF by Melville et al. (2002) which was originally designed for predicting the ratings. The last two methods achieve the best results on our reference data set.

3 Common and Hybrid Attribute-Aware CF Methods

3.1 Common CF Methods

In user-based CF (Sarwar et al. (2000)), recommendations are generated by considering solely the ratings of users on items, by computing the pairwise similarities between users, e.g., by means of vector similarity

$$\text{usim}^{\text{ratings}}(u, v) := \frac{\langle R_{u,.}, R_{v,.} \rangle}{||R_{u,.}||_2 ||R_{v,.}||_2}$$

where $u, v \in U$ are two users and $R_{u,.}$ and $R_{v,.}$ the vectors of their ratings. For each user, the k most-similar users are selected (neighborhood – N_u) and for predicting items for a target user u, items are ranked by decreasing frequency of occurrence in the ratings of his/her neighbors

$$p^{\text{cf}}(O_{u,i} = 1) := \frac{|\{v \in N(u) \,|\, O_{v,i} = 1\}|}{|N_u|}$$

A dualistic form of user-based CF is item-based CF (Deshpande and Karypis (2004)), where similarities are computed between each pair of items $i, j \in I$.

$$\text{isim}^{\text{ratings}}(i, j) := \frac{\langle R'_{.,i}, R'_{.,j} \rangle}{||R'_{.,i}||_2 ||R'_{.,j}||_2}$$

In content-based filtering, a naive Bayesian classifier is trained for the binary target variable $O_{u,.}$ depending on the binary predictors $D_{.,b}$ for all $b \in B$:

$$\hat{p}^{\text{cb}}(O_{u,.} = 1 \,|\, D_{.,b}, b \in B) := P(O_{u,.}) \cdot \prod_{b \in B} P(D_{.,b} \,|\, O_{u,.}) \tag{1}$$

3.2 Hybrid Attribute-aware CF Methods

The three existing hybrid methods in Tso and Schmidt-Thieme (2005) incorperate attributes into user-based and item-based CF.

Sequential CBF and CF is the adapted version of an existing hybrid approach, Content-Boosted CF, originally proposed by Melville et al. (2002) for predicting ratings. This method has been conformed to the predicting items problem here. It first uses CBF to predict ratings for unrated items and then filters out ratings with lower scores (i.e. keeping ratings above 4 on a 5-point scale) and applies CF to recommend topN items.

Joint Weighting of CF and CBF, first applies CBF on attribute-dependent data to infer the fondness of users for attributes. In parallel, user-based CF is used to predict topN items with ratings-dependent data. Both predictions are joint by computing their geometric mean. This mean combination is then used for performing the prediction:

$$\hat{p}(O_{u,i} = 1) := \hat{p}^{\text{cb}}(O_{u,i} = 1)^{\lambda} \cdot p^{\text{cf}}(O_{u,i} = 1)^{1-\lambda} \quad \text{with } \lambda \in [0, 1]. \qquad (2)$$

Attribute-Aware Item-Based CF extends item-based CF (Deshpande and Karypis (2004)). It exploits the content/attribute information by computing the similarities between items using attributes thereupon combining it with the similarities between items using ratings-dependent data.

$$\text{isim}^{\text{attributes}}(i, j) := \frac{\langle D_{i,.}, D_{j,.} \rangle}{||D_{i,.}||_2 ||D_{j,.}||_2}$$

$$\text{isim}^{\text{combined}} := (1 - \lambda)\ \text{isim}^{\text{ratings}} + \lambda\ \text{isim}^{\text{attributes}} \quad \text{with } \lambda \in [0, 1]$$

The last two methods use λ as weighting factor to vary the significance of CF or CBF.

3.3 Attribute selection

To our best knowledge, no similar analysis has been documented in literature affiliated with the sensitivity of attributes in RSs. As the number of attributes increases, quantitative measure for attribute selection are needed to filter the irrelevant ones. Thus, we define two quantitative measures (i) the total number of attribute occurrences (attribute frequency) and (ii) the χ^2 measure between item occurrences and attributes. The attribute frequency is simply the total number of occurrences of each item having a particular attribute. The more frequent an attribute is, the better it is judged.

This approach appears to be simple and clearly favors ubiquitous but attributes could eventually became non-informative. Thus, we also consider the χ^2 measure between item occurrences and attributes. It bases on the 2×2 table of all possible ratings $U \times I$ according to actual occurrence in the data $(O_{u,i})$ and having the attribute in question $(D_{i,b})$

	$D_{.,b} = 1$	$D_{.,b} = 0$
$O_{.,.} = 1$	$v_{1,1}$	$v_{1,2}$
$O_{.,.} = 0$	$v_{2,1}$	$v_{2,2}$

where

- $v_{1,1} := |\{(u,i) \in U \times I \,|\, O_{u,i} = 1 \text{ and } D_{i,b} = 1\}|$,
- $v_{1,2} := |\{(u,i) \in U \times I \,|\, O_{u,i} = 1 \text{ and } D_{i,b} = 0\}|$,
- $v_{2,1} := |U| \cdot |\{i \in I | D_{i,b} = 1\}| - v_{1,1}$,
- $v_{2,2} := |U| \cdot |\{i \in I | D_{i,b} = 0\}| - v_{1,2}$

computed by

$$\chi^2 = \sum_{i=0,1,j=0,1} \frac{(v_{i,j} - \hat{v}_{i,j})^2}{\hat{v}_{i,j}} \tag{3}$$

where $\hat{v}_{i,j} := v_{i,.} \cdot v_{.,j} / v_{.,.}$ represents the expected frequencies.

The stronger the dependency between item occurrence and an attribute, i.e. the higher the χ^2 value is, the better the attribute is judged.

4 Evaluation and Experimental Results

We have evaluated the three attribute-aware CF algorithms and have compared their performances with their corresponding non-hybrid base models, which do not integrate attributes.

Data set We evaluated the algorithms with the MovieLens datasets (*ml*; MovieLens (2003)), which contains approximately 1 million movie ratings of 6,040 users on 3,592 movies. The ratings are expressed on a 5-point rating scale. We looked at two different sets of movie attributes: (i) 18 genres that comes with the data set and (ii) Amazon taxonomy of 1074 different genres/classes provided by Ziegler et al. (2004). We will reference these two attributes sets as "18 genres" and "Amazon genres", respectively.

We took ten random subsets of the *ml* dataset with 1000 users and 1500 items each. Each dataset is split into 80% training set and 20% test set at random. The quality of the models are measured by comparing their top 10 recommendations computed from the training data against the actual items in the test set. We report the averages and standard deviations of the F1 values of the ten trials.

Metrics Our paper focuses on the item prediction problem, which is to predict a fixed number of top recommendations and not the ratings. Suitable evaluation metrics for item prediction problem are Precision, Recall and F1. Similar to Sarwar et al. (2000), our evaluations consider any item in the recommendation set that matches any item in the test set as a "hit".

Parameters We select optimal neighborhood sizes, by means of a grid search. Neighborhood size for user-based and joint weighting CF–CBF is 90, and 100 for item-based CF and attribute-aware item-based CF. Furthermore, λ parameters are set to 0.15 and 0.05 for joint weighting CF–CBF and attribute-aware item-based CF respectively. They are chosen from previous experiments (Tso and Schmidt-Thieme (2005)), which found to give reasonable results for the augmented attributes as well.

Experimental Results The results of our previous experiments (Tso and Schmidt-Thieme (2005)) on the 18 genres attribute set is summarized in Fig. 1. The attribute-aware methods enhance their respective base-models significantly, especially the joint weighting CF-CBF. Although Melville et al. (2002) reported that CBCF performed better than user-Based and CBF for ratings, it fails to provide quality topN recommendations for items in our experiments.

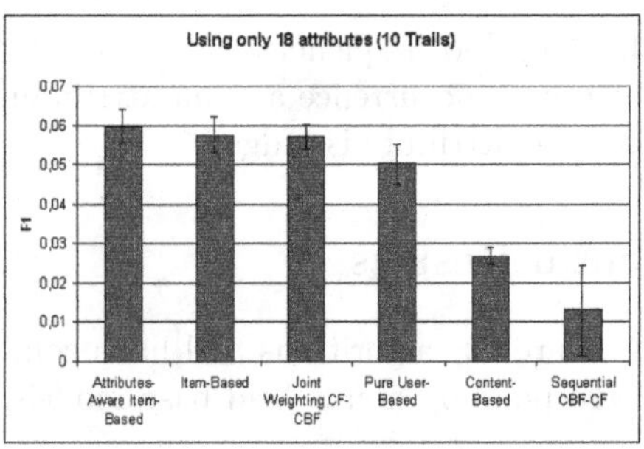

Fig. 1. F1 with 18 attributes

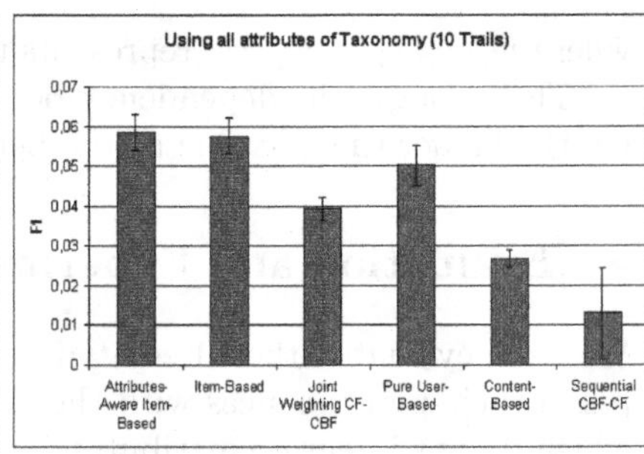

Fig. 2. F1 with all taxo attributes

We anticipate that the prediction quality could be improved by including more attributes. i.e. using the Amazon attribute set instead of the 18 genres. The results of the average of ten random trials using all attributes from the Amazon taxonomy are presented in Fig. 2.

Although attribute-aware item-based CF using all Amazon attributes still achieves the highest F1 value, the difference w.r.t. its base method is insignificant. It also can be observed that all attribute-aware methods perform worse for the 1074 Amazon attributes than for just the 18 genres. This indicates that the quality of attributes plays an important role in hybrid methods and that attribute selection should be performed. Since the results of Sequential CBF-CF scores way below the classical models, we therefore focus our discussion on the other two algorithms from now on.

Sensitivity of Attributes To analyze the impact of attributes on attribute-aware CF algorithms, we further partition the ten trials into subsets by varying the number of useful attributes by attribute frequency and χ^2 for each trial. The average sensitivity of attributes from the taxonomy of ten trials

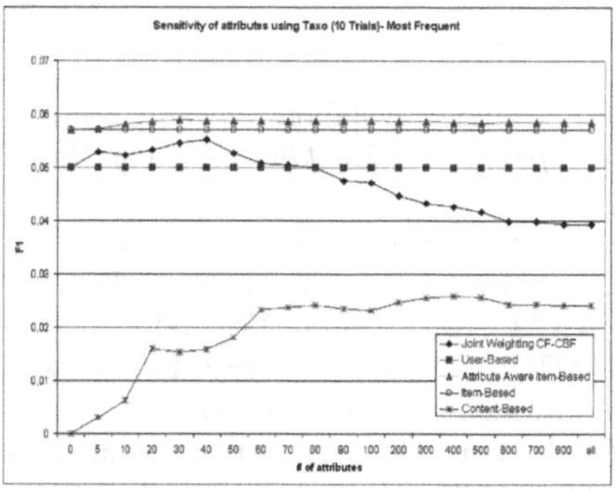

Fig. 3. Vary # of Most Freq. attributes

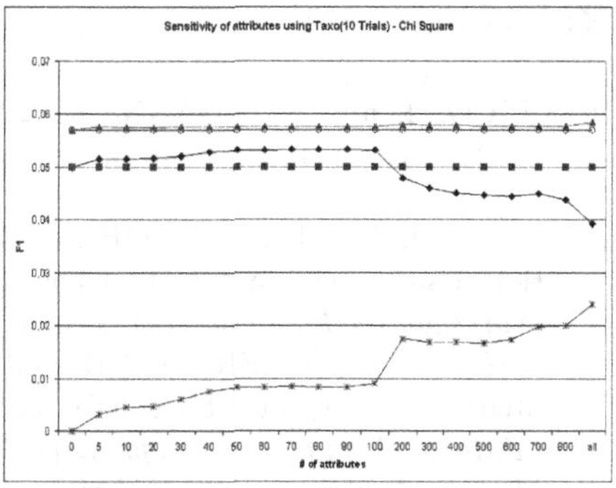

Fig. 4. Vary # of best χ^2 attributes

for each subset are presented in Fig. 3 and Fig. 4.

As shown in both figures, the selection of attributes does affect the quality of topN recommendation. In joint weighting CF-CBF, the quality increases gradually, reaches its peak and decreases dramatically as more irrelevant attributes are appended. In the case of attribute frequency measure, the algorithm reaches its peak at around 40 attributes, whereas in χ^2, the peak is reached in the range of 70-100 attributes. Taking the peaks of both attribute frequency and χ^2 measures, there is an increase of 10.4% and 6.7% respectively compared to its base models. On the other hand, in attribute-aware item-based CF, the quality of attributes has almost no effect on the quality of the recommendations. For attribute frequency, the F1 value quickly meets its peak and maintains rather constant as more irrelevant attributes are added to the algorithms, whereas for the χ^2 measure, the quality reaches the peak when most noise is presented. One of the reasons for these strange results could be due to the value of lambda being set too low as it controls the contribution of attributes to those algorithms.

5 Conclusions and Future Works

Our empirical analysis on state-of-the-art hybrid algorithms shows that the effectiveness of these methods depends on the selection of useful attributes. We have proposed two measures: attribute frequency and chi square. Joint weighting CF-CBF proves to be more effective and provides up to 10.4% gain in F1 than pure CF for movie taxonomy datasets.

As the quality of recommendations varies with the informativeness of the attributes, further studies on other attribute selection measures such as the information gain or the combination of various measures could be the future works.

References

BASILICO, J. and HOFMANN, T. (2004): Unifying collaborative and content-based filtering. In:*21st International Conference on Machine Learning.* Banff, Canada.

BASU, C., HIRSH H., and COHEN, W. (1998): Recommendation as classification: Using social and content-based information in recommendation. In: *1998 Workshop on Recommender Systems.* AAAI Press, Reston, Va. 11-15.

BREESE, J. S., HECKERMAN, D. and KADIE, C. (1998): Empirical analysis of predictive algorithms for collaborative filtering. In:*14th Conference on Uncertainty in Artificial Intelligence (UAI-98).* G. F. Cooper, and S. Moral, Eds. Morgan-Kaufmann, San Francisco, Calif., 43-52.

CLAYPOOL, M., GOKHALE, A. and MIRANDA T. (1999): Combining content-based and collaborative filters in an online newspaper. In: *SIGIR-99 Workshop on Recommender Systems: Algorithms and Evaluation.*

DELGADO, J., ISHII, N. and URA, T. (1998): Content-based Collaborative Information Filtering: Actively Learning to Classify and Recommend Documents. In: *Cooperative Information Agents II. Learning, Mobility and Electronic Commerce for Information Discovery on the Internet.* Springer-Verlag, Lecture Notes in Artificial Intelligence Series No. 1435.

DESHPANDE, M. and KARYPIS, G. (2004): Item-based top-N recommendation Algorithms. In: *ACM Transactions on Information Systems* **22/1**, *143-177.*

GOLDBERG, D., NICHOLS, D., OKI, B. M. and TERRY, D. (1992): Using collaborative filtering to weave an information tapestry. In: *Commun. ACM* **35**, *61-70.*

GOOD, N., SCHAFER, J.B., KONSTAN, J., BORCHERS, A., SARWAR, B., HERLOCKER, J., and RIEDL, J. (1999): Combining Collaborative Filtering with Personal Agents for Better Recommendations. In: *1999 Conference of the American Association of Artificial Intelligence (AAAI-99), pp 439-446.*

LI, Q. and KIM, M. (2003): An Approach for Combining Content-based and Collaborative Filters. In: *Sixth International Workshop on Information Retrieval with Asian Languages (ACL-2003) pp. 17-24.*

MELVILLE, P., MOONEY, R. J. and NAGARAJAN, R. (2002): Content-boosted Collaborative Filtering. In: *Eighteenth National Conference on Artificial Intelligence(AAAI-2002), pp. 187-192.* Edmonton, Canada.

MOVIELENS (2003): Available at http://www.grouplens.org/data.

PAZZANI, M. J. (1999).: A framework for collaborative, content-based and demographic filtering. In: *Artificial Intelligence Review 13(5-6):393-408.*

SARWAR, B. M., KARYPIS, G., KONSTAN, J. A. and RIEDL, J. (2000): Analysis of recommendation algorithms for E-commerce. In: *2nd ACM Conference on Electronic Commerce (EC'00).* ACM, New York. 285-295.

TSO, H. L. K. and SCHMIDT-THIEME, L. (2005): Attribute-Aware Collaborative Filtering. In:*29th Annual Conference of the German Classification Society 2005, Magdeburg, Germany. (to appear)*

ZIEGLER, C., SCHMIDT-THIEME, L. and LAUSEN, G. (2004): Exploiting Semantic Product Descriptions for Recommender Systems. In:*2nd ACM SIGIR Semantic Web and Information Retrieval Workshop (SWIR'04).* Sheffield, UK.

Part V

Multivariate Methods for Customer Satisfaction and Service Evaluation

Customer Satisfaction Evaluation: An Approach Based on Simultaneous Diagonalization

Pietro Amenta[1] and Biagio Simonetti[2]

[1] Department of Analysis of Economic and Social Systems, University of Sannio, Italy
amenta@unisannio.it

[2] Department of Mathematics and Statistics, University of Naples "Federico II", Italy
simonetti@unisannio.it

Abstract. Several methods have been proposed in literature for the service quality evaluation. These models measure the gap between customer's expectations for excellence and their perceptions of actual service offered. In this paper we propose an extension of a techniques which allows to analyze jointly the expectations and perceptions data.

1 Introduction

In literature there is a wide convergence on the evaluation models of the service performances regarding the typology of indicators which can be constructed with the aim to estimate the components in which the service is developed. Such classification previews three classes of indicators: structure, result and process. The Customer Satisfaction is a result indicator and its attainment is of great interest for the management.

In literature several methods have been proposed for the service quality evaluation. A lot of these models (Servqual, Servperf, Normed Quality, Qualitometro etc.) are based on the Gap Theory of Service Quality. These models measure the gap between customer's expectations for excellence and their perceptions of actual service offered, so service providers can understand both customer expectations and their perceptions of specific services. Several techniques have been proposed in literature for the description and the exploratory study of these data. Aims of these proposals are to estimate the multidimensional aspects of the investigated system and the introduction of criteria of judgment (Hogan et al. (1984), Cronin and Taylor (1992), Lauro et al. (1997), Vedaldi (1997), D'Ambra et al. (1999), D'Ambra and Amenta (2000), Amenta and Sarnacchiaro (2004)).

In this paper, starting from the statistical methods used to analyze data coming from evaluation service survey, we propose an approach based on the simultaneous diagonalization which allow to analyze jointly the data of expectations and perceptions of the customers respect to the quality evaluation of a service.

2 An Approach Based on Simultaneous Diagonalization

In order to evaluate the customer satisfaction, we consider a questionnaire taken on n statistical units at the beginning (expectation) and at the end (perception) of a given service. Data are represented in terms of matrices X (pre-service) and Y (post-service) of order $(n \times p)$. Let Q be positive defined symmetrical matrix (metric) of dimensions $(p \times p)$ and $D_n = diag(d_1, \ldots, d_n)$, $(\sum_{i=1}^{n} d_i = 1)$, weight matrix of statistical units. We obtain a couple of statistical triplets: (X, Q, D_n) and (Y, Q, D). Following Lafosse (1989), every triplet can be defined *fully matched* because they describe the same statistical units (n rows) by means the same variables (p columns). We suppose matrices X and Y centered as respective column means and, without lose of generality, we consider $Q = I_p$ with I_p identity matrix of order p. Moreover, let $A = X'X$, $B = Y'Y$ and $C = A + B$.

By using the classical tools of the multidimensional data analysis, in order to evaluate the quality service, several statistical approaches concentrate their attention on independent analyses of the perceived (Y) and expected (X) judgments of the quality service as well as they analyse their "gap" $(X - Y)$. Let I_X,I_Y and I_D the inertia matrices associated to Y, X and $X - Y$, respectively. In this context, other approaches study the symmetrical and non symmetrical relationships between the two blocks of variables by using techniques based on covariance criteria like, for example, Co-inertia Analysis (Tucker, 1958) and Partial Least Squares (Wold, 1966) and their generalizations. The co-inertia axes play a primary role in the customer evaluation framework and the independent analyses of the perceived and expected judgments as well as of the gap are included in the Co-Inertia of Fully Matched Tables (Co-Structure Analysis) by Torre and Chessel (1995).

Aim of our approach is to find common dimensions in the different sets of variables by means the simultaneous diagonalization (SD) of the A and B matrices. Joint diagonalization of square matrices is an important problem of numeric computation. Extensive literature exists on this topics topics (e.g. Noble and Daniel (1977), Van Der Vorst and Golub (1997)). The objective is to find a projection v that maximizes the generalized Rayleigh quotient $J(v) = \max_v [(v'Av)/(v'Bv)]$. The solution can be obtained analytically by solving a generalized eigenvalue problem: $Av = \lambda Bv$. Since the matrix $B^{-1}A$ is unnecessarily symmetric, the common vectors v are usually dependent.

It can be easily proved that the solutions maximizes also the quotient

$$J(v) = \max_v [(v'Av)/(v'Cv)]. \tag{1}$$

wth the same maximum value. Let v_1 the first axe obtained by the previous maximization problem. In order to obtain C-uncorrelated common vectors, we can reconsider the problem in a form that has been analyzed by Rao (1973) under the name of the restricted eigenvalue problem ($i > 1$):

$$\max_{v_i} [(v_i' A v_i)/(v_i' C v_i)] \tag{2}$$

subject to $v_i' C v_i = 1$ and $G_i v_i = 0_{i-1}$ where $G_i = V_i C$ and $V_i = [v_1, v_2, ..., v_{i-1}]$, where the last constraint is dropped when $i = 1$.

By applying the Lagrange method directly to problem (2) we obtain $C^{-1}(I - P_i) A v_i = \lambda_i v_i$ where $P_i = CV_i'(V_i C V_i')^{-1} V_i$ is such that $(I - P_i) G_i' = 0$ and $(I - P) C v_i = C v_i$ $(i = 1, ..., p)$. We obtain the constrained solutions (the eigenvector v_i corresponding to the largest eigenvalue λ_i) iteratively by diagonalizing, at each step, the following quantity

$$\left[C^{-1}(I - P_i)\right]^{1/2} A \left[C^{-1}(I - P_i)\right]^{1/2} u_i = \lambda_i u_i$$

with $v_i = \left[C^{-1}(I - P_i)\right]^{1/2} u_i$. We call this approach "C-algorithm".

By the Co-structure Analysis, we know that single analysis of matrices Y, X and $X - Y$, respectively, and Co-Inertia Analysis are linked by the following relationships

$$X'X + Y'Y = (X - Y)'(X - Y) + (X'Y + Y'X) \tag{3}$$

and $I_X + I_Y = I_D + 2tr(X' D_n Y Q)$. For the previous results, we have that $J(v) = \max_v [(v' A v)/(v' C v)]$ results to be also equal to

$$J(v) = \max_v \frac{v' X' X v}{v' \left[(X - Y)'(X - Y) + (X'Y + Y'X)\right] v}.$$

This implies that a single set of common vectors V simultaneously diagonalizes the following matrices $X'X$, $Y'Y$, $X'X + Y'Y$ and $[(X - Y)'(X - Y) + (X'Y + Y'X)]$ linked, respectively, to the single inertia analyses of expectations, perceptions, their sum (equivalent to the Principal Components Analysis of the row linked matrix $[X'|Y']'$) and co-structure with gap. The common vectors V are such that $V'X'XV = \Lambda_A$, $V'Y'YV = \Lambda_B$, $V'(X'X + Y'Y) V = I$ and $V' [(X - Y)'(X - Y) + (X'Y + Y'X)] V = I$.

We remark that by using this single set of common vectors V for all the components of identity (3), the previous Co-structure Analysis inertia decomposition $I_X^{(V)} + I_Y^{(V)} = I_D^{(V)} + 2tr(X' D_n Y Q)$ still holds.

We highlight that this approach leads to an optimal choice for the axes v in order to evaluate the expected (X) judgments by taking into account all the components of (3). On the contrary, Co-Structure Analysis obtains different system solutions for each component of identity (3).

Following the same approach, we can analyze the single other components of (3): perceptions, gap and co-structure by taking always into account all the others (see Table 1).

Component	Generalized Rayleigh quotient $J(v)$
Expectations	$\max_v \frac{v'X'Xv}{v'X'Xv+v'Y'Yv} = \max_v \frac{v'X'Xv}{v'[(X-Y)'(X-Y)+(X'Y+Y'X)]v}$
Perceptions	$\max_v \frac{v'Y'Yv}{v'Y'Yv+v'X'Xv} = \max_v \frac{v'Y'Yv}{v'[(X-Y)'(X-Y)+(X'Y+Y'X)]v}$
Gap	$\max_v \frac{v'(X-Y)'(X-Y)v}{v'(X-Y)'(X-Y)v+v'(X'Y+Y'X)v} = \max_v \frac{v'(X-Y)'(X-Y)v}{v'(X'X+Y'Y)v}$
Co-structure	$\max_v \frac{v'\left(X'Y+Y'X\right)v}{v'(X'Y+Y'X)v+v'(X-Y)'(X-Y)v} = \max_v \frac{v'\left(X'Y+Y'X\right)v}{v'(X'X+Y'Y)v}$

Table 1. Generalized Rayleigh quotient

In the same way, in order to obtain B and C-uncorrelated common vectors, we write $\max_{v_i} [(v_i'Av_i)/(v_i'Bv_i)]$ subject to $v_i'Bv_i = 1$ and $G_i v_i = 0_{i-1}$ where $G_i = V_i C$ and $V_i = [v_1, v_2, ..., v_{i-1}]$. The last constraint is dropped when $i = 1$. As before, by applying the Lagrange method we obtain $B^{-1}(I - P_i)Av_i = \lambda_i v_i$ where $P_i = CV_i'(V_i CB^{-1}CV_i')^{-1}V_i CB^{-1}$ with $(I - P_i)G_i' = 0$ and $(I - P)Bv_i = Bv_i$. We obtain the constrained solutions iteratively by diagonalizing, at each step, the following quantity

$$\left[B^{-1}(I - P_i)\right]^{1/2} A \left[B^{-1}(I - P_i)\right]^{1/2} u_i = \lambda_i u_i$$

with $v_i = \left[B^{-1}(I - P_i)\right]^{1/2} u_i$. We call this approach "BC-algorithm".

In this case, for the analysis of expectations, common vectors V are such that $V'X'XV = \Lambda_A$, $V'Y'YV = I$, $V'(X'X + Y'Y)V = (I + \Lambda_A)$ and $V'[(X - Y)'(X - Y) + (X'Y + Y'X)]V = (I + \Lambda_A)$.

Finally, we remark that it is possible to obtain the common axes V without computing the inverse of the matrix B and with the same "C-algorithm" results, by using the following sequential algorithm ("*Eigen*-algorithm") :

1. $L_A = diag\left[eigenvalues\,(A)\right]$
2. $U_A = eigenvectors\,(A)$
3. $H = U_A L_A^{-1/2}$
4. $L_C = diag\left[eigenvalues\,(H'CH)\right]$
5. $U_C = eigenvectors\,(H'CH)$
6. $V = HU_C\,(I + L_C)^{-1/2}$

3 Patient Satisfaction Data

We consider a day surgery patient satisfaction study in a Neapolitan Hospital for a sample of 511 patients. A questionnaire, developed according to the Servqual model (Parasuraman et al., 1985), with 15 items on 5 dimensions (Tangibility, Reliability, Responsiveness, Assurance, Empathy) (Table 2) and a five levels answer scale, has been given during a week at the beginning of the service (Expectations) and at the patient discharge (Perceptions). Due to

Dimension	Item	Label
Tangibility	The hospital has modern and efficient equipments	Var1
	The hospital structure is in a good and clean state	Var2
	The Hospital Personnel (HP) has a good appearance	Var3
Reliability	The HP supplies the promised service	Var4
	If there is a problem, the HP gives an understanding and reassuring attitude	Var5
	Doctors provide me with precise information about the disease	Var6
Responsiveness	The HP informs with precision when the service is given	Var7
	The HP supplies services promptly	Var8
	The HP is ready to give help to me	Var9
Assurance	The HP's behavior trusts me	Var10
	The HP has a good knowledge to answer to my questions	Var11
	The HP is kind to me	Var12
	The HP has the backing of management about the job	Var13
Empathy	The HP pays an individual attention to me	Var14
	The HPl takes my principal interests at heart	Var15

Table 2. Structure of the questionnaire.

the ordinal nature of data, we proceed to a transformation of them by using the "Rating Scale Models" (Wright and Masters, 1982).

We developed the Simultaneous Diagonalization and the Co-Inertia of Fully Matched Tables on the transformed data in order to highlight the main differences on the perception, on the expectations as well as on the gap evaluations.

Figure 1.a and figure 1.b display the coordinates of expectations variables on the first factorial plane given by SD and PCA, respectively. Typical aspect of the patient satisfaction studies, first PCA axe (figure 1.b) does not offer a remarkable information, by representing an overall information about the expectations for all variables considered in the questionnaire. First factorial axe obtained by SD method (figure 1.a) is a synthesis of the variables included in tangibility (Appearance of physical facilities, equipment, and communication material) and assurance areas (Competence, knowledge and courtesy of employees and their ability to convey trust and confidence). This leads to synthesize all the information relative to the evaluation of the physical structure of the hospital. In the specific context of the analyzed Hospital, the obsolete structure represents an influent negative factor in the overall patient satisfaction evaluation. Second PCA factorial axe seems to be not easily interpretable while the second one obtained by SD gives a better representation of the variables related to responsiveness (Willingness to help customers and provide prompt service) and to reliability (Ability to perform the promised service dependably and accurately). Related to psychological evaluation of the patients, it provides important information about the expectations connected to the human factor which is an essential component in a delicate sector like the health service.

More interesting information are given by the representation of the perceptions. Like the expectations variables, PCA does not provide consistent information for the first factorial plane (Figure 2.b).

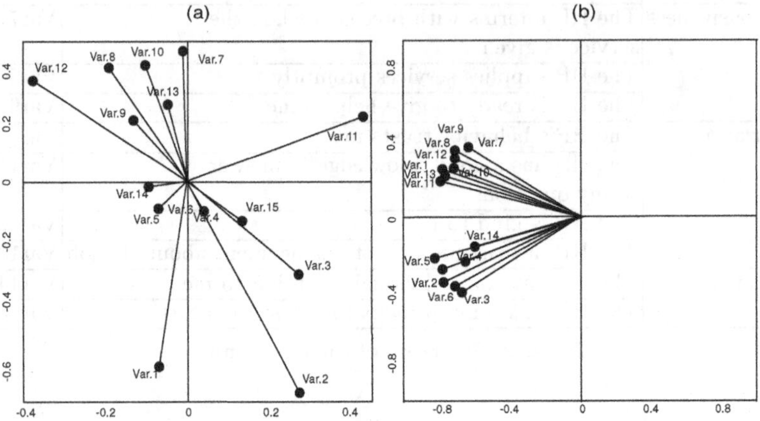

Fig. 1. Graphical representation of the perception variables on the first factorial plane computed by SD (a) and PCA (b).

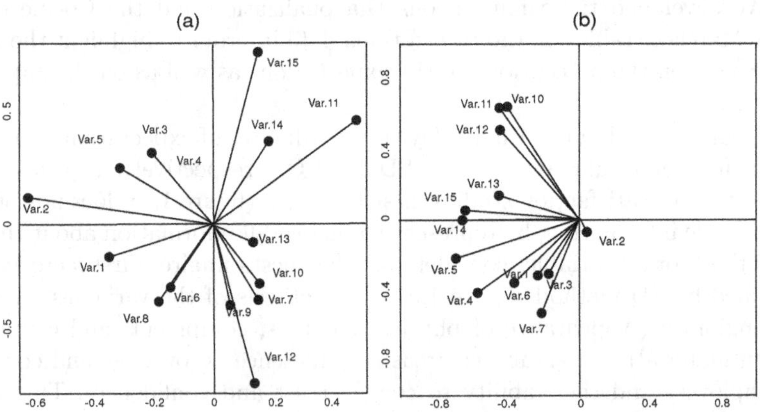

Fig. 2. Graphical representation of the expectation variables on the first factorial plane computed by SD (a) and PCA (b).

As before, first SD factorial axe gives an overview of the variables of the Hospital structure quality while the second one (figure 2.a) synthesizes all the variables linked to the evaluation of the hospital human aspect. We remark that there is a perfect interpretive correspondence between the first factorial planes generate for the expectations and the perceptions. We are able to summarize all information concerning the patients evaluation on Hospital physical structure and management (doctors and nurses) by the first factorial plane. Contextual factors like staff friendliness, treatment explanations and appearance of surroundings, are often used by patients to perform satisfaction. These two aspects represent useful underlining information for the hospital management in order to improve the service.

A gap score is the difference between a patient's level of satisfaction and level of importance on a specific service feature. To identify gaps between what is important to patients and their perception of the quality of received service, is a powerful activity in order to assist in targeting service improvement efforts and resource allocation. For this reason is of great interest the gap analysis generated by the difference between the perceptions and expectations matrices (Figure 3). Classical PCA (Figure 3.b) does not give, as in previous analysis, additional information about the hospital evaluation. First SD plane remarks the results obtained by perceptions analysis, by offering on the first axe a synthesis of the general difference between perceptions and expectations related to the hospital structure and the quality of management. A new element is the presence of the variables related to Empathy area (Caring, individualized, and professional attention the firm provides its customers) on the first SD factorial axe. Second SD axe highlights the quality relationships between the management (doctors and nurses) and the patients.

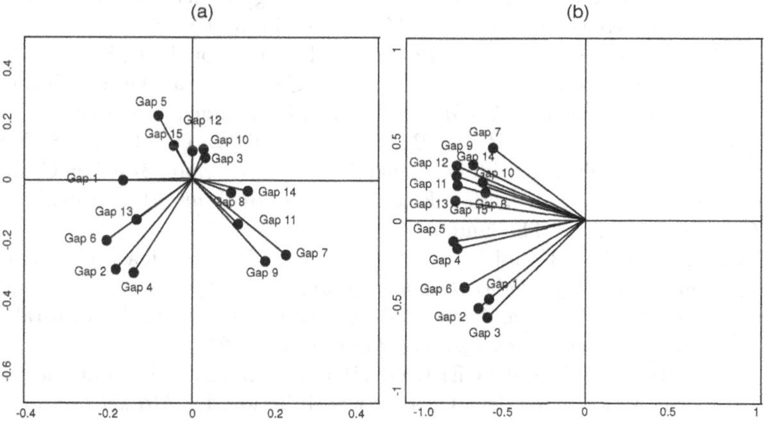

Fig. 3. Graphical Representation of Gap variables on the First factorial plane computed by SD (a) and PCA (b)

4 Conclusions

This paper introduces a new method to analyze jointly data structured in couple of matrices, which are related by logical relationship. Specifically in service quality evaluation the researcher is interested to investigate about the relationships between the expectations and perceptions in order to improve the offered service keeping in account the wishes of the customers. Other field of applications can be found in ecological as well as in chemometric frameworks. The proposed method is based on the simultaneous diagonalization theory. According to the results presented here, the method allows to enrich the interpretation of the factorial axes by supplying complementary or supplementary information to the classical dimensions reduction methods like the Co-Inertia of Fully Matched Tables.

Acknowledgement

This research has been supported with a PRIN 2004 grant (P. Amenta). Authors wish to thank the anonymous referees for their comments.

References

AMENTA, P. and SARNACCHIARO, P. (2004): Tensorial Co-Structure Analysis for the Full Multi-Modules Customer Satisfaction Evaluation. In: H. Bock, M. Chiodi and A. Mineo (Eds.): *Advances in Multivariate Analysis*, Springer, Berlin, 159-169.

CRONIN, J.J. and TAYLOR, SA. (1992): Measuring Service Quality: a Reexamination and Extension. *Journal of Marketing, 56, 55-68.*

D'AMBRA, L. and AMENTA, P. (2000): Multidimensional Statistical Methods Based on Co-Inertia for the 'Multi-Modules' Customer Satisfaction Evaluation. In: *Atti del Convegno CIMASI'2000 EHTP*, Casablanca, Marocco.

D'AMBRA, L., AMENTA, P., RUBINACCI, R., GALLO, M. and SARNACCHIARO, P. (1999): Multidimensional Statistical Methods Based on Co-Inertia for the Customer Satisfaction Evaluation. In: *Book of International Quality Conference IQC'99*, Thailandia.

HOGAN, J., HOGAN, R. and BUSH, C. M. (1984): How to Measure Service Orientation. *Journal of Applied Psychology, 69, 167-173.*

LAFOSSE, R.(1989): Ressemblance et différence entre deux tableaux totalement appariés, *Statistique et analyse des données, 14, 1-24.*

LAURO, N., BALBI, S. and SCEPI, G. (1997): L'Analisi Multidimensionale dei Dati per la Misurazione della Customer Satisfaction. In: *Atti del Convegno SIS "La Statistica per le Imprese"*, Torino.

NOBLE, B. and DANIEL, W. (1977): *Applied Matrix Algebra*, Prentice Hal, Inc., Englewood Cliffs, NJ.

PARASURAMAN, A., ZEITHAM, l.V. and BERRY, L. (1985): A Conceptual Model of Service Quality and its Implication for Future Research, *Journal of Marketing, 49, 41-50.*

RAO, C. (1973): *Linear Statistical Inference and Its Application*, New York, Wiley.

TORRE, F. and CHESSEL, D. (1995): Co-structure de deux tableaux totalement appariès, *Revue de Statistique Appliquèe, XLIII, 1, 109-121.*

TUCKER, L. (1958): An Inter-Battery Method of Factor Analysis, *Psycometrika, 23, 2, 111-136.*

VAN DER VORST, A. and GOLUB, G. (1997): 150 Years Old and Still Alive: Eigenproblems. In: *The state of the art in Numerical Analysis*, Oxford University Press,63, 93-120.

VEDALDI, R. (1997): Modelli Interpretativi per la Valutazione della Customer Satisfaction: l'impiego degli Strumenti Statistici, in: *Atti del Convegno SIS "La Statistica per le Imprese"*, Torino.

WOLD, H. (1966): Estimation of Principal Components and Related Models by Iterative Least Squares. In: *Multivariate Analysis*, Krishnaiah, New York.

WRIGHT, B.D. and MASTERS, G.N. (1982): *Rating Scale Analysis.* MESA press, Chicago.

Analyzing Evaluation Data: Modelling and Testing for Homogeneity

Angela D'Elia and Domenico Piccolo

Dipartimento di Scienze Statistiche,
Università di Napoli Federico II, Italy
{angela.delia,domenico.piccolo}@unina.it

Abstract. In the evaluation process of a given service, different issues are worth of analysis. In first instance, it is interesting to assess how the evaluation responses changes over the time and whether there is an effect of the raters' features. Secondly, when the service is made up by different items, it is important to verify if the satisfaction feelings of the users/consumers are the same with respect to all the dimensions. At this scope, the paper proposes a modelling approach for analyzing and testing ordinal/rating data. Some evidence from University services evaluation shows the usefulness of this procedure in a real case-study.

1 Introduction

In evaluation contexts, raters are often required to express their judgements about a set of different items, which are assumed to contribute to the whole performance of a specific service (Parasurman, *et al.*,1988; Peña, 1997). Besides, in most of cases, the raters have to give also an overall rating expressing their global satisfaction towards the service.

As a consequence, for a given service made up by k different items and evaluated by n raters, usually we observe a $(n \times k+1)$ matrix of rank data $\{r_{ij}\}$, $(i = 1, 2, ..., n;\ j = 1, 2, ..., k, k+1)$ expressing the satisfaction feelings of the service's users towards the k items and, in the last column, towards the service as a whole. Given this setting, it is relevant to assess

- how the global satisfaction changes over the time and/or over different group of raters;
- if the raters' feelings towards the k different dimensions of the service are substantially the same.

These tasks can be addressed by means of a mixture model for ordinal/rating data originally proposed by D'Elia (2003), Piccolo (2003), and D'Elia and Piccolo (2005a): in particular, this paper highlights how the peculiar model's parametrization may be usefully exploited for dealing with both the issues. In section 2, we motivate why the model is sensible for evaluation data and we briefly recall some related results; then, in section 3, first results from a real evaluation case study are shown. In section 4, likelihood ratio tests among nested models are developed and, finally, further evidence from the evaluation data set is discussed.

2 The Model

The degree of satisfaction $r\,(=1,2,...,m)$ assigned to a given item may be analyzed as a realization of a discrete ordinal random variable R. In fact, the judgement process is intrinsically continuous and, since the basic feeling (for instance, liking or disliking) depends on several causes, it could be thought to follow a Gaussian distribution.

Indeed, the underlying variable approach for the analysis of ordinal data often assumes that the observations are generated by an unobserved continuous variable (say R^*) normally distributed (Joreskog and Moustaki, 2001), so that $(r = a) \;\Leftrightarrow (\tau_{a-1} < r^* < \tau_a)$, where $\tau_0 = -\infty$, $\tau_m = +\infty$, and τ_a, $a = 1,2,...,m-1$, are the ordered threshold parameters to be estimated. This kind of analysis is widely discussed by Cagnone *et al.* (2004), Moustaki (2000) and Moustaki *et al.* (2004).

Following this idea, but maintaining the discrete nature of the observations, a suitable model for achieving the mapping of the unobserved continuous variable R^* into a discrete random variable defined on the values $r = 1,2,...,m$, may be the shifted Binomial distribution.

However, it seems sensible to assume that each evaluation process is also characterized by an uncertainty component, that adds up to the expression of the basic feeling: this component can be adequately described by means of the discrete Uniform distribution, since this random variable is known to maximize the entropy on a finite support.

As a consequence, we assume that the level of satisfaction r is the realization of a Mixture of a Uniform and a shifted Binomial distribution. Thus, we define $R \sim MUB(m,\pi,\xi)$ if:

$$Pr(R=r) = \pi \binom{m-1}{r-1} (1-\xi)^{r-1}\xi^{m-r} + (1-\pi)\frac{1}{m}, \qquad r = 1,2,\ldots,m;$$

where $\pi \in [0,1]$ and $\xi \in [0,1]$.

The π parameter is inversely related to the uncertainty component of the probabilistic model: thus, the estimate of $(1-\pi)$ is a measure of the uncertainty in the ratings. On the other hand, both π and ξ are related to the liking feeling towards the item, since

$$E(R) = \pi(m-1)\left(\frac{1}{2} - \xi\right) + \frac{m+1}{2}.$$

The exact meaning of ξ changes with the setting of the analysis and, being the MUB model reversible, it depends on how the ratings have been codified (the greater the rating the greater the satisfaction, or viceversa). In the case study considered in this paper, higher ratings are associated to higher satisfaction: thus, the estimate of $(1-\xi)$ can be thought of as a measure of liking. Finally, the ratio $\pi/(1-\pi)$ measures the relative weights of the basic feeling and of the uncertainty components, respectively, in the evaluation of a given service.

Given the observed frequencies vector $(n_1, n_2, \dots, n_m)'$, where n_r is the frequency of $(R = r)$, and letting $p_r(\pi, \xi) = Pr(R = r \mid \pi, \xi)$, $r = 1, 2, \dots, m$, the log-likelihood function for the MUB model is:

$$\log L(\pi, \xi) = \sum_{r=1}^{m} n_r \log\{p_r(\pi, \xi)\}. \tag{1}$$

As it is common for mixture models (McLachlan and Peel, 2000), the maximum likelihood estimates of the parameters π and ξ can be obtained by means of an E-M algorithm. Computational details for the estimation algorithm and for obtaining the asymptotic variance and covariance matrix of the maximum likelihood estimates are given by D'Elia and Piccolo (2005a).

The MUB model is an extremely flexible tool for fitting several real data sets and for explaining different judgement choices (ratings, preferences, agreement, perceptions, etc.) also when the raters' covariates are considered, as discussed in D'Elia (2004) and D'Elia and Piccolo (2005b). However, in this paper we limit our attention to an evaluation data context.

3 The Evaluation of Orientation Activities in the University

The 13 Faculties of the University of Naples Federico II, during the years 2002-2004, have provided an Orientation program to their students. At the end of each year a survey was carried out, in order to check for the students' satisfaction towards this service: it involved n = 2179, 2536, 3183 students for the years 2002, 2003 and 2004, respectively.

The survey was based on a 5-items questionnaire: each student was asked to give a score in [1, 7] expressing his/her *overall satisfaction* towards the Orientation service, where 1 means "completely unsatisfied" and 7 means "completely satisfied". Besides, four dimensions of the service were investigated: *willingness* and *competence* of the Orientation staff, *clearness* of the information, adequateness of *timetable*.

In first instance, we fitted a MUB model to the rating data expressing the global assessment of the overall satisfaction in the 13 Faculties, through the time. By means of the estimated $(\hat{\pi}, \hat{\xi})$ we got a display of the estimated parameters in the parameter space (Figure 1).

Thus, looking at the ξ axis, it emerges that almost all the Faculties exhibit high values of $(1 - \hat{\xi})$, that means high ratings for their Orientation service, and this behavior has became more homogeneous during the years. On the other hand, if we consider the π axis, we can notice that all the Faculties get a very low estimate of $(1 - \pi)$, meaning that there is a very low weight of the uncertainty component in the assessment of the global satisfaction. Moreover, this behavior becomes more evident in the last year (2004): indeed, through the years, there is a shift towards higher values of π (that is, lower weight of uncertainty). Such a pattern may depend on an increasing skill -for both the students and the interviewers- in filling in the evaluation questionnaire.

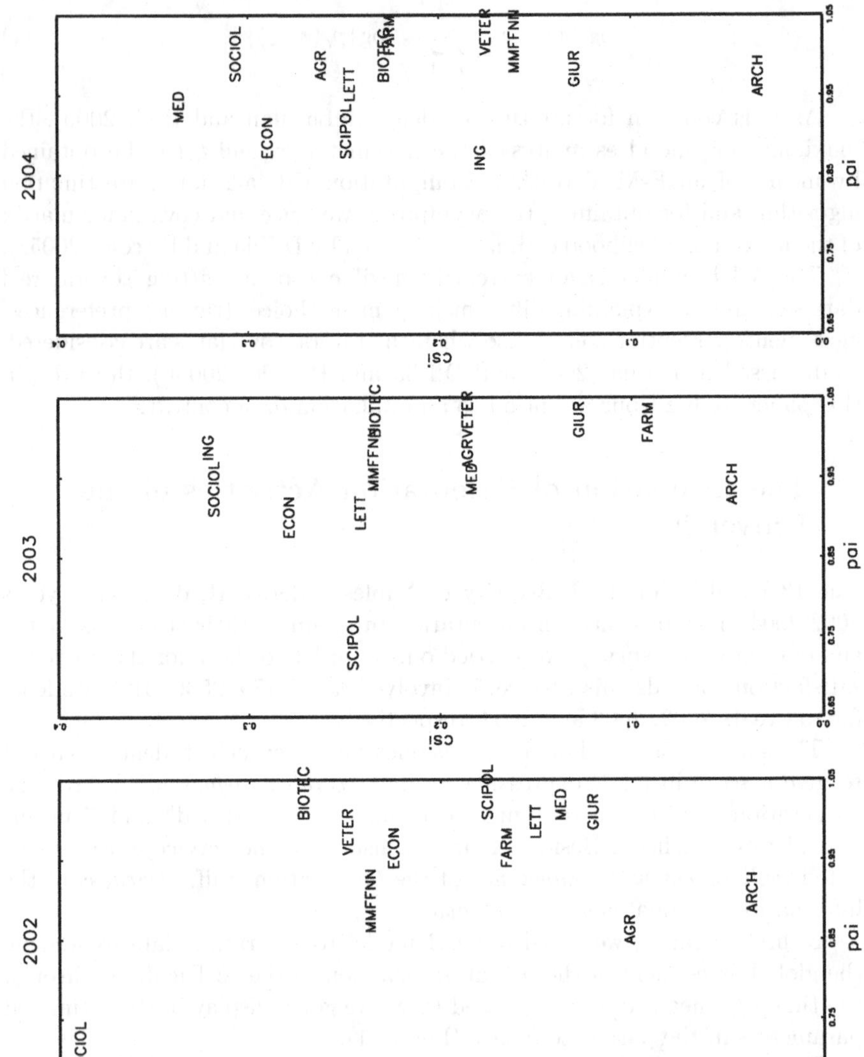

Fig. 1. Estimated parameter space for the overall satisfaction

Of course, in the analysis of evaluation/rating data it is also important to detect if and how the un/satisfaction changes with different group of raters (e.g. students). In the following Figure 2 we display the estimated parameter space (through the years), when we considered the students clustered on the basis of the high school (CL="classico", SC="scientifico", TE="tecnico", PR="professionale", LI="linguistico", AL="other") they have attended before starting their University career.

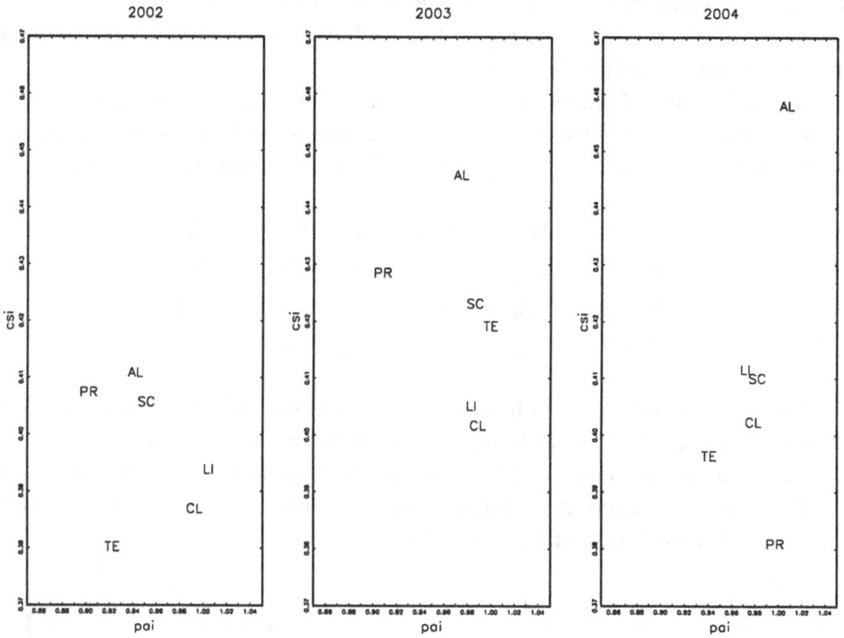

Fig. 2. Estimated parameter space for different high schools

In fact, we cannot detect a sharp distinction among the different group of students as far as it concerns both the ξ and the π estimated values, meaning that the kind of high school the students attended has not influenced their evaluation of the Orientation service. But, if we look at the behavior of each group (type of high school) through the 3 years, we are able to notice that the estimated ξs in most cases exhibit the following pattern: with respect to the year 2002 (beginning of the Orientation service), the $\hat{\xi}$s increase in 2003 and they decrease in 2004. This means that for most of the students there has been first an enthusiastic judgement of the service, followed by a more critical assessment and finally by a new increase in the liking. The only exception are the students from "LI" and "AL" -accounting for 14% of the sample- for whom there has been a decreasing satisfaction during the years[1].

[1] A more analytic approach is also possible, by including explicitly the raters' covariates in the specification of the MUB model, as discussed in D'Elia (2004) and D'Elia and Piccolo (2005b).

4 Testing Among Models

From the previous analysis, we noticed a decreasing role of the uncertainty component in the assessment of the satisfaction, but an alternating behavior of the parameter ξ: then, since the overall satisfaction for the service is the result of the feelings towards its different dimensions, it seems interesting to study the pattern of the ratings for the component items.

More in general, let us suppose that we have a multi-items service, made up of k different specific dimensions.

Then, for each of them we get: $R_j \sim MUB(m, \pi_j, \xi_j)$, $j = 1, 2, ..., k$.

In order to assess if the raters' feelings towards the k different dimensions of the service are the same, the following hypotheses can be specified:

$$
\begin{aligned}
&H_0 : \pi_1 = \pi_2 = ... = \pi_k; && \xi_1 = \xi_2 = ... = \xi_k; \\
&H_1 : \pi_1 = \pi_2 = ... = \pi_k; && \exists \xi_j \neq \xi_h,\ j \neq h; \\
&H_2 : \exists \pi_j \neq \pi_h,\ j \neq h; && \xi_1 = \xi_2 = ... = \xi_k; \\
&H_3 : \exists \pi_j \neq \pi_h,\ j \neq h; && \exists \xi_j \neq \xi_h,\ j \neq h.
\end{aligned}
$$

These hypotheses yield a hierarchical sequence of MUB models, since H_0 is nested with respect to both H_1 and H_2, that are nested into H_3.

Then, in the more general hypothesis (H_3), assuming that the k judgements towards k different dimensions of a service are made independently, the log-likelihood function turns out to be:

$$
\begin{aligned}
\log L(\pi_j, \xi_j) &= \sum_{j=1}^{k} \sum_{i=1}^{n} \log \left\{ \pi_j \binom{m-1}{r_{ij}-1} \xi_j^{m-r_{ij}} (1-\xi_j)^{r_{ij}-1} + \frac{1-\pi_j}{m} \right\} \\
&= \sum_{j=1}^{k} \sum_{r=1}^{m} n_r^{(j)} \log\{p_r(\pi_j, \xi_j)\},
\end{aligned}
$$

where we let $n_r^{(j)} = \#\{r_{ij} = r,\ i = 1, 2, ..., n\}$, $j = 1, 2, ..., k$; $r = 1, 2, ..., m$.

In the previous expression, under the hypotheses H_1 and H_2, we let $\pi_j = \pi$, $\forall j$, and $\xi_j = \xi$, $\forall j$, respectively; under the hypothesis H_0, equation (1) holds, as specified in Section 2.

Thus, given the nested structure of the above hypotheses, the testing procedure can be based upon the likelihood-ratio test:

$$
\lambda = 2[\log L(\hat{\boldsymbol{\theta}}_M) - \log L(\hat{\boldsymbol{\theta}}_{Mr})] \overset{a}{\sim} \chi^2_{p-q},
$$

where $\boldsymbol{\theta} = (\boldsymbol{\pi}, \boldsymbol{\xi})'$ are the parameters to be estimated under the different hypotheses, and $\log L(\hat{\boldsymbol{\theta}}_M)$ and $\log L(\hat{\boldsymbol{\theta}}_{Mr})$ represent, respectively, the log-likelihood functions for a given model M (with p unconstrained parameters), and a restricted model Mr (with $q < p$ unconstrained parameters).

5 Testing for Homogeneity in the Evaluation of Orientation Activities

In addition to the overall satisfaction towards the Orientation service, the survey investigated also the feelings of the students towards 4 specific dimensions of the service which concern, respectively, the *willingness* and the *competence* of the Orientation staff, the *clearness* of the information, and the adequateness of *timetable*.

Thus, it seems interesting to verify whether the satisfactions expressed by the students are the same with respect to the 4 items of the analyzed service. In particular, we can check the homogeneity among the $k = 4$ different evaluations by means of the hypotheses on the MUB model's parameters, with the following meaning:

- H_0 : uncertainty and satisfaction are equal for all the items;
- H_1 : satisfaction changes with the specific j-th item;
- H_2 : uncertainty changes with the specific j-th item;
- H_3 : both the satisfaction and the uncertainty change with the item.

Then, the likelihood ratio test among nested MUB models was performed, yielding the results in Table 5 where, for each year of the survey and for each Faculty, the hypothesis that increases the likelihood more significantly is displayed.

Faculty	*2002*	*2003*	*2004*	**Faculty**	*2002*	*2003*	*2004*
Agraria	H_1	H_1	H_1	Medicina e Chirurgia	H_1	H_1	H_1
Architettura	H_2	H_2	H_2	Medicina Veterinaria	H_1	H_3	H_3
Economia	H_3	H_1	H_1	Scienze Biotecnologiche	H_1	H_1	H_1
Farmacia	H_1	H_1	H_1	Scienze MM. FF. NN.	H_1	H_3	H_3
Giurisprudenza	H_1	H_1	H_3	Scienze Politiche	H_1	H_1	H_1
Ingegneria	H_1	H_3	H_1	Sociologia	H_1	H_1	H_1
Lettere e Filosofia	H_3	H_1	H_1				

Table 1. Likelihood ratio test results for homogeneity

Thus, it emerges that most of the evaluations are homogeneous with respect to the uncertainty, but the satisfaction level changes with the dimensions of the service (hypothesis H_1). This behavior is constant through the years for 6 Faculties, while it alternates with the hypothesis H_3 (both satisfaction and uncertainty changing) in other 6 cases.

The only "anomalous" behavior has been found for the Faculty of Architettura, where it appears to be an inhomogeneous uncertainty, but an homogeneous satisfaction (H_2); in fact, the related data deserve some suspicion (that could be noticed also by the inspection of the previous Figure 1, where the Faculty of Archittettura appears always isolated with respect to the others).

6 Concluding Remarks

Two points arise from the previous analysis. Firstly, the assumption of independent judgements in the evaluation of different items in some circumstances could be untrue; however, subtracting from each evaluation data a global measure of the satisfaction for the whole service, we found that the net differences were almost uncorrelated: thus, it seems that possible dependencies do not arise from real interrelationships among the items themselves, but they are the results of a global feeling towards the Orientation services.

Secondly, it emerges a quite different role of the π and ξ parameters, and then of the component they represent, in the satisfaction assessment. Indeed, the uncertainty appears homogeneous with respect to the items and decreasing in the years; on the other hand, the liking is heterogeneous towards the dimensions and it exhibits an alternating pattern in the years; at this regard, the inclusion of raters/items covariates in the model might supply a deeper insight for explaining such a behavior.

References

CAGNONE, S., GARDINI, A. and MIGNANI, S. (2004): New developments of latent variable models with ordinal data. In: *Atti della XLII Riunione Scientifica SIS*. CLEUP, Padova, 221–232.

D'ELIA, A. (2003): A mixture model with covariates for ranks data: some inferential developments. *Quaderni di Statistica, 5, 1–25.*

D'ELIA, A. (2004): New developments in ranks data modelling with covariates. In: *Atti della XLII Riunione Scientifica SIS*. CLEUP, Padova, 233–244.

D'ELIA, A. and PICCOLO, D. (2005a): A mixture model for preferences data analysis. *Computational Statistics & Data Analysis, 49, 917–934.*

D'ELIA, A. and PICCOLO, D. (2005b): Un modello statistico per l'analisi dei giudizi di gradimento. In: Metodi, Modelli e Tecnologia delle Informazioni a Supporto delle Decisioni: *Atti del Convegno MTISD2004*. Franco Angeli, Milano, in press

JÖRESKOG, K. and MOUSTAKI, I. (2001): Factor analysis of ordinal variables: a comparison of three approaches. *Multivariate Behavioral Research, 36, 347–387.*

McLACHLAN, G., and PEEL, G.J. (2000): *Finite mixture models.* Wiley, New York.

MOUSTAKI, I. (2000): A latent variable model for ordinal data. *Applied Psychological Measurement, 24, 211-223.*

MOUSTAKI, I., JÖRESKOG, K. G. and Mavridis, D. (2004): Factor models for ordinal variables with covariate effects on the manifest and latent variables: a comparison of LISREL and IRT approaches. *Structural Equation Modeling Journal, 11, 487–513.*

PARASURMAN, A., ZEITHAML, V.A. and Berry, L.L. (1988): SERVQUAL: a multiple item scale for measuring consumer perceptions of service quality. *Journal of Retailing, 64, 12–40.*

PEÑA, D. (1997): Measuring service quality by linear indicators. In: P. Kunst, and J. Lemmink (Eds.): *Managing service quality.* Chapman Publishing Ltd., London, 35–51.

PICCOLO, D. (2003): Computational issues in the E-M algorithm for ranks models estimation with covariates. *Quaderni di Statistica, 5, 27–48.*

Archetypal Analysis for Data Driven Benchmarking

Giovanni C. Porzio[1], Giancarlo Ragozini[2], and Domenico Vistocco[1]

[1] Dipartimento di Scienze Economiche,
Università di Cassino, Italy
porzio@eco.unicas.it, vistocco@unicas.it

[2] Dipartimento di Sociologia "G. Germani",
Università di Napoli Federico II, Italy
giragoz@unina.it

Abstract. In this work, adopting an exploratory and graphical approach, we suggest to consider archetypal analysis as a basis for a data driven benchmarking procedure. The procedure is aimed at defining some reference performers, at understanding their features, and at comparing observed performances with them. Being archetypes some extreme points, we propose to consider them as reference performers. Then, we offer a set of graphical tools in order to describe these archetypal benchmarks, and to evaluate the observed performances with respect to them.

1 Introduction

Managing and running complex organizations requires the measurement and the analysis of their performances. Monitoring the ongoing processes, and comparing results of different activities lead organizations to continuously improve their jobs. With this goal, hence, organizations collect and analyze a large number of performance indicators measured on large sets of performers. Such data sets are exploited for performance analysis, that mainly aims at benchmarking and studying relationships among indicators (Camp, 1989). Through benchmarking organizations evaluate various aspects of their processes with respect to some standards, and many statistical methods, either confirmatory or exploratory, numerical or graphical, can be profitably exploited with this aim.

In this work, adopting an exploratory and graphical approach, we address some statistical issues related to the standards definition, their interpretation, and their use for comparison. We propose to consider archetypal analysis (Cutler and Breiman, 1994) as a method for selecting some points as best or worst performers, detecting any outlying performances, comparing and clustering performances.

Archetypes represent a sort of "pure individual types", that synthesize multivariate data through few points lying on the boundary of the data scatter. In our opinion, this feature fits well with the benchmarking aims. Being archetypes extreme pure performers, we propose first to use them to define some "pure benchmark" points. Exploiting geometric properties of archetypes, we design exploratory tools to visually analyze them, and to understand benchmark characteristics through the inspection of archetypes composition. Finally, we propose to use parallel coordinates to compare observed performances among them and with respect to the benchmarks.

The paper is organized as follows. We present basic archetypal analysis, introducing a computational geometry point of view in Section 2. Then we discuss our interpretation of archetypes as benchmarks, and we offer a benchmarking procedure organized in three different steps (Section 3). An illustrative example on 209 computer CPU performances will be used along the whole paper. Some concluding remarks follow.

2 Archetypal Analysis

Archetypal analysis is a statistical method aiming at synthesizing a set of multivariate observations through few points not necessarily observed. These points, the archetypes, can be considered a sort of "pure" types as all the data points must be a mixture of them. In addition, to ensure that these "pure" points are the nearest to the observed data, archetypes must be also a convex combination of the data points.

Archetypal analysis has found applications in spatio-temporal dynamics and cellular flames (Stone and Cutler, 1996; Stone, 2002), in astronomy (Chan *et al.*, 2003), and in market researches (Anderson and Weiner, 2004). In performance analysis archetypes have been used for ordering multivariate performances (D'Esposito and Ragozini, 2004; 2006).

Formally, the archetypes $\{\mathbf{a}_j\}_{j=1,\ldots,m}$ should be those points in the p-dimensional Euclidean space such that:

$$\mathbf{x}_i' = \alpha_i'\mathbf{A} \tag{1}$$

with

$$\alpha_{ij} \geq 0 \quad \forall i, j \qquad \alpha_i'\mathbf{1} = 1 \quad \forall i, \tag{2}$$

where $\{\mathbf{x}_i\}_{i=1,\ldots,n}$ are the observed data, $\mathbf{A}$ is the archetype matrix with $\mathbf{a}_j'$ its j-th row, and α_i' is the vector of the convex combination coefficients with elements $\{\alpha_{ij}\}_{j=1,\ldots,m}$.

At the same time, all the archetypes should be also a mixture of the observed data:

$$\mathbf{a}_j' = \beta_j'\mathbf{X} \tag{3}$$

with

$$\beta_{ji} \geq 0 \quad \forall j, i \qquad \beta_j'\mathbf{1} = 1 \quad \forall j, \tag{4}$$

where $\mathbf{X}$ is the observed data matrix, and the convex combination coefficient β_{ji}'s are the n elements of the β_j' vectors.

By definition of convex hull, the eqn.s (1) and (2) imply that all the data belong to the convex hull of the archetypes, that is the archetypes could be the vertices of any convex p-polytope including the data scatter. On the other hand, eqn.s (3) and (4) imply that archetypes belong to the convex hull of the data. Consequently, archetypes are the vertices of the data convex hull.

However, in practice, the number of the data convex hull vertices is generally too large to synthesize data through few pure types. For this reason, looking for a smaller number of pure types, and wishing to preserve their closeness to the data (eqn.s 3 and 4), Cutler and Breiman (1994) defined the archetypes as those m points that fulfill as much as possible eqn. (1), satisfying at the same time eqn.s (2), (3) and (4).

More precisely, let us rewrite eqn. (1) as $\mathbf{x}_i' - \alpha_i'\mathbf{A} = \mathbf{0}$. For the discussion above, if the number of archetypes is less than the number of the data convex hull vertices, then the eqn (1) does not hold. In particular, for the points i^* lying outside the convex hull of the archetypes, we have that $\|\mathbf{x}_{i*}' - \alpha_{i*}'\mathbf{A}\| > 0$, where $\|\cdot\|$ is the L_2 norm of a vector. The archetypes, given m, have been then defined as the points $(\mathbf{a}_1, \ldots, \mathbf{a}_m)$ minimizing

$$\sum_{i=1}^{n} \|\mathbf{x}_i' - \alpha_i'\mathbf{A}\| , \tag{5}$$

holding equations (2),(3) and (4).

The solution to this minimization problem depends on m, and solutions are not nested as m varies. That is, denoting with $\mathbf{a}_j'(m)$ the j-th archetype for a given m, $\mathbf{a}_j'(m) \neq \mathbf{a}_j'(l)$, with $m \neq l$.

As for the choice of m, Cutler and Breiman (1994) suggest to look at the quantity:

$$RSS(m) = \sum_{i=1}^{n} \|\mathbf{x}_i' - \tilde{\mathbf{x}}_i'(m)\| \tag{6}$$

where $\tilde{\mathbf{x}}_i'(m) = \alpha_i'(m) \cdot \mathbf{A}(m)$ is the best approximation of $\mathbf{x}_i'$ through the m archetypes. The residual sum of squares $RSS(m)$ is then the sum of the euclidean distances of the observed data from their best approximation, and therefore it measures to what extent the m archetypes synthesize the data.

3 Benchmarking Through Archetypes

Given the geometrical and statistical properties discussed above, we propose to use archetypes for a three step data driven benchmarking procedure. The procedure is aimed at defining some reference performers, to understand their features, and to compare observed performances with them.

Usually, reference performers (or benchmarks) are defined on the basis of some expert knowledge or some field agreement. For example, financial dealers agree that the stock index Nasdaq may be used as a benchmark against which a performance of a technology stock is compared. Depending on the application context, the benchmark has two main different meanings.

It can be intended as some average performance, e.g. the Nasdaq index, or as some "best" performer, such as a sector leader. In the simplest case of a single performance indicator, the benchmark can be respectively some weighted averages or the maximum observed value.

If many performance indicators are measured, it is somehow harder to identify the benchmark. Looking for an average standard, in our opinion some multivariate centrality measures can be adopted, say the simple mean or some more appropriate location depth measures (Liu *et al.*, 1999; Mizera and Müller, 2004). Searching for the best standard, two different issues have to be addressed. As such benchmark is a multivariate extreme value, first it is necessary to define it. Secondly, as a data scatter presents many multivariate extremes, it is necessary to identify the "best practice" among them.

In the following, we present details on our procedure, and with an illustrative purpose we introduce a real data example. The data we use come from a study on the performance of central processing units (CPU), and consist of a set of 209 CPU's to be compared considering some performance indicators (Ein-Dor and Feldmesser, 1987). The six indicators we consider are: Cycle time (ns), Minimum memory (kb), Maximum memory (kb), Cache size (kb), Minimum channels, and Maximum channels of the CPU's. The goal is first to identify some benchmarks for the CPU's, and then to evaluate these latters with respect to the identified benchmarks.

3.1 Looking for Extremes Through Archetypes

Geometrically, the multivariate extremes of a data scatter are the points lying on its boundary. Using a center-outward ordering based on some depth notion, the extremes are the points with empirical depth equal to zero, and they lie on the furthest empirical center-outward quantile contour. Following Liu *et al.* (1999), these extreme points are the convex hull boundary.

Extreme observed points correspond then to the vertices of the data scatter convex hull. However, there is a non-trivial problem of defining one or few benchmarks, as the convex hull vertices cardinality could be extremely large. As known, the number of vertices tends quickly to the number of observed data as the number of variables increases. In our case, the data set on the CPU performance indicators present 104 convex hull vertices, not an adequate number for a benchmarking purpose.

For this reason, given an observed performance data set, we suggest to use their archetypes to select a reduced number of extremes to be used as benchmarks. As for their number, we suggest to select it looking at the graphical devices we are going to illustrate in an exploratory and iterative approach. As a matter of fact, the sole analysis of the RSS behaviour (as suggested by Cutler and Breiman, 1994) does not provide enough information to decide the number of archetypes to be used as benchmarks. With respect to our example, the $RSS(m)$ function suggest to select five archetypes, being

$RSS(5) \cong 0$. However, after some analysis that we discuss shortly, we decided to select three benchmark-archetypes ($m = 3$).

3.2 Describing Archetypes to Identify Benchmarks

By their nature, multivariate extremes cannot be expressed in terms of "maxima" or "minima". In other words, having selected few extreme values through the archetypes, the second issue to be addressed is to evaluate which of them can be considered as the "best" and which as the "worst" pure performances.

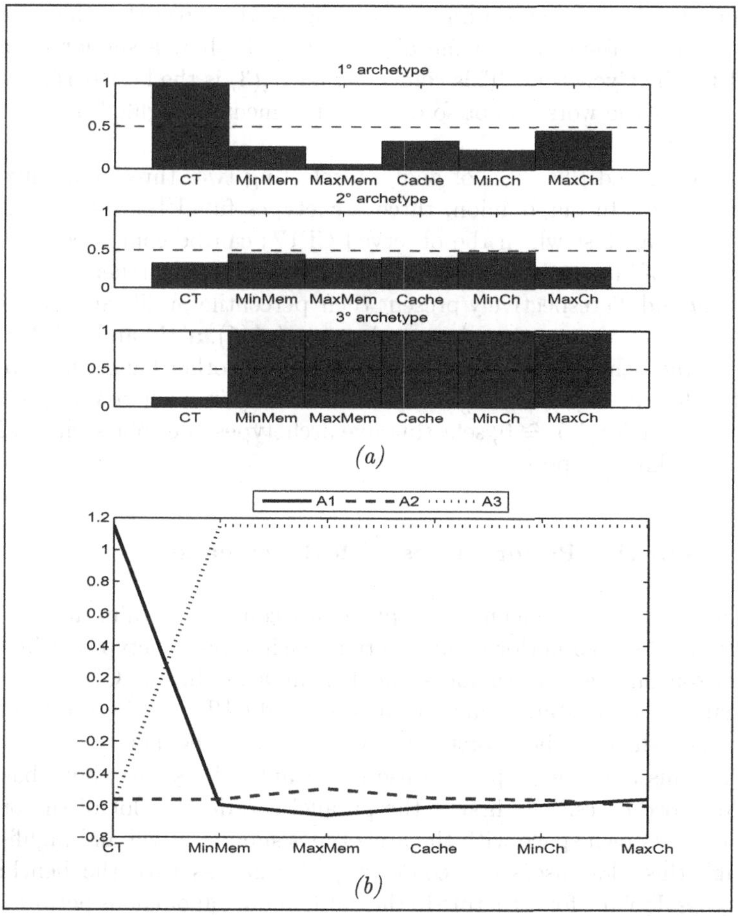

Figure 1. Describing archetypes: *(a)* percentile profile plot, and *(b)* parallel coordinate plot for the CPU's data when three archetypes are chosen ($m = 3$).

With this goal, we propose to visually analyze them in their own space, using iconic plots, such as the percentile profiles or stars (Hartigan, 1975), or a parallel coordinate plot (Inselberg, 1985; Wegman, 1990; Milioli and Zani, 1996). We note that these graphical representations are adequate for this aim because of the reduced number of points to be visualized.

With respect to the CPU's data set, we decided to select three archetypes, and in Figure 1*a* and 1*b* we respectively present their percentile profile and parallel coordinate plots. Visual inspection of Figure 1*a* highlights that the first archetype $\mathbf{a}_1'(3)$ is the "worst type" as it performs badly on the Cycle time (being the highest percentile) and has substantially poor performances on the other variables (being generally among the lower percentiles). The second selected archetype $\mathbf{a}_2'(3)$ represent a pure type slightly under the median performer, whilst the third archetype $\mathbf{a}_3'(3)$ is clearly the pure positive CPU benchmark. The parallel coordinate plot (Figure 1b) shows that the third type is further from the others, while $\mathbf{a}_1'(3)$ and $\mathbf{a}_2'(3)$ show a similar behaviour except for the Cycle time. This confirms that $\mathbf{a}_3'(3)$ is the best performer and suggests that the worst is not so far from the median, highlighting a skewed data scatter.

We performed this kind of analysis selecting two, three, four, five, and six archetypes. In our opinion, three archetypes fulfill the interesting pure benchmarks against which the observed CPU's can be compared. However, for the sake of illustration, we discuss also the case of five archetypes ($m = 5$). Figure 2*a* and 2*b* respectively present their percentile profile and parallel coordinate plots. We note that three archetypes ($\mathbf{a}_1'(5)$, $\mathbf{a}_2'(5)$ and $\mathbf{a}_3'(5)$) closely resemble the benchmarks described above. On the other hand, the two additional archetypes $\mathbf{a}_4'(5)$ and $\mathbf{a}_5'(5)$ have not an easy immediate interpretation. Hence, even if $RSS(5) \cong 0$, selecting five archetypes does not seem useful for a benchmarking purpose.

3.3 Comparing Performances with Benchmarks

The third step in our benchmarking procedure consists of evaluating the closeness of the observed performances to the benchmark-archetypes, achieving a comparison among performances and benchmarks. In our CPU's example, the goal is to understand which of the observed CPU's is close to the "best" CPU type, which to the "worst", and which to the "median".

With this aim, we propose to use two graphical displays, both based on parallel coordinates. The first is the parallel coordinate plot of the original data set in its own space with the archetypes superimposed and highlighted. Through this plot, users can compare performances with the benchmark-archetypes, looking for structure in the data, finding anomalous performances, identifying which variable mostly determine the results obtained by the performers. A dynamic parallel coordinate plot (Wegman and Qiang, 1997), incorporating selection, deletion and colouring, could allow users to better

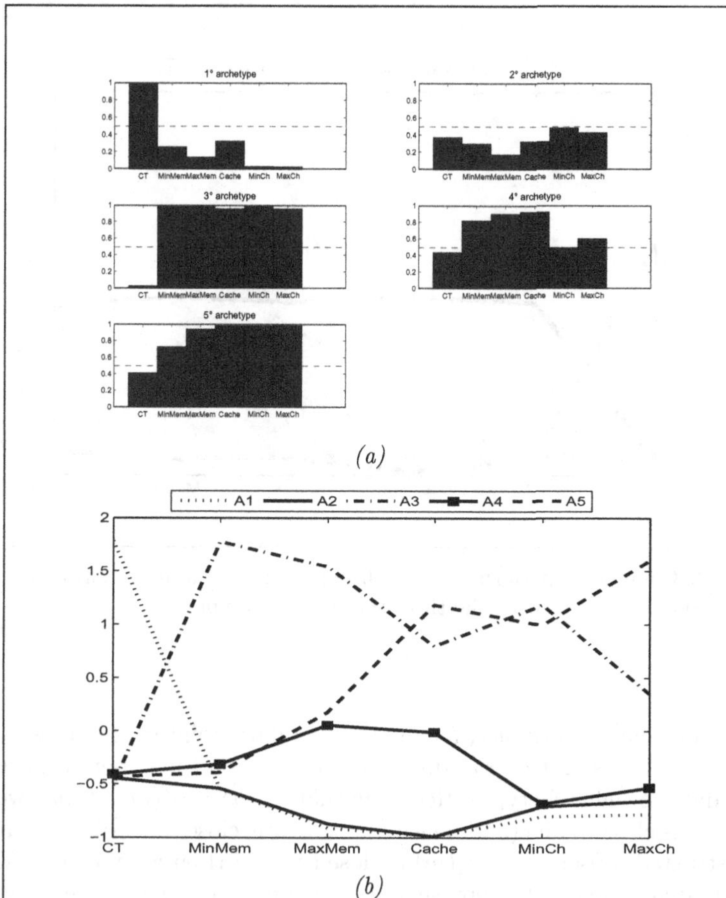

Figure 2. Describing archetypes: *(a)* percentile profile plot, and *(b)* parallel coordinate plot for the CPU's data when five archetypes are chosen ($m = 5$).

understand benchmarks, and to classify performers considering their closeness to the archetypes.

Figure 3 represents the suggested plot for the CPU data set along with the three selected archetypes. Many data points belong to a narrow band, close to both $\mathbf{a}_1'(3)$ and $\mathbf{a}_2'(3)$. The best pure performer $\mathbf{a}_3'(3)$, that lies in the data tails, clearly synthesizes CPU's widely scattered around it. It appears that these latters perform better on some variables and worse on some others with respect to the pure benchmark $\mathbf{a}_3'(3)$. Dynamic selection of the curves could make easier to identify such performers.

The second display we suggest is the parallel coordinate plot of the convex combination coefficient $\alpha_i'(m)$'s. In such a case, we represent the n data

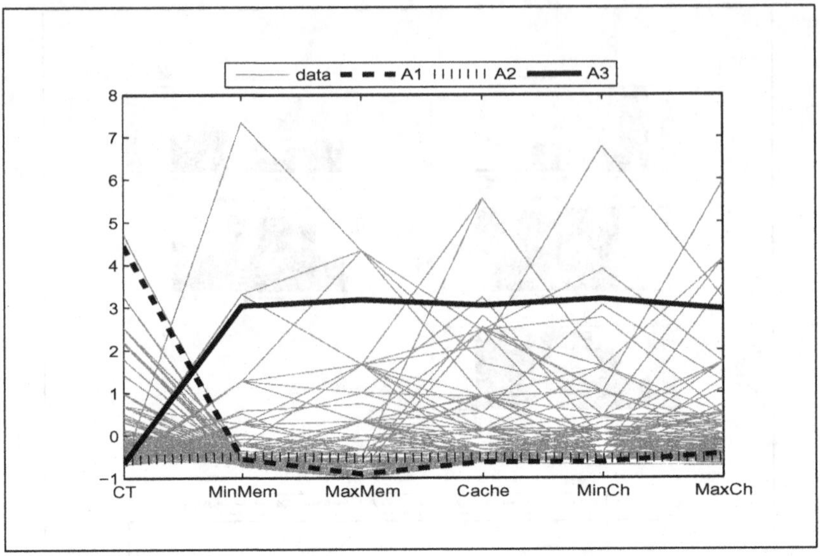

Figure 3. Comparing performances with benchmarks: parallel coordinate plot of the CPU data set along with the three selected archetypes.

points on the m α_j variables, having the i-th data point coordinates $\alpha_{ij}(m)$, $j = 1, \ldots, m$. This plot carries direct information on the closeness of the observed data to the archetypes. Recalling that $\tilde{\mathbf{x}}_i'(m) = \alpha_i'(m) \cdot \mathbf{A}(m)$, we have that if α_{ij} is close to one (and hence α_{ik}, $k \neq j$, is close to zero), then the i-th (reconstructed) observation $\tilde{\mathbf{x}}_i'(m)$ is close to the j-th archetype. In addition, the j-th archetype can be represented in this plot by the curve having coordinates $\alpha_j = 1$ and $\alpha_k = 0$ for $k \neq j$. To highlight the archetypes, we superimpose then dots in correspondence of the coordinates $\alpha_j = 1$, $j = 1, \ldots, m$.

Cutler and Breiman (1994) alternatively suggest to display the same information through a triangular plot based on the $\alpha_i'(m)$'s. However, unlike our proposed graph, the triangular plot is unfeasible for $m > 3$. Furthermore, we note that the plot we suggest represents "reconstructed" performances. Hence, whenever $\|\mathbf{x}_i' - \tilde{\mathbf{x}}_i'(m)\| > 0$, some points can be badly represented. For this reason, we recommend to use this plot jointly with the previous parallel coordinate plot, possibly exploiting some dynamic linking tools.

In our CPU's example this coefficient plot (Figure 4) suggests that a relatively small group of performers tends to resemble the worst benchmark behaviour $\mathbf{a}_1'(3)$, a large majority resembles the median $\mathbf{a}_2'(3)$, while few are somehow similar to the best pure type $\mathbf{a}_3'(3)$. We note that this plot on the reconstructed data avoids to display some noise present in the original data set (i.e. the data scattered around $\mathbf{a}_3'(3)$ in Figure 3). It seems that this feature makes easier to associate performers to benchmarks.

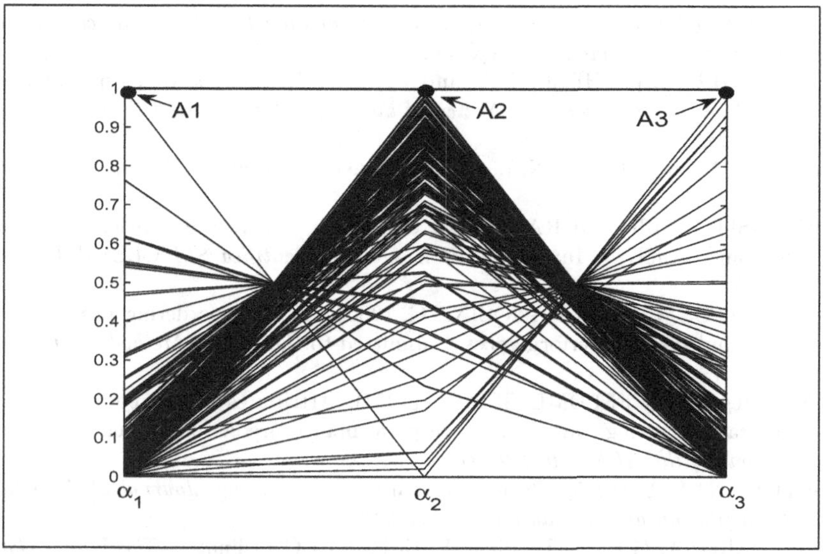

Figure 4. Comparing performances with benchmarks: parallel coordinate plot of the convex combination coefficient $\alpha_i'(m)$'s for the CPU's data when three archetypes are chosen ($m = 3$).

4 Some Concluding Remarks

Archetypal analysis is a powerful tool to select few extreme points belonging to the boundary of the data convex hull. This feature allowed us to design a data driven benchmarking procedure that we believe may benefit users in many application fields.

As further work, we aim at measuring to what extent the benchmark-archetypes synthesize the data, and to evaluate the goodness of the reconstruction for each performer. In addition, we plan to provide users with an *R* function that perform the benchmarking procedure we suggested.

Acknowledgments

Authors have been supported by M.I.U.R. PRIN2003 national grant number 2003130242: "Multivariate Statistical and Visualization Methods to Analyze, to Summarize, and to Evaluate Performance Indicators".

References

ANDERSON, L. and WEINER, J.L. (2004): *Actionable Market Segmentation Guaranteed (Part Two).* Knowledge Center Ipsos-Insight (*www.ipsos.com*).

CAMP, R.C. (1989): *Benchmarking, the search for industries best practice that lead to superior performance.* ASQC Quality Press, Milwaky.

CHAN, B.H.P., MITCHELL, D.A. and CRAM L.E. (2003): Archetypal analysis of galaxy spectra. *Monthly Notice of the Royal Astronomical Society, 338, 3, 790–795.*

CUTLER, A. and BREIMAN, L. (1994): Archetypal Analysis. *Technometrics, 36, 338–347.*

D'ESPOSITO, M.R. and RAGOZINI, G. (2004): Multivariate Ordering in Performance Analysis. In: *Atti XLII Riunione Scientifica SIS.* CLEUP, Padova, 51–55.

D'ESPOSITO, M.R. and RAGOZINI, G. (2006): A new r-ordering procedure to rank multivariate performances. Submitted to: *Statistical Methods and Applications.*

EIN-DOR, P. and FELDMESSER, J. (1987): Attributes of the performance of central processing units: a relative performance prediction model. *Communications of the ACM, 30, 308–317.*

HARTIGAN, J. A. (1975): Printer Graphics for Clustering. *Journal of Statistical Computation and Simulation, 4, 187–213.*

INSELBERG, A. (1985): The Plane With Parallel Coordinates. *The Visual Computer, 1, 69–91.*

LIU, R.Y., PARELIUS, J.M. and SINGH, K. (1999): Multivariate Analysis by Data Depth: Descriptive Statistics, Graphics and Inference, *The Annals of Statistics, 27, 783-858.*

MILIOLI, M.A. and ZANI, S. (1996): Alcune varianti del metodo delle coordinate parallele per la rappresentazione di dati multidimensionali, *Statistica Applicata, 8, 397-412*

MIZERA, I., and MÜLLER, C.H. (2004): Location-Scale Depth (with discussion), *Journal of American Statistical Association, 99, 981–989.*

STONE, E. (2002): Exploring Archetypal Dynamics of Pattern Formation in Cellular Flames. *Physica D , 161, 163–186.*

STONE, E. and CUTLER, A. (1996): Introduction to Archetypal Analysis of Spatio-temporal Dynamics. *Physica D, 96, 110–131.*

WEGMAN, E.J. (1990): Hyperdimensional data analysis using parallel coordinates. *Journal of the American Statistical Association, 85, 664–675.*

WEGMAN, E.J. and QIANG, L. (1997): High dimensional clustering using parallel coordinates and the grand tour, *Computing Science and Statistics, 28, 352–360.*

Determinants of Secondary School Dropping Out: a Structural Equation Model

Giancarlo Ragozini[1] and Maria Prosperina Vitale[2]

[1] Dipartimento di Sociologia "G. Germani",
Università di Napoli Federico II, Italy
giragoz@unina.it

[2] Dipartimento di Scienze Economiche e Statistiche,
Università di Salerno, Italy,
mvitale@unisa.it

Abstract. In this work we present the main results of a research program on dropping out in secondary school, carried out for the Labor Bureau of Campania Region in Italy. We exploited structural equation modeling to identify determinants of the phenomenon under study. We adopt a social system perspective, considering data coming from official statistics related to the 103 Italian Provinces. We provide some details for the model specification and the estimated parameters. Some relevant issues related to the estimation process due to the small sample size and the non-normality of the variables are also discussed.

1 Introduction

Recently, Western countries' policy makers have devoted much interest to the quality of teaching and to education system performances. As higher qualifications are now considered the essential baselines for successful entry into the labor market and as dropouts number in secondary school is increasing in European countries, a lot of researches have focused the attention on these phenomena. Besides the straight measurement of dropping out rates, the analysis of their characteristics and factors affecting them is a crucial point for designing education and employment policies.

In Italy, recent reports documented that the dropping out rate in secondary school is still 4.62%, up to 11.30% for the first year in some technical curriculum (Miur, 2002). This phenomenon represents a social and economic problem, as dropouts, that do not achieve higher levels of educational attainments, are more unlikely to participate in the labor market. However, most of the studies have been concerned only with the primary school where, instead, the dropping out rate has been decreasing to "physiological" levels.

In this work, we present the main results of a study on dropping out in secondary school that has been carried out for the Labor Bureau of Campania Region in Italy (A.R.Lav.), and was oriented to regional welfare policy planning. The research program aimed at understanding the main social and economic features affecting the education system and dropping out in secondary school. In particular, we introduce the structural equation modeling approach to explain the dropping out phenomenon in terms of social and economic context, education system, family environments, and student performances in primary school. We adopt a different perspective with respect to usual researches that focus the attention on individual characteristics of dropouts, relying on questionnaires or interview data. In our case, instead, we consider the national social system as a whole, and the 103 Italian Provinces as statistical units. Hence, our data sources are the available official statistics of Italian National Statistical Institute (Istat) and Department of Education (Miur).

In this paper, we provide the theoretical model specification based on dropping out literature (Section 2), and we discuss some issues related to data sources and to the use of exploratory data analysis tools for the variable selection (Section 3). In Section 4 the estimation problems due to nonnormality are addressed and the final model estimates are presented. Some concluding remarks are in Section 5.

2 Theoretical Hypotheses and Structural Model Specification

To understand secondary dropping out causes, to define the dimensions affecting it, and to specify the theoretical model we first consider the sociological literature (Benvenuto *et al.*, 2000; Jimerson *et al.*, 2000). Both foreign and domestic surveys usually highlighted as relevant features the individual traits and psychological measures, such as behavior problems, poor peer relationship (Farmer *et al.*, 2003). Some demographic status variables and family factors like parental school involvement, quality of parent child interactions, family lifestyles and values are also pointed out (Liverta Sempio *et al.*, 1999; Clarizia and Spanó, 2005). Other few studies consider the impact of welfare policies, like activities for family support or parental employment opportunities, on dropping out rates (Orthner and Randolph, 1999; Prevatt and Kelly, 2003).

Following the literature and paying more attention to features at country system level instead of the individual one, we define four main factors: the social, economic and demographic contexts, the education system, the family structures, and student personal traits. In particular, considering the research perspective, we intend the last factor related to the primary school careers and student performances. While it is clear that all the previous factors affect directly dropping out, other causal relationships among them can be also hypothesized. Indeed, the first three dimensions can yield effects on students careers in primary school. At same time the social and economic context could exert some influence on all the other factors. To describe such a complex phenomenon and its causal patterns, we introduce structural equation modeling for the relationships among the defined factors, each of them measured by one or more observed indicators.

Structural Equation Models (SEM's), developed by Jöreskog in 1970, have increasingly been applied in psychological and social studies, as well as in marketing researches during the last decade. They have been applied also in education to explain individual dropping out behavior (Kaplan *et al.*, 1995). To briefly recall, SEM's combine the path analysis among observed variables and latent factors, and their measurement models. Formally, given a set of observed exogenous variables $\mathbf{x}$ and endogenous variables $\mathbf{y}$, let us denote with $\boldsymbol{\xi}$ and $\boldsymbol{\eta}$ the latent exogenous and endogenous factors. The measurement parts of SEM's can be described by the following two equations:

$$\mathbf{x} = \mathbf{\Lambda}_x \xi + \delta,$$
$$\mathbf{y} = \mathbf{\Lambda}_y \eta + \epsilon,$$

where $\mathbf{\Lambda}_x$ and $\mathbf{\Lambda}_y$ are the coefficient matrices relating the observed variables to the latent factors, while $\boldsymbol{\delta}$ and $\boldsymbol{\epsilon}$ are the associated residual errors. The model includes also a third equation describing the structural part

$$\eta = \mathbf{B}\eta + \Gamma\xi + \zeta,$$

where $\boldsymbol{\Gamma}$ and $\mathbf{B}$ are the regression coefficient matrices among the latent factors and $\boldsymbol{\zeta}$ is the related error vector.

One of the crucial phase in the SEM construction consists in the specification of latent factors and their relationship pattern, as well as in the choice of observed variables to measure the factors. With regard to the former step, we define a model to describe causal relationships among the four exogenous latent constructs and the endogenous one, the dropping out. Although the four factors could be all considered exogenous, the primary school careers actually depends also on the context, the education system, and on family structures, as well as the school system is affected by the social, economic and demographic contexts. For these reasons, the specified model consists of two completely exogenous factors and three endogenous factors, two of them mediators. Figure 1 reports the corresponding theoretical path diagram.

3 Data Sources and Measurement Models' Specification

Once the theoretical structural model have been defined, the observed indicators for the measurements parts should be selected to completely specify the model. Due to the complexity of the phenomena and to the lack of information in some areas, the specification of measurement models required more attention than usual. As an example, consider that in the case of secondary school it is quite difficult to find detailed measures of dropping out in the official statistics (the Miur research of 2004 on this phenomenon reports data only for compulsory education).

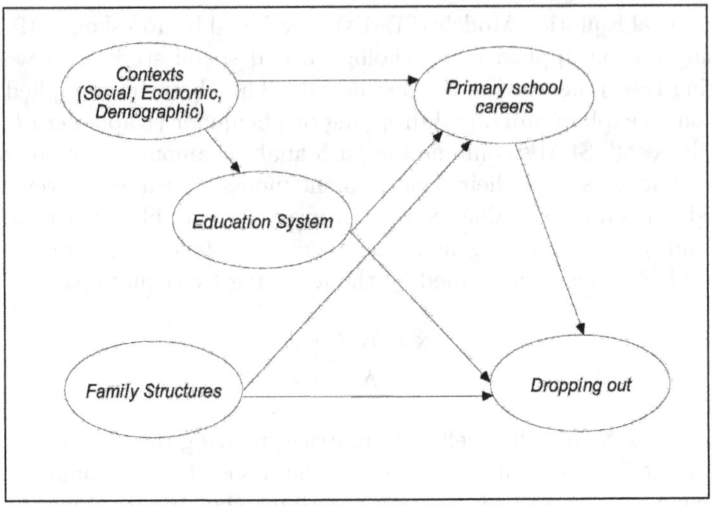

Fig. 1. Theoretical path diagram for dropping out in secondary school

Taking into account only Italian official statistics available to provincial level, given the 103 geographical units (i.e. the Italian Provinces), we selected 62 variables from the Social Indicator System and the 11th Italian Census from Istat, the Miur database and the Quality of Life Indicator System of Sole24ore. Hence, many indicators were available to define the latent factors.

However, given the relative small number of units, exploratory analyses were carried out to select a smaller variable set (for details see Ragozini and Vitale, 2005). In a first step, the bunch of indicators was split in coherent groups and principal component analyses were performed on the subsets in order to look at the global correlation structure and to select a reduced number of variables. In this phase, some expert knowledge about sociological theories on dropping out was also exploited.

In a second phase, we looked at correlation and scatter plot matrices to go deeper in the relationship structures among the phenomena under study. Following the results of the exploratory analyses, and aiming to select very few variables to gain parsimony, we defined seven factors for the final models. The measurement models' specification process took into account the structural model identification constrains.

For the social and economic context ($\boldsymbol{\xi}_1$) we introduced two economic indicators of province wealth and one for the presence of high cultural level population. Moreover, to better describe the social context we consider also the crime rate as a factor ($\boldsymbol{\xi}_2$). The family structures ($\boldsymbol{\xi}_3$) are measured through indicators related to separations and divorces. The education system is described by two mediator factors related to the teaching characteristics (η_1) and to the student enrollment in different curricula (η_2). We considered also one endogenous factor for dropping out in primary schools (η_3) measured by students that failed some years. Finally, concerning the target endogenous factor η_4, we used three indicators to describe it: the rate of students with one failure, the rate of students with more than one failure during all the course of study, and the rate of students that did not enroll to second grade.

The observed indicators selected for the final measurement models, along with their labels, are reported in Table 1.

Indicators for latent constructs	Labels	Model
Social and Economic Contexts	**Context**	ξ_1
Percentage of graduates	Cont1	X_1
Per Capita Added Value	Cont2	X_2
Employment rate	Cont3	X_3
Crime Rate	**Crime**	ξ_2
Crime Rate	Crime	X_4
Family structures	**Family**	ξ_3
Separation rate	Family1	X_5
Rate of new families after a divorce	Family2	X_6
Percentage of female single-parent families	Family3	X_7
Education System	**EduSyste**	η_1
Rate of teachers with tenureship less then 5 yrs in pr. sch.	Edu1	Y_1
Rate of teachers with tenureship less then 5 yrs in s. sch.	Edu2	Y_2
Ratio of teachers with and without tenureship	Edu3	Y_3
School Curricula	**TSchool**	η_2
Rate of enrollment of students in technical schools	TSchool	Y_1
Primary school career	**PSCareer**	η_3
Rate of students with more than 1 failure in primary school	Primary	Y_5
Dropping out in secondary school	**Dropout**	η_4
Rate of students with 1 failure	Drop1	Y_6
Rate of students with more than 1 failure	Drop2	Y_6
Rate of students that did not enroll to second grade	Drop3	Y_7

Table 1. Measurements models: observed indicators (*italic*) for exogenous and endogenous factor (**bold**) along with their labels and role in the model.

Given the selected indicators and the hypotheses discussed in the previous section, skipping the measurement equations, the structural model represented in Figure 1 can be expressed through the following four equations:

$$
\begin{aligned}
\eta_1 &= \gamma_{11}\xi_1 + \zeta_1 \\
\eta_2 &= \gamma_{21}\xi_1 + \beta_{21}\eta_1 + \zeta_2 \\
\eta_3 &= \gamma_{31}\xi_1 + \gamma_{32}\xi_2 + \gamma_{33}\xi_3 + \beta_{31}\eta_1 + \zeta_3 \\
\eta_4 &= \gamma_{41}\xi_1 + \gamma_{42}\xi_2 + \gamma_{43}\xi_3 + \beta_{41}\eta_1 + \beta_{42}\eta_2 + \beta_{43}\eta_3 + \zeta_4
\end{aligned}
$$

4 The Estimated Structural Equation Models

Before estimating the specified model, univariate and multivariate normality of variables should be verified, as SEM's methodology assumes the **x** and **y** variables to be normal. We assessed distributional assumptions through Jacque-Berra tests, and almost all the variables turned out to be not normal, with severe skewness and kurtosis, with some extreme values.

Under non-normal conditions, as in our case, asymptotically distribution free (ADF) estimators should be used, since they perform relatively better than maximum likelihood (ML) and generalized least squares (GLS) estimators (Browne, 1984; Satorra, 1990). However, the ADF method is not adequate to deal with small sample size. For this reason, we decided to adopt an alternative approach: first we used power transformations to obtain approximate univariate normality for all the variables, and then applied ML estimators on the transformed indicators.

It has to be noted that, even if after transformations univariate normality holds, multivariate normality is still not verified. Hence, to asses the estimated model goodness of fit, we looked at Normed Fit Index (NFI), Non-Normed Fit Index (NNFI) and Comparative Fit Index (CFI), because they perform better with respect to other indices for small sample size and under non-normality conditions (Lei and Lomax, 2005). Moreover, being chi-square statistics the least robust fit index with respect to deviation from normality, it was used only to compare different models, and to evaluate the effect of model parameter modifications (Bollen, 1989).

Finally, the structural equation model was estimated starting from the correlation matrix of the transformed measured variables through the LISREL 8.54 software. In Table 2, the estimated **B** and $\boldsymbol{\Gamma}$ coefficients are reported, while to compare the relative influence of factors, the standardized coefficients for dropping out phenomenon are shown on the path diagram in Figure 2.

	Γ coefficients			B coefficients		
	Context	*Crime*	*Family*	*TSchool*	*EduSyste*	*PSCareer*
TSchool	*n.s*	–	–	–	.62 (*3.22*)	–
EduSyste	.26 (*3.89*)	–	–	–	–	–
PSCareer	-.50 (*-3.16*)	.23 (*2.30*)	.45 (*2.87*)	–	.61 (*3.10*)	–
Dropout	-.46 (*-3.45*)	.30 (*3.81*)	.22 (*1.75*)	.18 (*2.76*)	.42 (*2.67*)	.26 (*3.53*)

Table 2. Γ and B unstandardize estimated coefficients (direct effects) for the specified model and related t-values (*italic*).

The analysis of the t-values of the estimated parameters reveals that all the $\boldsymbol{\Lambda}_y$ and $\boldsymbol{\Lambda}_x$ coefficients among the measurement variables and the latent constructs, and almost all the hypothesized regression coefficients between latent constructs $\boldsymbol{\Gamma}$ and **B** are statistically significant at least at 0.01 level. The γ_{21} coefficient of economic context on rate of enrollment in technical curriculum was removed from the model, turning out completely not significant ($t - value = -0.05$). On the other hand, the γ_{43} coefficient, even if not significant ($t - value = 1.75$), have been retained in the model for two reasons: first, the $t - value$ is quite close to the significance threshold, and second, it contributes to the significance of total effects we consider shortly. Then, the structural equations with the standardized estimated parameters are:

$$\begin{aligned}
EduSyste &= .47\,Context \\
TSchool &= .35\,EduSyste \\
PSCareer &= -.49\,Context + .23\,Crime + .45\,Family + .34\,EduSyste \\
Dropout &= -.59\,Context + .39\,Crime + .28\,Family + .24\,TSchool + \\
&\quad + .30\,EduSyste + .33\,PSCareer
\end{aligned}$$

Furthermore, to enrich the interpretation we consider the total effects that represent the sum of the direct effects of the simple path and the indirect effects of the compound paths (Table 3). They are all significant at least at 0.05 level. In particular, the factors related to the social and economic context (-.41) and the education system (.69) exert the stronger influence on dropping out . The crime rate (.36) and the family structure (.34) are also determinant, while the performances in primary school (.26) and the enrollment in technical schools (.18) present lower effects. The total effects on the other endogenous factors are also of interest: the economic context heavily determines the education system (.26) and the dropping out in primary school (-.34), while it has some influence on the choice of technical curriculum (0.16). High rates of legal separations and divorces, the education system, and the crime rate affect the primary school dropping out (with total effects respectively equal to .45, .62 and .23).

	$\boldsymbol{\Gamma}$ coefficients			**B** coefficients		
	Context	*Crime*	*Family*	*TSchool*	*EduSyste*	*PSCareer*
TSchool	.16 (*2.91*)	–	–	–	.62 (*3.22*)	–
EduSyste	.26 (*3.89*)	–	–	–	–	–
PSCareer	-.34 (*-2.24*)	.23 (*2.30*)	.45 (*2.87*)	–	.61 (*3.10*)	–
Dropout	-.41 (*-3.09*)	.36 (*4.25*)	.34 (*2.54*)	.18 (*2.76*)	.69 (*3.81*)	.26 (*3.53*)

Table 3. $\boldsymbol{\Gamma}$ and $\boldsymbol{B}$ unstandardize total effects for the specified model and related t-values (*italic*).

Summarizing then, the negative coefficients of the context indicates that in rich and developed areas the risk of pushing out students from secondary

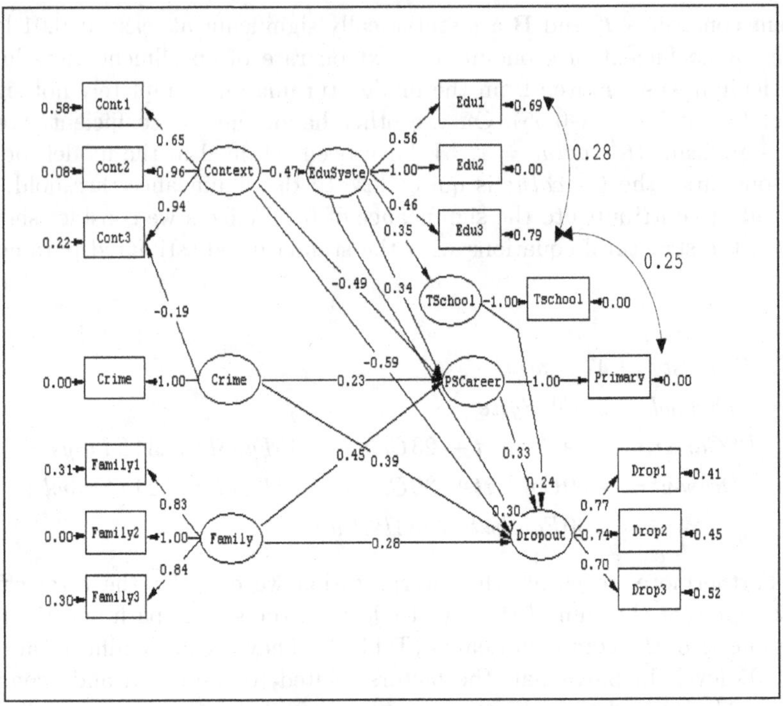

Fig. 2. Standardized coefficients of final structural model predicting dropping out in secondary school. Paths significant at least the .01 level. $\chi^2 = 144.47$ (77 *df*); $RMSEA = 0.094$; $CFI = .93$; $NFI = .88$; $NNFI = .91$.

schools is reduced. The other relationships of the model were all positive indicating that an education system poor in terms of high occupational mobility of teachers, an high presence of students enrolled in technical schools, an high crime rate, a family with internal relational problems, and, finally, high rate of failures in primary school increase the dropping out in secondary school.

5 Further Developments and Concluding Remarks

Although the proposed model do not exhaust the possible causal relationships, its results offered useful interpretations to the policy makers involved in the research program.

By a statistical point of view, besides the issues related to small sample size and non-normal distributions, we want to stress the role of preliminary exploratory data analyses. In our case, they allowed us the variable selection and showed the presence of subregional differences. As an example, considering only Southern Italy Provinces, the scatterplot matrices highlighted that the average family size and the incidence of large families present stronger relationships with dropping out with respect to the factors included in the national model. It would have been interesting to develop a model including regional submodels. However, since there are only 36 Southern Provinces, any structural submodel could be carried out. Probably, in such a case path analysis or PLS path modeling (Tenenhaus *et al.*, 2005) could be more feasible and effective.

Furthermore, even if the parameters in our structural model are identifiable and the solutions are coherent with the sociological hypothesis, the problem of indeterminacy of the factor scores of the latent variables still remains. This issue have been investigated by several authors starting from Guttman (1955) and different methods are described in literature to obtain unique solutions (among the others Schönemann and Steiger, 1976; Nooan and Wold, 1982 ; Vittadini and Haagen, 1991).

On the other hand, the following three sufficient conditions for the uniqueness under factor rotation are investigated in Bollen and Jöreskog (1985): *i)* $\boldsymbol{\Phi}$, the covariance matrix of $\boldsymbol{\xi}$ is a symmetric positive definite matrix with diag($\boldsymbol{\Phi}$)=$\mathbf{I}$; *ii)* $\boldsymbol{\Lambda}$ has at least $k-1$ fixed zeros in each column, with k is the number of latent variables; *iii)* $\boldsymbol{\Lambda_s}$ has rank $k-1$, where $\boldsymbol{\Lambda_s}$, $s = 1, \ldots, k$ is the submatrix of $\boldsymbol{\Lambda}$ consisting of the rows of $\boldsymbol{\Lambda}$ which have fixed zero elements in the s column. In our model, the first two condition are satisfied for both the $\mathbf{x}$ and $\mathbf{y}$ measurement models, while the third are not realized.

In order to fulfill also the condition *iii)*, modifications of the final model introducing new λ coefficients are under study, as well as the possibility of measuring the indeterminacy of the proposed model through some appropriate index (see e.g. Vittadini, 1989).

Acknowledgment

Authors had support from the research program: "Analysis of inequalities cultural capital and social inclusion due to differences in social classes and subregional areas", Action 3.6 a of P.O.R. CAMPANIA 2000-2006 (P. Clarizia and A. Spanó managers), and by Miur PRIN 2003 national grant #2003130242: "Multivariate Statistical and Visualization Methods to Analyze, to Summarize, and to Evaluate Performance Indicators" (M.R. D'Esposito manager).

References

BENVENUTO, G., RESCALLI, G. and VISALBERGHI, A. (2000): *Indagine sulla dispersione scolastica.* La Nuova Italia, Firenze.

BOLLEN, K.A. (1989): *Structural Equation with Latent Variables.* John Wiley, New York.

BOLLEN, K.A. and JÖRESKOG, K.G. (1985): Uniqueness Does Not Imply Identification, *Sociological Methods and Research, 14, 155–163.*

BROWNE, M.W. (1984). Asymptotically distribution-free methods for the analysis of covariance structures *British Journal of Mathematical and Statistical Psychology, 37, 1–21.*

CLARIZIA, P. and SPANÓ, A. (2005): La survey sulle famiglie. In: A. Spanó and P. Clarizia (Eds): *Dentro e fuori la scuola. Percorsi di abbandono e politiche di contrasto.* Giannini, Napoli, *163–255.*

FARMER, T.W., ESTELL, D.B., LEUNG, M.C., TROTT, H., BISHOP, J. and CAIRNS, B.D. (2003): Individual characteristics, early adolescent peer affiliations, and school dropout: an examination of aggressive and popular group types, *Journal of School Psychology, 41, 217–232.*

GUTTMAN, L. (1955): The determinacy of factor scores matrices with implications for five other basic problems of common factor theory, *British Journal of Statistical Psychology, 8, 65–81.*

HAAGEN, K. and VITTADINI, G. (1991), Regression Component Decomposition in Structural Analysis, *Communications in Statistics, 20, 1153–1161.*

JIMERSON, S., EGELAND, B., SROUFE, L.A. and CARLSON, B. (2000): A Prospective Longitudinal Study of High School Dropouts Examining Multiple Predictors Across Development, *Journal of School Psychology, 38, 525–549.*

JÖRESKOG, K.G. (1970): A General Method for Analysis of Covariance Structures, *Biometrika, 57, 239–251.*

KAPLAN, D.S., PECK, B.M. and KAPLAN, H.B. (1995): A Structural Model of Dropout Behavior: A Longitudinal Analysis, *Applied Behavioural Science Review, 3, 177-193.*

LEI, M. and LOMAX, R.G. (2005): The Effect of Varying Degrees of Nonnormality in Structural Equation Modeling, *Structural Equation Modeling, 12, 1–27.*

LIVERTA SEMPIO, O., GONFALONIERI E. and SCARATTI G. (Eds.) (1999): *L'abbandono scolastico, aspetti culturali, cognitivi, affettivi.* Raffaello Cortina Editore, Milano.

M.I.U.R. (2002): Indagine campionaria sulla dispersione scolastica nelle scuole statali elementari, medie e secondarie superiori: Anno scolastico 2001-2002. Roma, www.istruzione.it.

NOOAN, R. and WOLD, H. (1982): PLS Path Modeling with indirectly observed variables, in *Systems under indirect observation: causality, structure, prediction, JÖRESKOG K.G and WOLD H. Eds.*, North-Holland, Amsterdam, *263–270.*

ORTHNER, D.K. and RANDOLPH, K.A. (1999): Welfare Reform and High School Dropout Patterns for Children, *Children and Youth Services Review, 21, 881–900.*

PREVATT, F. and KELLY, F.D. (2003): Dropping out of school: A review of intervention programs, *Journal of School Psychology, 41, 377–395.*

RAGOZINI, G. and VITALE, M.P. (2005): L'analisi della dispersione scolastica in italia attraverso i modelli causali. In: A. Spanó and P. Clarizia (Eds): *Dentro e fuori la scuola. Percorsi di abbandono e politiche di contrasto.* Giannini, Napoli, *9–91.*

SATORRA, A. (1990): Robustness issues in structural equation modeling: a review of recent developments, *Quality & Quantity, 24, 367–386.*

SCHÖNEMANN, P.H. and STEIGER, J.H. (1976): Regression component analysis, *British Journal of Mathematical and Statistical Psychology, 29, 175–189.*

TENENHAUS, M., ESPOSITO VINZI, V. , CHATELIN, Y. and LAURO, C. (2005): PLS path modelling, *Computational Statistics and Data Analysis, 48, 159–205.*

VITTADINI, G. (1989): Indeterminacy Problems in the Lisrel Model, *Multivariate Behavioral Research, 24, 397–414.*

Testing Procedures for Multilevel Models with Administrative Data

Giorgio Vittadini[1], Maurizio Sanarico[2], and Paolo Berta[3]

[1] Dipartimento di Statistica,
Università Milano-Bicocca, Italy
giorgio.vittadini@unimib.it
[2] Noustat s.r.l.
m.sanarico@noustat.it
[3] C.R.I.S.P.
crisp.statistica@unimib.it

Abstract. Recent Relative Effectiveness studies of the Health Sector have strongly criticized hierarchical ranking in hospitals. As an alternative, they propose a multifaceted approach which evaluates the quality and characteristics of Hospital services. In this direction, the use of administrative data has proven highly useful. This data is less precise than clinical data but performs more effectively in describing general situations. The numerosity of the population renders all the parameters significant in linear model tests. We must therefore utilize resampling schemes in order to verify the hypotheses concerning the significance of the parameters in opportunely drawn subsamples.

1 Hospital Effectiveness with Administrative Data

Several recent statistical papers deal with risk-adjusted comparisons on the basis of mortality or morbidity outcomes corrected by means of Multilevel models in order to take into account different case-mix of patients (Goldstein, H. and Spiegelhalter (1996)). These papers, accurate from the methodological point of view, are all based on small samples of patients with particular pathologies. Other medical papers propose risk-adjusted comparisons as a method for evaluating effectiveness of health structures (Iezzoni (1997)). Moreover, in some countries private or public External Health Agencies gather, ad-hoc, larger data sets and use linear and logistic models in order to validate quality indicators (AHRQ (2003), JCAHO(2004)). In other cases, they benchmark health structures by means of risk-adjusted comparisons (CIHI(2003), NHS(2004)). Recently, this use of risk adjusted comparisons for benchmarking health structures has been strongly criticized (Lilford et al. (1994)). In particular, it has been stated that: "The sensitivity of an institution's position in league tables to the method of risk adjustment used suggests that comparisons of outcomes are unlikely to tell us about the quality of care". Therefore it has been suggested that "The agencies should facilitate the development and dissemination of a database for best practice and improvement based on the results for primary and secondary research."(Lilford et al.

(1994)). In this direction, the use of administrative data, used by payors to pay bills and manage operations, can be very useful. In fact: "These data are typically computerized, making it easy to collect and use large quantities of information Administrative data have been used to examine geographic variation in utilization of surgical and medical procedures, monitor the use of health services, assess the effects of a policy change on health expenditures, evaluate the relationships between hospital death rates and hospital characteristics"(Damberg et al.(1998)). Which problems do linear models and, in particular, multilevel models involve when they are used with administrative data? Besides problems connected with the accuracy of data (i.e. coding accuracy, timing of diagnoses uncertainty etc.) (Damberg (1998)), there is a relevant methodological problem. It is known that when there are large data sets the significance tests associated with linear models refuse the null hypothesis in all cases. In fact the sample size influences the results, and beyond a certain threshold, it is the only determining factor of the test. Every explicative variable seems to be significant for explaining the outcomes, and this result is particularly misleading for the topics mentioned above. Therefore we need appropriate testing procedures able to verify hypotheses regarding the significance of explicative variables in samples drawn from the population associated with administrate data. In general terms, we must devise inference methods in heterogeneous samples collected from large data sets (Duncan(1998)). The conclusion obtained for tests connected with the Multilevel Model used for the evaluation of healthcare institutions, can be generalized for Linear models.

2 The Model

Let us consider a number of outcomes obtained from hospital discharge forms. These outcomes are binary variables and due to the hierarchical structure of **n** observations in a Logistic Multilevel Model. Therefore given the variable $\mathbf{Y}_{ij}$ = Bin (n, π_{ij}), we fit the following model

$$\begin{aligned} Logit(\pi_{ij}) = f(\mathbf{X}, \mathbf{Z}) & \\ &= \gamma_{00} + \gamma_{10}\mathbf{X}_{ij} + \gamma_{01}\mathbf{Z}_j \\ &+ \gamma_{11}\mathbf{X}_{ij}\mathbf{Z}_j + u_{1j}\mathbf{X}_{ij} + u_{0j} + \varepsilon_{ij} \end{aligned} \tag{1}$$

$$\pi_{\mathrm{ij}} = \frac{1}{1 + \exp(-f(\mathbf{X}, \mathbf{Z}))} \tag{2}$$

Before the estimation of the model, we applied a process aimed at optimising the univariate relationship between the outcomes and the predictors. Such a process consisted of a discretization based on the automatic identification of the linear intervals in each relationship, assigning an indicator variable to each interval. In this way, non-linearities were captured and modelled, thereby enhancing the expressive power of the independent variables. The model building initiates with a model including only the dummies; in this phase we facilitated a chunk elimination; the second step consisted in adding all the remaining predictors. A backward elimination completed the model building process. A positive side effect of the transformation process was the weakening of the evidence calling for the inclusion of interactions. A possible explanation of this fact is that most of the variability is retained by the main effects after transformation, and that non linearity and second-order interaction compete for the same information.

3 Problems Involved in Tests with Large Data Sets

Because of the high number of degrees of freedom, all first level effects turned out to be significant, whereas this is not true in the case of the hospital level effects. This result is inherent in the classical estimation procedure, designed for small to medium samples (at most some thousands), and it is due to the indefinite narrowing of the standard errors as sample increases. The imbalance between the first and second level analysis has been overcome by utilizing an empirical testing procedure looking first at the size of effects, then at their relative variability (gauged by confidence intervals) and finally the statistical significance of the effects. This procedure adheres to the following logic: first evaluate the practical importance, then the extent of statistical variability associated with the effect (uncertainty), then the significance is used as a standard measure of the deviation to a null effect. The next step in the present work tries to go beyond what is described above. Statistical inference based on very large sample (> 100.000), containing many heterogeneous groups, leads to irrelevance of statistical testing because of the exceeding power. We think that the real interest is to disentangle the complex data structure. In summary "failing to reject the null hypothesis" is not the same as "accepting the null hypothesis" or as "rejecting the alternative hypothesis" because of the large size of the sample, the null hypothesis is rejected but this does not mean that the alternative hypothesis of significance is accepted.

4 The New Proposal

We propose a scheme of analysis in which we first attempt to discover and represent the heterogeneity, then we model the detailed data incorporating the structure which has emerged, performing the inference using a number of competing approaches: conducting the analysis within each sub sample, patching together the results using and comparing a standard approach, a Bayesian approach, resampling-based approaches. The standard approach has many drawbacks: in this paper it is used to provide a reference point. The Bayesian approach, by modelling the probabilities directly, seems to be immune to the problems discussed above (Albert and Chib(1951)). However, apart from being computationally very expensive, it is not clear how this approach performs in the presence of large samples. The structured resampling-based approach is an attempt to overcome the problems described by combining resampling-based techniques (for example, various versions of Bootstrap or Boosting) and a representation of the heterogeneity in the sample which can be obtained either by a data-driven approach (useful for achieving a statistically representative analysis of the heterogeneity) or by a knowledge-driven approach (useful to test hypotheses or to explore specific well-identified sub-samples)(Di Ciccio and Efron (1996), Efron (1996)). Multilevel models represent the statistical relationships existing between a given dependent variable or response function and a set of predictors, taking into account the objects of different size to which such predictors are associated and the relationships (most of the time hierarchical) between these objects. This allows taking into account the different sources of variability in the data correctly. However the multilevel paradigm is not able to capture all of the variability and heterogeneity in the data. For example, it is not able to explain the heterogeneous behaviour of a given agent (hospital) considered from the response function conditional to the set of predictor, relative to other similar agents. Being alike after modelling out the variability associated to the multi-level model amount, they are probably alike from a managerial point of view. Given this, we considered a more complementary approach in which no predefined structure was super-imposed on the observations. This methodology has been called cluster-weighted modelling or soft-clustering. The idea is to combine the results in order to highlight cases that behave in a well-characterized way (belonging to a single domain of influence or cluster) and cases whose response has characteristics partially shared by more than one cluster. In the final part of this section, we define soft-clustering methodology in further detail. The clusters do not interact or describe the data locally with respect to the maximum of the joint probability. There is no prior information (an arbitrary cost function E_{ij}, is used to express the energy associated to z_i in the cluster $\mathrm{C}_j$with centre μ_j^z); an iterative process with many clusters utilized to achieve a satisfactory partitioning of the data space through a sequential fusion of the clusters. The probability that point $x_i \in C_j$ belongs to cluster C_j is expressed by $\mathrm{p}_{ij.}$. The total average cost is therefore:

$$< E >= \sum_{j=1}^{M} \sum_{i=1}^{N} p_{ij} E_{ij} \tag{3}$$

Equation (3) acts as a boundary condition to the data distribution. To find a stable distribution we follow the maximum entropy principle during each step of the iterative process. The p_{ij} which maximizes entropy is:

$$H = -\sum_{j=1}^{M}\sum_{i=1}^{N} p_{ij} \log (p_{ij}) \qquad \begin{cases} \sum_{i=1}^{N} p_{ij} = 1 \\ \sum_{i=1}^{N}\sum_{j=1}^{M} p_{ij}E_{ij} =< E >= cost \end{cases} \tag{4}$$

the Boltzmann distributions are:

$$p_{ij} = \frac{\exp\left[-\beta\, E_{ij}\right]}{Z_j} \qquad Z_j = \sum_{i=1}^{N} \exp\left(-\beta\, E_{ij}\right) \tag{5}$$

where the distribution function β is a Lagrange multiplier. Using a thermodynamic analogy, if $\beta \propto 1/T$, where T is the a "temperature" of the system, with increasing β, the system tends to be frozen and only the closer points influence each other. With decreasing β we have a more disordered system (observations have a higher degree of interaction). The assumption of independence between the clusters and the p_{ij} of different clusters allows us to define the free energy F_j for the cluster C_j:

$$F_j = -\frac{1}{\beta} \log Z_j \qquad \frac{\partial F_j}{\partial \mu_j^2} = 0 \qquad \forall j \tag{6}$$

Considering the squared euclidean distance:

$$E_{ij} = |z_i - \overrightarrow{\mu_i}|^2 = |y_i - \mu_i^y|^2 + |x_i - \mu_i^x|^2 \tag{7}$$

we obtain:

$$\mu_j^z = \sum_{i=1}^{N} \frac{z_i \exp\left[-\beta\left(z_i - \mu_j^z\right)^2\right]}{\sum_{i=1}^{N} \exp\left[-\beta\left(z_i - \mu_j^z\right)^2\right]} \tag{8}$$

Equation (8) cannot be solved analytically. A solution can be obtained through fixed-point iteration of the following formula:

$$\mu_j^z(n+1) = \sum_{i=1}^{N} \frac{z_i \exp\left[-\beta\left(z_i - \mu_j^z(n)\right)^2\right]}{\sum_{i=1}^{N} \exp\left[-\beta\left(z_i - \mu_j^z(n)\right)^2\right]} \tag{9}$$

which is typically iterated until a stable μ_j^z is obtained. The process converges to a local minimum with respect to specified initial conditions and β, which reflects the number of clusters used to represent the data. This topic, currently under investigation, is the application of the cluster-weighted modelling of all data, taking into account the patient as well as the hospital level, constructing a multi-level cluster weighted framework of analysis, able to answer some very interesting questions without the necessity of using ad-hoc procedures. Below, we give a brief account of the theory. Let us consider the data vectors as $\{\mathbf{y}_n, \mathbf{x}_n, \mathbf{z}_n\}$, where $\mathbf{y}$ is the response function, $\mathbf{x}$ the level-1 variables and $\mathbf{z}$ the level-2 variables. We then infer $p(\mathbf{y}, \mathbf{x}, \mathbf{z})$ as the joint probability density. This density is expanded over a sum of cluster C_k, each cluster containing an input distribution, a local model, and an output distribution. The input distribution is:

$$
\begin{aligned}
p(\mathbf{y},\mathbf{x},\mathbf{z}) &= \sum_{k=1}^{K} p(\mathbf{y},\mathbf{x},\mathbf{z},c_k) = \sum_{k=1}^{K} p(\mathbf{y},\mathbf{x},\mathbf{z}|c_k)p(c_k) = \\
&= \sum_{k=1}^{K} p(\mathbf{y}|\mathbf{z},\mathbf{x},c_k)p(\mathbf{x}|\mathbf{z},c_k)p(\mathbf{z}|c_k)p(c_k)
\end{aligned} \tag{10}
$$

with the normalization condition $\Sigma_n p(c_k)=1$. In the presence of both discrete and continuous predictors we must further partition them accordingly. The next point is to associate a specific density to each of the terms in the formula. Normally, the conditional distributions $p(\mathbf{x}|\mathbf{z},c_k)$ and $p(\mathbf{z}|c_k)$ are taken to be Gaussian distributions with appropriate covariance matrices (for example, diagonal or structured). The output distribution $p(\mathbf{y}|\mathbf{x},\mathbf{z},c_k)$ depends on whether $\mathbf{y}$ is continuous-valued or discrete, and on the type of local model connecting $\mathbf{x},\mathbf{z}$ and y: $f(\mathbf{x},\mathbf{z},b_k)$. In most cases a linear function is enough, given the composition of many local functions (as many as required by the data distribution), to represent complex nonlinear functions.

5 An Application

The study is based on the administrative data provided by the Lombardy Regional Health Care Directorate regarding 1.152.266 admissions to 160 hospitals. The data consists of: regional population anagraphical records, Administrative Hospital Discharge records and hospitals' structural characteristics. Response variables are: in-hospital and post-discharge mortality, patient's discharges against medical advice, transfers to other hospitals, unscheduled hospital re-admissions, unscheduled returns to operating room. Patients' case mix and hospitals' characteristics are also collected from the same sources. We use a logistic Multilevel model to investigate best and worse practices of hospitals connected with their characteristics (i.e.: size, private vs public status, general vs specialized, etc.). The test procedures mentioned above are used in order to evaluate the significance of parameters related to explicative variables in the context of large populations. Multilevel models produce a variety of useful results, and in the Health Care Effectiveness Evaluation context level-2 residuals are particularly important. In the present case, we have 160 residuals, one for each hospital, with a considerable level of heterogeneity, indicating either a possible difference in managerial effectiveness, or some other source of variability. Is there further information, perhaps at a higher level, not-well defined at the sampling design stage, that can reasonably account for a significant portion of the level-2 residual variation? Is it possible to individualize combinations of conditions associated with specific portions of the level-2 residual distribution? In order to answer to these questions, a simple but effective strategy identifies a set of variables able to determine the aforementioned patterns. Then, we apply a multiple comparison Bonferroni adjusted procedure to identify their statistical significance. Three criteria are used: the effect size, the adjusted p-value and the standard errors (confidence intervals) of the effect. In the analysis we also include second-order interaction effects. At this point we pursue an integrated approach in combining the Multilevel and Cluster weighted models. Therefore we present an outline of a practical application of the association between Multilevel modelling and cluster-weighted clustering, namely, to apply the soft-clustering to level-2 residuals to obtain automatically the main patterns of variation in the joint distribution of residuals. As a further interesting result, we obtain the probability of each hospital to belong to each cluster. This allows us to discern well-characterized hospitals from other less-clearly defined ones. The case study concerns the soft-clustering analysis for mortality rate within 30 days after discharge outcome. The procedure detected 4 clusters. All the observations were well-characterized by a single cluster with one exception, as already mentioned. The tables below show an analysis of variance on these clusters and their composition.

Cluster	N Obs	Variable	Mean	Minimum	Maximum	Std Dev	Median
cluster0	55	PRIV	1.00	1.00	1.00	0.00	1.00
		IRCCS	0.00	0.00	0.00	0.00	0.00
		DEAS	0.16	0.00	1.00	0.37	0.00
		PRS	0.15	0.00	1.00	0.36	0.00
		RR	1.00	0.24	3.05	0.59	0.06
cluster1	24	PRIV	0.00	0.00	0.00	0.00	0.00
		IRCCS	0.25	0.00	1.00	0.44	0.00
		DEAS	0.00	0.00	0.00	0.00	0.00
		PRS	0.00	0.00	0.00	0.00	0.00
		RR	1.21	0.24	0.175	0.05	1.10
cluster2	49	PRIV	0.02	0.00	1.00	0.14	0.00
		IRCCS	0.08	0.00	1.00	0.28	0.00
		DEAS	1.00	1.00	1.00	0.00	1.00
		PRS	0.00	0.00	0.00	0.00	0.00
		RR	1.16	0.27	2.17	0.41	1.11
cluster3	32	PRIV	0.03	0.00	1.00	0.18	0.00
		IRCCS	0.03	0.00	1.00	0.18	0.00
		DEAS	0.00	0.00	0.00	0.00	0.00
		PRS	1.00	1.00	1.00	0.00	1.00
		RR	1.25	0.39	2.56	0.55	1.23

Parameter		Estimate	Standard Error	t Value	Pr > |t|
Intercept		-.3277484344	0.20144526	-1.63	0.7347222
Cluster	cluster0	0.0340194876	0.37985182	0.09	6.45
Cluster	cluster1	0.5136519229	0.22580980	2.27	0.16875
Cluster	cluster2	-.1232972155	0.25478708	-0.48	4.36875
Cluster	cluster3	0.0000000000	.	.	.
DEAS		0.6099373653	0.17460280	3.49	0.0006
PRS		0.4559875042	0.18319776	2.49	0.0965278
IRCCS		-.4917529255	0.16183465	-3.04	0.0028
UNI		-.1586361703	0.14348553	-1.11	1.8798611
PRIV		-.0270973074	0.34221433	-0.08	6.5069444

The risk within clusters presents a significant heterogeneity. The presence of DEAS and emergency units remain significant in explaining the heterogeneity. Another interesting result is that most of the hospitals with a given profile have similar patterns of residual variation (hospital-specific effectiveness), whereas some are deviate from this norm. This is a significant finding, allowing the proposal of innovations for the improvement in Quality for specific health care facilities.

References

AHRQ (2003): *Guide to Inpatient Quality Indicators.* www.ahrq.gov/dat/hcup.

ALBERT, J. and CHIB, S. (1951): Bayesian Tests and Model Diagnostics in Conditionally Independent Hierarchical Models. *Journal of the American Statistical Association, 92, 916–925.*

CIHI (2003): *Hospital Report 2002, Acute Care Technical Summary.* secure.cihi.ca/cihiweb/splash.html.

DAMBERG, C. and KERR, E. A. and McGLYNN, E. (1998): Description of data Sources and Related Issues. In: E. A. McGlynn, C. L. Damberg, E. A. Kerr and R.H. Brook (Eds.): *Health Information Systems, Design Issues and Analytical Applications.* RAND Health Corporation, 43–76.

DI CICCIO, T. J. and EFRON, B.(1996): Bootstrap Confidence Intervals. *Statistical Science, 11, 189–212.*

DUNCAN, C., JONES, K. and MOON (1998): Context, Composition and Heterogeneity: Using Multilevel Models in Health Research. *Social Science and Medicine, 46, 97–117.*

EFRON, B.(1996): Empirical Bayes Methods for Combining Likelihoods. *Journal of the American Statistical Association, 91, 538–565.*

GOLDSTEIN, H. and SPIEGELHALTER, D. J. (1996): League Tables and Their Limitations: Statistical Issues in Comparisons of Institutional Performance. *Journal of the Royal Statistical Society A, 159, 385–443.*

IEZZONI, L.I. (1997): The risk of Risk Adjustment. *JAMA, 278, 1600–1607.*

JCAHO (2004): www.jcaho.com.

LILFORD, R., MOHAMMED, M.A., SPIEGELHALTER, D. J. and THOMSON, R. (1994): Use and Misuse of Process and Outcome Data in Managing Performance of Acute Medical Care: Avoiding Institutional Stigma. *The Lancet, 364, 1147–1154.*

NATIONAL HEALTH SERVICE, CHI (2004): *Commentary on Star Ratings 2002-2003.* www.chi.nhs.uk./ratings.

Multidimensional Versus Unidimensional Models for Ability Testing

Stefania Mignani[1], Paola Monari[1], Silvia Cagnone[1], and Roberto Ricci[1]

Dipartimento di Scienze Statistiche "Paolo Fortunati", Italy
{mignani,monari,cagnone,rricci}@stat.unibo.it

Abstract. Over last few years the need for an objective way of evaluating student performance has rapidly increased due to the growing call for the evaluation of tests administered at the end of a teaching process and during the guidance phases. Performance evaluation can be achieved busing the Item Response Theory (IRT) approach. In this work we compare the performance of an IRT model defined first on a multidimensional ability space and then on a unidimensional one. The aim of the comparison is to assess the results obtained in the two situations through a simulation study in terms of student classification based on ability estimates. The comparison is made using the two-parameter model defined within the more general framework of the Generalized Linear Latent Variable Models (GLLVM) since it allows the inclusion of more than one ability (latent variables). The simulation highlights that the importance of the dimensionality of the ability space increases when the number of items referring to more than one ability increases.

1 Introduction

In the Italian educational system, the raised level of formative requirement needs a particular consideration in the assessment and evaluation field. Over the last few decades, the assessment issue has acquired new facets such as the self-evaluation, the measurement of the level of a skill and the effectiveness of a teaching process. The need for statistical tools for developing large-scale testing had greatly increased.

If one is interested in evaluating the abilities acquired by a student in a given phase of a learning process it may be opportune to focus attention on the problem of assessing individual performance. That is, one has to conceive it as the manifest expression of a set of latent traits underlying the cognitive process. Several disciplines are involved here: psychology, investigation topics, statistics, computer science, and so on. Each expert contributes to fine-tuning the evaluation tool, which cannot be efficient if certain aspects are neglected. The evaluation of students' learning and competence can be typically carried out by administering a questionnaire containing a set of items relating to the analyzed skills. The responses to the items can be used to estimate the student's ability that is assumed to be a latent trait, namely, a facet that is not directly observable. To this aim, a model that expresses the latent traits in function of the answers to the items can be determined (Mignani *et al.*,

2005; Cagnone *et al.*, 2005). What requires investigation is the question of whether each item involves only one or more abilities. In literature, the latent space is referred to as complete if all latent traits influencing the student's performance have been specified. In real situations there are always several cognitive factors affecting test results. These factors might include different levels of motivation, anxiety, readings skills, and so on. However, under the assumption of unidimensionality, one factor is taken as dominant and is referred to as the ability evaluated by the test. Items are often related to a set of abilities and their evaluation in a unidimensional skill space can limit questionnaire validity. This aspect has to be carefully analyzed in order to assess whether the loss of information following a reduction in the ability space dimension is not excessively large. A unidimensional latent space is often assumed as it enhances the interpretability of a set of test scores.

The aim of this work is to compare a multidimensional model with a unidimensional one. The former is defined on a multidimensional ability space and allows us to take into account the complexity of the learning process. Nevertheless, the latter is more feasible in practical terms, although it depicts the phenomenon analyzed in less detail. The concordance and discordance of the two models are discussed in terms of student classification.

2 Measuring Students Abilities

The methodology used to evaluate the individual ability in this paper is the Item Response Theory (IRT) (Lord and Novick, 1968), which was introduced into the educational test field in order to deal with the case in which student ability is measured using a questionnaire. The main feature of IRT is that it simultaneously permits evaluation the characteristics of the item (difficulty, discrimination power and guessing) and student performance. As ability is not directly observable and measurable, it is referred to as a latent trait. An important feature of the IRT models is the possibility of using estimates obtained to compare students that have not necessarily answered to same items. The only important theoretical requirement is that the examinees receive questionnaires that measure the same ability (Van der Linden and Hambleton, 1997).

Unidimensional IRT models are usually useful when tests are designed to measure only one trait. However, cognitive processes are typically more complex and therefore it may be appropriate to introduce IRT defined in the multidimensional ability space. Indeed, students demonstrate that they have a wide variety of cognitive skills. Furthermore, some test tasks are sensitive to different kind of skills.

In principle, it is possible to introduce several mathematical models that are used to describe the relationship between an examinee's psychological trait and his or her responses. As reported in the literature, when a multidimensional phenomenon is analyzed using a unidimensional model, incorrect inferences regarding a student's proficiency in a given subject may arise (Reckase, 1997). The effects of multidimensionality are even more important when the student classification is obtained by referring to ability estimates. The classification criteria have to be connected to the aim of evaluation in different steps of the formative process: guidance, entrance, and the end of a course. In this work, the impact of the multidimensionality on a classification criterion is explored through a simulation study. The IRT model considered is located within the GLLVM (Bartholomew and Knott, 1999), a general framework that makes it possible to simultaneously treat more abilities.

3 The Model

The GLLVM represents a general framework within which different statistical methods are conveyed. In particular, if the observed variables are categorical, as in the case of the items of a questionnaire, it is restricted to the IRT approach. However, a substantial feature of the GLLVM that makes it more appealing than the IRT, is that the latent trait is treated as a multidimensional random normal variable so that more abilities underlying the learning process can be investigated (Cagnone and Ricci, 2005).

This different way of considering the latent abilities leads also to a different way of estimating the model. More details are given in (Bartholomew and Knott, 1999; Moustaki and Knott, 2000).

In order to describe the GLLVM for binary data, consider $\mathbf{y} = (y_1, \ldots, y_p)$ as the vector of p observed binary variables (binary items) and $\boldsymbol{\theta} = (\theta_1, \ldots, \theta_q)$ as the vector of q independent latent traits. We define $P(y_i = 1|\boldsymbol{\theta}) = \pi_i(\boldsymbol{\theta})$, for $i = 1, \ldots, p$ and $y_r = (0, 1, 1, 0, \ldots, 1)$ the r-th complete p-dimensional response pattern, that is the set of the possible responses given to the p items.

The probability of y_r can be expressed as follows

$$\pi_r = \int_{-\infty}^{+\infty} \cdots \int_{-\infty}^{+\infty} g(\mathbf{y} \mid \boldsymbol{\theta}) f(\boldsymbol{\theta}) \mathrm{d}\boldsymbol{\theta}, \tag{1}$$

where $f(\boldsymbol{\theta})$ is assumed to be standard multivariate normal distributed and $g(\mathbf{y} \mid \boldsymbol{\theta})$ is assumed to be Bernoulli distributed

$$g(\mathbf{y} \mid \boldsymbol{\theta}) = \prod_{i=1}^{p} g(y_i \mid \boldsymbol{\theta}) = \prod_{i=1}^{p} \pi_i(\boldsymbol{\theta})^{y_i} [1 - \pi_i(\boldsymbol{\theta})]^{1-y_i}, \tag{2}$$

The second equality in (2) is obtained by assuming the conditional independence of the observed variables given the latent traits.

The relation between the observed variables and the latent traits is expressed through a logit link function so that:

$$\text{logit}(\pi_i) = \alpha_{i0} + \sum_{j=1}^{q} \alpha_{ij}\theta_j. \tag{3}$$

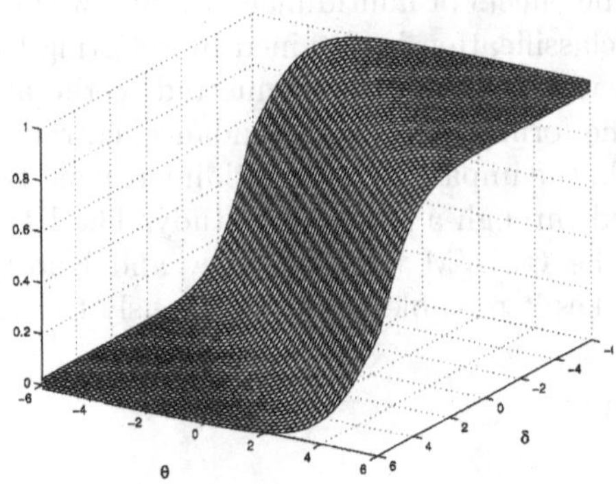

Fig. 1. The two dimensional model

where the parameter α_{i0} can be interpreted as difficult parameters, representing the trait level necessary to respond above 0.50 probability. α_{ij} is the discrimination parameter since the bigger α_{ij} the easier it will be to discriminate between a pair of individuals a given distance apart on the latent scale. Figure 1 represents a two dimensional model, for α_{i0} and α_{ij} fixed.

The model thus defined can be viewed as a generalized linear model with latent variables, $g(\mathbf{y}|\boldsymbol{\theta})$ being a member of the exponential family and the link function being the logit of the probability associated to a correct answer.

Model parameters are estimated using the maximum likelihood estimation by an E-M algorithm. At step M of the algorithm, a Newton-Raphson iterative scheme is used to solve the non-linear maximum likelihood equation.

To score the individuals on the latent dimension we can refer to the mean of the posterior distribution of $\boldsymbol{\theta}$ defined as:

$$\hat{\boldsymbol{\theta}}_h = E(\boldsymbol{\theta}|\mathbf{y}_h) = \int \boldsymbol{\theta} f(\boldsymbol{\theta}|\mathbf{y}_h)\mathrm{d}\boldsymbol{\theta} \tag{4}$$

where $h = 1, \ldots, n$ and n is the sample size.

4 Unidimensional Versus Bidimensional Psychometric Models

The choice of either a unidimensional or a multidimensional scheme depends on the aims of the evaluation. From a psychometric point of view, situations representing different learning scheme can occur.

In this paper we consider three conditions of a bidimensional structure that, although not exhaustive, can be viewed as a starting point of a more complex and more complete analysis than in the unidimensional case. More specifically, a structure referring to three possible alternatives has been considered:

1. each items is causally related to both the abilities (bidimensional items)
2. some items are related to one ability, some items are related to both the abilities (bidimensional and unidimensional items)
3. each item is related to one and only one ability (unidimensional items, the so called 'simple structure')

The three theoretical schemes listed above correspond to three different models that, from a methodological point of view, are differently specified and, from an empirical point of view, refer to different testing situations.

The three models are represented in Figure 2.

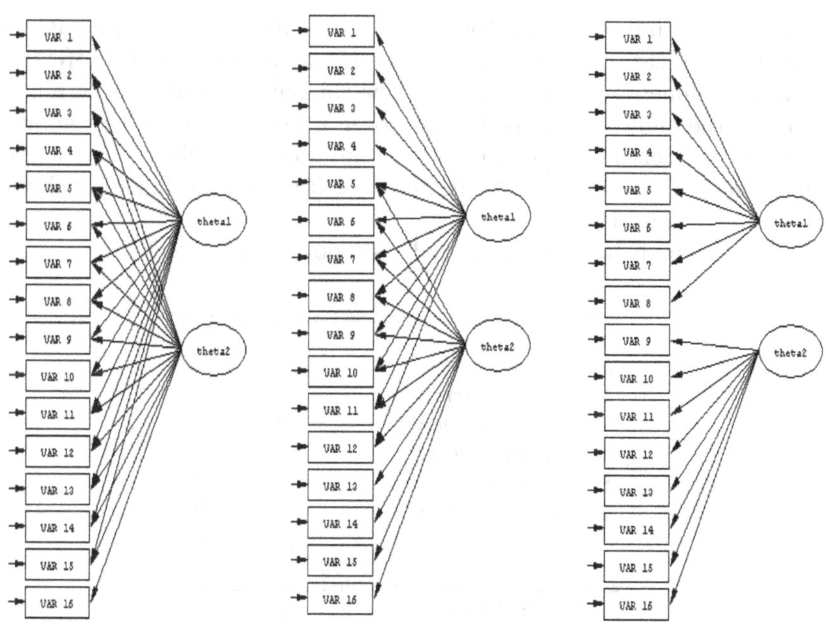

Fig. 2. Path diagrams representing the dimensionality of the items: model A, B, C

The first path diagram (model A) shows the case in which almost all the items are bidimensional (one of the parameters α_{ij} has been fixed to avoid identification problems). In the second case (model B) only eight items are bidimensional whereas the remanent 8 are unidimensional. In the last path diagram (model C) all the items are unidimensional.

The aim of this work is to analyze the effect of fitting a unidimensional model when the bidimensional models described above are assumed in the population by performing a simulation study. More specifically, we propose a classification criterion, based on the ability estimates, used to evaluate the loss of information that occurs when a unidimensional model is fitted.

The classification criterion refers to a level of ability that makes it possible to discriminate between smart student performance from poor student performance. This level of ability is determined by the object of evaluation. Classification is more severe if the aim is measuring the ability of a student who wants to access to a formative iter, it is moderate if it provides a guideline in the choice of a formative iter (self-evaluation tests).

5 The Simulation Study

A simulation study was carried out in order to evaluate the performance of the student classification based on a unidimensional model when data are known to be multidimensional. In literature, few simulation studies concerning the effects of the multidimensionality have been proposed due to the complexity of the experimental design and the computational problems. Furthermore, empirical studies need to assess the applicability of a multidimensional model to diverse testing situations.

The design of the simulation study is reported in Table 1.

N. of factors	2
N. of binary items	16
Discrimination parameters	$\mathcal{U}(1.4, 1.8)$
Difficulty parameters	$\mathcal{N}(0, 1)$
Sample size	500
Number of samples	50
Software	Fortran 95

Table 1. Design of study

The data have been fitted through a two-factor and a one-factor model. As pointed out above, the analysis of the results is focused on ability estimates. As for the bidimensional case, we assume a compensatory approach by which an individual is able to offset a low ability on one dimension with a high ability on the second dimension. It makes it possible to obtain an ability measurement ($\hat{\theta}_{av}$) by averaging $\hat{\theta}_1$ and $\hat{\theta}_2$. For both models student classification is obtained by considering the criterion that a value of ability greater than 0 indicates an adequate level of knowledge. In Table 2 the percentages of classification agreement and disagreement of the two-factor model versus the one-factor model is reported. As for Model A, the percentage of the overall disagreement is 5.3. This means that the two classification criteria give coherent results. This holds for both positive and negative abilities. The percentage of disagreement doubles for Model B and it is the highest in Model C. In this latter case the two criteria give different classification results for a quarter of the students analyzed.

Model A	*2 factors*	
1 factor	$\hat{\theta}_{av} \geq 0$	$\hat{\theta}_{av} < 0$
$\hat{\theta} \geq 0$	45.9	2.8
$\hat{\theta} < 0$	2.5	48.8

Model B	*2 factors*	
1 factor	$\hat{\theta}_{av} \geq 0$	$\hat{\theta}_{av} < 0$
$\hat{\theta} \geq 0$	45.1	5.1
$\hat{\theta} < 0$	5.7	44.1

Model C	*2 factors*	
1 factor	$\hat{\theta}_{av} \geq 0$	$\hat{\theta}_{av} < 0$
$\hat{\theta} \geq 0$	40.9	11.1
$\hat{\theta} < 0$	12.9	35.1

Table 2. Results of the simulation study: model A, B, C

In general, passing from two factors to one factor leads to substantial percentages of disagreement. This behavior becomes more evident as the number of unidimensional items increases. That is, the one-factor model seems to be not recommended when the structure of the items is simple. However, these results are strictly related to the choice of a compensatory approach.

6 Concluding Remarks

In recent years in educational testing multidimensional models have became increasingly popular due to the awareness that a variety of cognitive skills are required to solve the items correctly. This has generated the need to measure a multiple ability from a single exam. It is therefore important to understand how best to interpret results of such tests in order to classify students accurately.

In this paper we carried out a simulation study in order to evaluate the performance of a unidimensional IRT model and a bidimensional IRT model when data are generated from the latter. Three model structures, representing different test situations, were considered. Analysis was based on a classification criterion according to which a student demonstrates an adequate level of knowledge if his estimated ability is higher or equal to zero. In the bidimensional case, a compensatory approach was assumed. The results highlighted that the choice of the simpler one factor model leads to good results, i.e. little loss of information occurs and hence agreement in the classification for the one-factor and the two-factor models exists, if all the items are bidimensional. The results become as worse as the number of unidimensional items increases. This means that if we refer to a simple structure only the bidimensional model would appear preferable. However, particularly important is a more in depth understanding of how test composition affects the ability. Furthermore different classification criteria should be analyzed in order to evaluate whether results change. For example, different ways of combining abilities in the bidimensional case should be explored. Finally, since feasibly latent traits are related, correlations among them should be included in the model.

References

BARTHOLOMEW, D. and KNOTT, M. (1999): *Latent Variable Models and Factor Analysis.* Kendall's Library of statistics, London.

CAGNONE, S. MIGNANI, S. RICCI, R. CASADEI, G. and RICCUCCI, S. (2005): Computer-Automated Testing: an evaluation of undergraduate student performance. In: Courtiat, J.P., Davarakis, C. and Villemur, T. (Eds.): *Technology Enhanced Learning.* Springer, Berlin, 59–68.

CAGNONE, S. and RICCI, R. (2005): Student Ability Assessment based on two IRT Models. *Metodoloski Sveski, 65, 391–411.*

LORD, F.M. and NOVICK, M.E. (1968): *Statistical theories of mental test scores.* Addison-Wesley Publishing Co., New York.

MIGNANI, S. CAGNONE, S. CASADEI, G. and CARBONARO, A. (2005): An item response theory model for students ability evaluation using computer-automated test results. In: M. Vichi, P. Monari, S.Mignani and A. Montanari (Eds.): *Studies Classification, Data Analysis and Knowledge organization.* Springer, Berlin, 322–332.

MOUSTAKI, I. and KNOTT, M. (2000): Generalized latent trait models. *Psychometrika, 65, 391–411.*

RECKASE, M. D (1997): A Linear Logistic Multidimensional Model for Dichotomous Item Response Data. In: Van der Linden, W. J. and Hambleton, R.K. (Eds.): *Handbook of Modern Item Response Theory.* Springer-Verlag, New York, 271–286.

VAN der LINDEN, W.J. and HAMBLETON, R.K. (1997): *Handbook of Modern Item Response Theory.* Springer-Verlag , New York.

Part VI

Multivariate Methods in Applied Science

A Spatial Mixed Model for Sectorial Labour Market Data

Marco Alfó[1] and Paolo Postiglione[2]

[1] Dipartimento di Statistica, Probabilitá e Statistiche Applicate
Universitá "La Sapienza" di Roma
[2] Dipartimento delle Scienze Aziendali, Statistiche, Tecnologiche e Ambientali
Universitá "G. d'Annunzio" di Chieti-Pescara

Abstract. A vast literature has been recently concerned with the analysis of variation in overdispersed counts across geographical areas. In this paper, we extend the univariate semiparametric models introduced by Biggeri et al. (2003) to the analysis of multiple spatial counts. The proposed approach is applied to modeling the geographical distribution of employees by economic sectors of the manufacturing industry in Teramo province (Abruzzo) during 2001.

1 Introduction

During last decades a wide class of models has been introduced to analyze univariate counts observed over a set of adjacent regions with particular emphasis on mapping disease or mortality counts. Both parametric and semiparametric approaches have been proposed for the univariate case, while the multivariate one has only recently known an increasing interest, see among others Leyland et al. (2000) or Jin et al. (2005). A similar interest can be found in the econometrics literature; see, among others, Munkin and Trivedi (1999) and Chib and Winkelmann (2000). Alfó and Trovato (2004) give a recent and detailed review of this topic. In this paper, we propose a method for mapping several dependent counts recorded over a set of contiguous areas; the multivariate model is defined through a set of conditionally independent univariate models, which are linked through outcome- and area- specific latent effects. We show how model parameters and the distribution of the random effects can be straightforwardly estimated using a standard EM algorithm for ML in finite mixtures. The paper is structured as follows. Details of the theoretical framework for the univariate case is reported in section 2. Section 3 entails the extension to the multivariate context. In section 4 we analyze the results obtained by mapping the geographical distribution of employees by economic sectors (clothing and leather goods) in the Teramo province. Finally, section 5 gives some concluding remarks and outlines potential future research agenda.

2 Univariate Modeling

Let us suppose that the analyzed region is composed by n adjacent areas and that counts of employees o_i in a given economic sector have been recorded for each area. We denote with $N_{(i)} = \{j : c_{ij} \neq 0\}$ the neighborhood of the $i-th$ area, where $c_{ij} = 1$ if areas i and j are adjacent, $c_{ij} = 0$ otherwise. Observed counts o_i are considered as realizations of n conditionally independent Poisson random variables:

$$O_i \, | \lambda_i \sim \text{Poisson} \, (E_i \lambda_i) \tag{1}$$

where E_i, $i = 1, \ldots, n$ represent the expected number of employees in the $i-th$ area obtained by standardization. Therefore, λ_i is a parameter measuring the departure of the expected standardized ratio (in a mortality setting, $SMR_i = O_i \, / E_i$) from 1 (the overall average value); it represents a propensity parameter for the population of the $i-th$ area to be employed in a given economic sector as compared to the whole area. Estimates based on crude ratios are often misleading since the corresponding variance is greater for less populated areas; simple homogeneous Poisson models are not adequate to give reliable estimates of model parameter. The goal is to filter the extra-Poisson variation in order to give reliable estimates of the underlying relative risks (propensities). Once, point estimates are obtained, a related interest is in classifying areas or in mapping results, usually through thresholding techniques. The goal is to identify areas with extremely high (low) relative propensities to the event of interest. Therefore, statistical analyses generally focus on the parameter vector $\boldsymbol{\lambda} = (\lambda_1, ..., \lambda_n)^{\mathsf{T}}$, which can be modeled using a canonical link as:

$$\log (\lambda_i) = \beta_0 + u_i + v_i \tag{2}$$

where β_0 represent the overall intercept (i.e. $\beta_0 = \log(\lambda_0)$ is the average log-propensity in the entire region) while u_i and v_i are mean zero random components measuring, respectively, individual heterogeneity and spatial dependence with respect to a specified neighborhood. This formulation is based on the conjecture that heterogeneity due to *local* features can be separated from spatial dependence.

This formulation has been introduced by Besag et al. (1991) and is usually referred to as the *convolution model*, see Mollié (1996). Various parametric approaches are based on this model; for a recent review, see Alfó and Vitiello (2003). In this paper we follow the semiparametric approach described by Alfó and Postiglione (2002) for binary data and by Biggeri et al. (2003) for counted outcomes. They modeled the term v_i, $i = i, \ldots, n$ using an autoregressive-type component:

$$v_i = \rho_i v_i^* = \rho_i \log \left(\sum_{j \in N(i)} o_j \Bigg/ \sum_{j \in N(i)} E_j \right) \tag{3}$$

The term ρ_i can be considered a *smoothness* parameter measuring the strength of spatial dependence. The pseudolikelihood function is obtained by integrating out the random vector $\mathbf{d}_i = (u_i, \rho_i)^\mathsf{T}$ as follows:

$$L(\cdot) = \prod_{i=1}^{n} \left\{ \int_{\mathcal{D}} f\left(o_i \,| O_j, j \in N(i), \mathbf{d}_i\right) \mathrm{d}G(\mathbf{d}_i) \right\} \tag{4}$$

where $\mathcal{D}$ represents the support for $G(\mathbf{d}_i)$, the distribution function of $\mathbf{d}_i$. Even when normality is assumed, the above integral has not analytical solution. Ordinary or Adaptive Gaussian Quadrature or other numerical techniques give a possible solution but they can be computationally demanding and far from accurate; this problem is substantial when the random effects distribution is far from being symmetric and unimodal. Avoiding a parametric specification for the random term $\mathbf{d}_i = (u_i, \rho_i)$, $i = 1, \ldots, n$ we can proceed to estimation by nonparametric maximum likelihood through an EM-type algorithm. Laird (1978) and Lindsay (1983a, 1983b) showed that, under certain general conditions, the log-likelihood function is maximized by at least a discrete distribution on a finite number of mass-points, say K. Adopting this approach the pseudolikelihood function can then be rewritten as:

$$L(\cdot) = \prod_{i=1}^{n} \left\{ \sum_{k=1}^{K} f\left(o_i \,| O_j, j \in N(i), \mathbf{d}_k\right) \pi_k \right\} \tag{5}$$

The discrete nature of the solution helps detecting clusters of areas characterized by homogeneous values of *local* random effects u_i. Estimated locations u_k, $k = 1, \ldots, K$ represent mean zero random deviations from β_0, $l = 1, \ldots, p$, conditionally on the observed effect of the autoregressive component, measured by $v_i = \rho_k v_i^*$. With respect to the k-th component, we may write:

$$\lambda_i = \exp\left(\beta_0 + u_k + v_i\right) = \exp(\beta_0)\exp(u_i)\exp(\rho_k v_i^*) \tag{6}$$

where $\exp(u_k)$, $k = 1, \ldots, K$, represent the relative propensity for the $k-$th component as compared to $\exp(\beta_0)$. Therefore, a positive value of u_k (i.e. $\exp(u_k) > 1$) indicates an increase in the relative propensity with respect to the whole region, while a negative value of this term (i.e. $\exp(u_k) < 1$) implies a decrease in the relative propensity. Locations $\mathbf{d}_k$ and corresponding masses π_k represent unknown parameters that are estimated as follows. Denoting for compactness with $\boldsymbol{\delta}$ the parameter vector, we have:

$$\frac{\partial \log[L(\boldsymbol{\delta})]}{\partial \boldsymbol{\delta}} = \frac{\partial \ell(\boldsymbol{\delta})}{\partial \boldsymbol{\delta}} = \sum_{i=1}^{n} \sum_{k=1}^{K} w_{ik} \frac{\partial \log f_{ik}}{\partial \boldsymbol{\delta}} \tag{7}$$

where w_{ik} is the posterior probability that the $i - th$ area comes from the $k - th$ component of the mixture. These score equations are weighted sums of those for an ordinary spatial GLM with weights w_{ik}. Solving score equations for fixed weights and updating the weights from current parameter estimates is an EM algorithm. Using posterior probabilities $w_{ik}, i = 1, \ldots, n, k = 1, \ldots, K$, we could cluster areas by employing a Maximum a Posteriori (MAP) approach.

3 Extension to the Multivariate Case

In the multivariate context, we observe counts o_{il} of number of employees in L economic sectors. Let $\mathbf{o}_i = (o_{i1}, o_{i2}, ..., o_{iL})^{\mathsf{T}}$ denote the vector of observed counts for the $i-th$ borough, with $i = 1, ..., n$. As noted by Jin et al. (2005), when we have several measurements recorded at each spatial location (for example, information on $p \geq 2$ diseases from the same population groups or regions), we need to consider multivariate areal data models in order to handle the dependence among the multivariate components as well as the spatial dependence between sites. Due also to efficiency reasons we propose to define a multivariate model for all the analyzed outcomes considered as jointly determined. Since longitudinal data represent a particular case of multivariate data, a first choice for modeling dependence among counts could be that of using, as they are, random coefficient models developed for the longitudinal case; these are often referred to as uni-factor models, see e.g. Munkin and Trivedi (1999) or Chib and Winkellman (2000). The common latent structure represents the only form of association among responses recorded in the same subregion. But this approach lacks generality, since it implies unit correlation between random terms in the L equations. We propose rather to use correlated random effects which are specific for area and outcome. Let $\mathbf{u}_i = (u_{i1}, ..., u_{iL})$ denote the set of random effects; using for simplicity only canonical links, the corresponding propensity parameters can be modeled as:

$$\log(\lambda_{il}) = \eta_{il} = \beta_{0l} + u_{il} + v_{il} \qquad i = 1, \ldots, n \quad l = 1, \ldots, L \tag{8}$$

where u_{il} and v_{il} have the same meaning as before, but refer only to the $l-$th outcome; they represent random terms measuring respectively *local discontinuities* and spatial dependence. We leave $G(\cdot)$ as completely unspecified, and proceed as in the univariate case. Using the definition for the spatial component discussed above we have:

$$\log(\lambda_{il}) = \beta_0 + u_{il} + \rho_{il} \frac{\sum_{j \in N(i)} o_{jl}}{\sum_{j \in N(i)} E_{jl}} \qquad i = 1, \ldots, n \quad l = 1, \ldots, L \tag{9}$$

Using this assumption, the multivariate outcomes can be treated by using the Poisson regression model described above for the univariate case and parameter estimation can be carried out by slightly adapting the EM algorithm of the univariate case. In fact, integrating out the random effects and approximating the mixing distribution $G(\cdot)$ with a discrete distribution, we obtain for the log-pseudolikelihood function the following expression:

$$\ell(\cdot) = \sum_{i=1}^{n} \left\{ \log \left[\sum_{k=1}^{K} f\left(\mathbf{o}_i \,|\mathbf{O}_j, j \in N(i), \mathbf{d}_k\right)\pi_k \right] \right\} \tag{10}$$

where $\pi_k = \Pr(\mathbf{d}_k)$, $k = 1, ..., K$ represent the joint probability of location $\mathbf{u}_k$. Denoting with $\boldsymbol{\delta}$ the parameter vector, and proceeding as before, we have:

$$\frac{\partial \ell\left(\boldsymbol{\delta}\right)}{\partial \boldsymbol{\delta}} = \sum_{i=1}^{n}\sum_{k=1}^{K} w_{ik} \frac{\partial \log f_{ik}}{\partial \boldsymbol{\delta}} = \sum_{i,k} w_{ik} \sum_{l=1}^{L} \frac{\partial \log f_{ilk}}{\partial \boldsymbol{\delta}} \tag{11}$$

Equations (11) are weighted sums of likelihood equations for L GLMs with weights w_{ik}, where:

$$w_{ik} = \frac{\pi_k f_{ik}}{\sum\limits_{k=1}^{K} \pi_k f_{ik}} = \frac{\pi_k \prod_{l=1}^{L} f_{ilk}}{\sum\limits_{k=1}^{K} \pi_k \prod_{l=1}^{L} f_{ilk}}$$

The EM algorithm is again defined by two steps: solving the score equations for fixed weights (M step), and updating the weights for fixed parameter estimates (E step). The standard EM algorithm for finite mixtures of univariate distributions applies, with the crucial difference that w_{ik} are now computed by considering jointly the counted responses, i.e. by considering the joint distribution of the random effects. The E- and M-steps are repeatedly alternated until the log pseudolikelihood relative difference changes by an arbitrarily small amount. Since $\ell^{(r+1)} \geq \ell^{(r)}$ $r \in \mathbb{N}$, convergence is obtained with a sequence of likelihood values which are bounded from above. A formal choice of the number of components may be based on penalized likelihood criteria (such as AIC, CAIC or BIC, see e.g. Keribin, 2000) or on bootstrapping the LRT (see Feng and McCulloch, 1996). Also in this case, subregions can be clustered by using a MAP approach, with groups defined by homogeneous values of a $L-$variate relative propensity parameter.

4 Working Example

The proposed approach is applied to mapping the geographical distribution of employees by economic sectors of the manufacturing industry in Teramo province during 2001. Analyzed data are drawn from the 8th Census of Industry and Services by Istat. It is worth recalling that, as they are defined, the industrial districts are agglomerations of firms, primarily small or medium sized, which operate in the same sector and are localized in a territorially delimited area. In the present context, our analysis refers to the district of Vibrata-Tordino-Vomano, which is specialized in the textile and clothing industry, but also contains a good number of firms in the leather making sector.

The territorial extension includes 20 boroughs for a total of 627,56 km^2. We aim at defining clusters of administrative areas which are homogeneous with respect to the proportion of total labor forces employed in the analyzed economic sectors. Our aim is to compare the industrial district, as defined from the *administrative* perspective, with the *observed* clusters of areas with a similar structure of labour market.

For this purpose, we employed model (9) with respect to the sectors of clothing and leather goods. The number of components $K = 3$ has been chosen using the BIC criterion, as suggested by Keribin (2000). Due to the analysis of model fit changes, the spatial component coefficients have been kept fixed across area, i.e. $\rho_k = \rho$, $k = 1, \ldots, K$ and $\mathbf{d}_k = u_k$. Parameter estimates are reported in Table 1. As can be observed, the overall intercept β_0 is close to zero, indicating that the overall propensity to employment in both analyzed economic sectors is set to unit. The strength of spatial dependence as measured by the ρ parameter, is quite different in the two sectors; in particular, while it seems to be negligible for Clothing ($\rho = 0.06$), it seems to play a substantial role for Leather ($\rho = 0.53$). Moreover, also the strength of local discontinuities, as measured by the variance of *local* random effects, is different, but with a greater value for the first analyzed process (clothing sector, $\sigma_1^2 = 0.77$) rather than for the second one (leather sector, $\sigma_1^2 = 0.47$). These results suggest that local heterogenity has a stronger effect in the first sector than in the second one, while spatial dependence has a primary role in the latter. Therefore, uni-factor models would not be effective in mapping component membership, and the multivariate approach with correlated random effects is more appropriate.

	Clothing		Leather	
Parameter	Estimate	s.e.	Estimate	s.e.
β_0	-0.18	0.13	-0.12	0.09
ρ	0.06	0.02	0.53	0.02
ℓ	-2191.67			
σ_1^2	0.77			
σ_2^2			0.47	
$\text{corr}(u_1, u_2)$	0.75			

Table 1. Spatial analysis of employees by two economic sectors in Teramo province, 2001. Parameter estimates.

As can be easily noticed, the correlation between the random effects in the two equations is high and positive, meaning that sources of heterogeneity due to local discontinuities are substantially related. However, care is needed when interpreting correlation coefficients since we are working on a finite distribution which is, in this sense, not as accurate as a continuous one.

Corresponding estimates of to the correlation coefficient are often higher than expected. Table 2 reports the mixing distribution estimates.

$$\widehat{G}(\cdot) = \begin{pmatrix} \pi & 0.25 & 0.43 & 0.32 \\ u_1 & -1.27 & 0.01 & 1.07 \\ u_2 & -1.19 & 0.42 & 0.34 \end{pmatrix}$$

Table 2. Spatial analysis of employees by two economic sectors in Teramo province, 2001. Mixing distribution estimates.

The estimated components can be easily interpreted using the corresponding locations estimates, and the posterior classification of boroughs (through a MAP approach), detailed in Figure 1. The first component (white coloured in Figure 1) is characterized by low propensities to both clothing and leather sectors; the second component (light grey coloured) shows an average propensity to clothing sector and a significant propensity to the leather sector; while, at last, the third component (dark grey coloured) has a high propensity to clothing sector and a significant propensity to the leather sector.

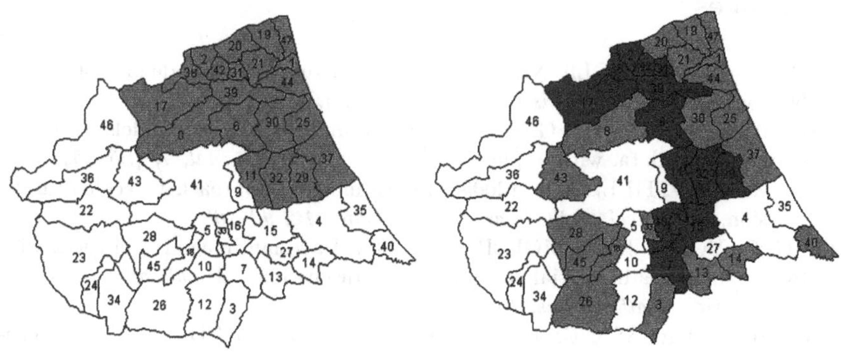

Fig. 1. Actual district (left) and estimated class membership (right)

As it is shown by Figure 1, all boroughs belonging to the industrial district of Vibrata-Tordino-Vomano are correctly identified by the model. They are all classified in the second or in the third class, which are the classes with higher propensities to the textile and leather sector. It is however interesting noticing that other areas belong to components 2 and 3, since they show a similar economic structure with respect to the sectors under investigation. As shown by Figure 1, the propose method identifies empirically a wider "industrial district", which is composed of 33 boroughs of Teramo province.

With respect to the *administrative* industrial district, it is wider and extends towards the south. In this perspective, our model can be used to evaluate the effective coherence of industrial districts (as defined by regional laws) to the economic structure of the analyzed region. These results can be therefore used to take policy actions regarding the economic integration of the analyzed regions.

5 Discussion and Further Developments

The suggested approach allows the mapping of L-variate counts in a simple and straightforward way, which represents a natural extension of standard methods for univariate mapping. It is simple to be interpreted and to be implemented, and avoids inconsistent estimates due to tight assumptions upon the parametric mixing distribution. Nevertheless, the definition of the term representing spatial dependence is quite unsatisfactory and further research is needed in order to adopt a proper Gibbs prior distribution, following the proposal of Green and Richardson (2002). This method has been applied to a dataset concerning employees distribution by economic sector in Teramo Province (Abruzzo) during 2001.

References

ALFÓ, M. and POSTIGLIONE, P. (2002): Semiparametric modelling of spatial binary observations, *Statistical Modelling, 2, 123–137.*

ALFÓ, M. and TROVATO, G. (2004): Semiparametric mixture models for multivariate count data, with application, *Econometrics Journal, 7, 426–454.*

ALFÓ, M. and VITIELLO, C. (2003): Finite mixture approach to ecological regression, *Statistical Methods and Applications, 12, 93–108.*

BESAG, J., YORK, J. and MOLLIÉ, A. (1991): Bayesian image restoration, with two applications in spatial statistics (with discussion), *Annals of the Institute of Statistical Mathematics, 43, 1–59.*

BIGGERI, A., DREASSI E., LAGAZIO, C. and BÖHNING D. (2003): A transitional non-parametric maximum pseudolikelihood estimator for disease mapping, *Computational Statistics and Data Analysis, 41, 617–629.*

CHIB, S. and WINKELMANN, R. (2001): Markov chain monte carlo analysis of correlated count data, *Journal of Business and Economic Statistics, 19, 428–435.*

FENG, Z. and McCULLOCH, C. (1996): Using bootstrap likelihood ratios in finite mixture models, *Journal of Royal Statistical Society B, 58, 609–617.*

GREEN, P.J. and RICHARDSON, S. (2002): Hidden Markov Models and Disease Mapping, *Journal of the American Statistical Association, 97, 1–16.*

JIN, X., CARLIN, B.P., and BANERJEE, S. (2005): Generalized hierarchical multivariate CAR models for areal data, *Biometrics, 61, 950–961.*

KERIBIN, C. (2000): Consistent estimation of the order of mixture models, *Sankhyā: Indian Journal of Statistics, 62, 49–66.*

LAIRD, N. (1978): Nonparametric maximum likelihood estimation of a mixing distribution, *Journal of the American Statistical Association, 73, 805–811.*

LANGFORD, I.H., LEYLAND, A.H., RASBASH, J. and GOLDSTEIN, H. (1999): Multilevel modelling of the geographical distribution of diseases, *JRSS C, Applied Statistics, 48, 253–268.*

LAWSON, A.B. et al. - Disease mapping collaborative group - (2000): Disease mapping models: an empirical evaluation, *Statistics in Medicine, 19, 2217–2241.*

LEYLAND, A.H., LANGFORD, I., RASBASH, J. and GOLDSTEIN, H. (2000): Multivariate spatial models for event data, *Statistics in Medicine, 19, 2469–2478.*

LINDSAY, B. (1983a): The geometry of mixture likelihoods: a general theory, *Annals of Statistics, 11, 86–94.*

LINDSAY, B. (1983b): The geometry of mixture likelihoods, part ii: the exponential family, *Annals of Statistics*, 11, 783–792.

MOLLIÈ, A. (1996): Bayesian mapping of disease. In: W. Gilks, S. Richardson and D. Spiegelhalter (Eds.): *Markov Chain Monte Carlo in Practice.* Chapman & Hall, London.

MUNKIN, M. and TRIVEDI, P. (1999): Simulated maximum likelihood estimation of multivariate mixed-poisson regression models, with application, *Econometrics Journal, 2, 29–48.*

The Impact of the New Labour Force Survey on the Employed Classification

Claudio Ceccarelli[1], Antonio R. Discenza[2], and Silvia Loriga[2]

[1] Servizio Condizioni economiche delle famiglie, Istat, Italy
clceccar@istat.it
[2] Servizio Formazione e lavoro, Istat, Italy
{discenza; siloriga}@istat.it

Abstract. Regulation n. 577/1998 of the European Council gives the rules to be used by the Community countries to design and conduct the Labour Force Survey (*LFS*). In order to apply this regulation the Italian *LFS* has been completely revised regarding several aspects of the survey such as frequency, definitions, questionnaire, survey design, interviewers network. All these changes caused a break in the time series of the main labour force estimates. The aim of this work is to describe and evaluate these differences and their impact on the employed classification.

1 Introduction

In order to apply the Regulation of the European Council n. 577/1998 on the organisation of harmonised Labour Force Surveys (*LFS*) in the European Member States, the Italian *LFS* has been completely revised. First of all, the new labour force survey has to be a continuous survey (in the following we'll refer to it as *CLFS*): it means that the reference weeks are spread uniformly throughout the whole year (the previous survey had a quarterly frequency; we'll refer to it as *QLFS*). The following EU Regulations n. 1575/2000 and n. 1897/2000 introduce other innovations regarding the set of variables to collect and their definitions, explore in details the definitions of employed and unemployed persons and give further methodological guidelines for the formulation of the questions on the labour status.

In order to face up these three regulations and owing to the greater complexity of the set of variables to collect, an electronic questionnaire has been developed. It allows to simplify the interview and to improve data quality. Moreover, a new data collection strategy has been adopted (a combination of different Computer Assisted Interviewing techniques) and a new professional interviewer network has been created by Istat. The new frequency of the survey, the new data collection strategy, the new definitions of employed and unemployed persons represent the core of the transformation of the Italian *LFS* and introduce breaks in the time series of the main labour force estimates.

In the following paragraphs, the sample design, the weighting procedure and the survey design of the *CLFS* will be synthetically described; in paragraph 5 the main changes from the *QLFS* to the *CLFS* - in particular those related to the employed definition - will be described. In paragraph 6 we compare the estimates of employed persons produced by the *QLFS* and by the *CLFS*, taking into account the main changes among the two surveys, and evaluating their impact on the final estimates. These results refer to the first quarter 2004, when both the surveys were carried on simultaneously.

2 Sample Design of the *CLFS*

The new Italian labour force survey is a continuous survey (*CLFS*): it means that all the weeks of the year are reference weeks, so that the sample splits into 52 sub-groups with the same size.

The main characteristics of the sample design, such as the frequency of the survey, the territorial representativeness, the allowed levels for sampling errors, the auxiliary information to be used to improve accuracy of the estimates, satisfy Eurostat Regulation n. 577/98. The sample has two stages of selection. Primary units are the Municipalities which are stratified into 1,238 strata at the provincial level, taking into account their demographic size. One primary unit is selected from each stratum, according to inclusion probabilities proportional to the Municipality demographic size. Secondary units are the households and they are selected from the Municipalities' population registers; 76,900 households are interviewed each quarter (that is i.e. 307,600 each year, with a yearly sampling rate approximately equal to 1.4%). The sample design follows a rotation scheme 2-2-2 of secondary units. Each household has to be interviewed four times during a 15 months period; households participate to the survey for 2 consecutive quarters, then they temporally exit for the following 2 quarters, and then come back in the sample for 2 quarters, after which they definitely exit. This sample scheme improves the efficiency of the estimates of quarterly and yearly differences of labour market indicators.

The *CLFS* sample has been designed to guarantee reliable yearly estimates of the main indicators of labour market at provincial level (NUTS III), quarterly estimates at the regional level (NUTS II) and monthly estimates at the national level[1].

3 Weighting Procedure of the *CLFS*

The estimates of the labour force survey are obtained through a calibration estimator (Deville and Särndal, 1992). Grossing weights for each sample unit are computed as follows:

[1] The *QLFS* sample design had similar characteristics to the new one: two selection stages, stratification of primary units, rotation scheme of secondary units, sample size, but only one reference week for each quarter.

- Initial weights are obtained for all selected households as the inverse of the inclusion probability;
- The initial weights are adjusted by a correction factor for total non-response in order to reduce the biasing effect. This factor is worked out as the inverse of the response ratio for sub-groups of households with similar characteristics identified by rotation group, region, household size, sex and age of the head of the household;
- Final weights are obtained by a further adjustment using post-stratification. They are obtained as solution of a minimisation problem under constraints, which requires the equalisation of the sampling estimates of certain auxiliary variables with their respective totals obtained from independent demographic sources based on population registers. The procedure uses the following constraints:
  1. distribution of population by sex and fourteen 5-year age groups at NUTS II region level (0-14, 15-19, ..., 70-74, 75 and more years);
  2. distribution by sex at NUTS II region level for each of the three months of the quarter;
  3. distribution by sex and five age groups at NUTS III province level (0-14, 15-29, 30-49, 50-64, 65 and more years);
  4. distribution by sex and five age groups for thirteen large municipalities with more than 250.000 inhabitants (0-14, 15-29, 30-49, 50-64, 65 and more years);
  5. number of households at NUTS II region level for each rotation group;
  6. non-national population at NUTS II region level classified in male, female, EU citizens of the 25 countries, NonEU citizens[2].

4 Survey Design of the *CLFS*

The whole survey process is based on several complex operations which have different impacts on the final outcomes; for the *CLFS* a complex monitoring system has been set up in order to keep under control all the activities, and to guarantee good levels of quality.

The electronic questionnaire simplifies the interview process reducing the duration and allowing for a significant improvement in the quality; this is mainly due to the use of automatic branching, help on-line, interactive codification of open items through search engines, root and coherence rules, confirmation items for re-interviews[3].

---

[2] In the *QLFS* the weights were obtained by a similar procedure applying most of the constraints used for the *CLFS* (excluded 2 and 6); but these constraints were not applied simultaneously and different non-response correction factors were computed.

[3] The *QLFS* adopted a paper questionnaire; the data quality check was conducted only ex-post.

The first interview is carried out by CAPI technique, the three following interviews are generally carried out by CATI technique. This mixed technique allows to make up for the advantages and the disadvantages of respectively face to face and by phone interviews, given the budget and organization constraints. To carry out CAPI interviews, a network of 311 professional interviewers operating on the national territory has been created, directly trained and managed by Istat experts.

5 Main Changes from *QLFS* to *CLFS* on the Employed Classification

Regulations n. 1575/2000 and n. 1897/2000 give a detailed description of the definitions of employed and unemployed persons which have to be adopted by all the member states to produce comparable estimates on the labour market by means of a *LFS*. With the transition from the Italian *QLFS* to the *CLFS* some changes occurred to the definitions adopted, so that definitions used by the *CLFS* are exactly those established by the regulations. In the following we focus on the definition of employed persons.

According to the *QLFS* had to be classified as Employed all those who are aged 15 years or more, perceiving themselves as employed or reporting to have been at work (at least one hour) during the reference week no matter for their perceived labour force status.

On the contrary, the definition established by the community regulations which has been adopted by the *CLFS* is the following[4]: have to be classified as Employed all those who are aged 15 years or more, reporting to have been at work (at least one hour) during the reference week or they were not working but had a job or business from which they were absent during the reference week; precise conditions regarding the absence period are specified: self-employed persons must maintain their business while for employees the total absence period from work has to be no longer than three months or he/she continues to perceive at least 50% of the wage or salary from the employer. Permanently disabled people and conscripts on compulsory military service are excluded from the reference population by both surveys.

Due to both the new survey technique and the new definition of employed persons, several changes were also made to the questionnaire even if the overall structure didn't change (a first section to collect social-demographic characteristics of all members of the household, and several individual subsections that identify homogeneous thematic areas). Referring to the employed definition, the main difference among the *QLFS* and the *CLFS* questionnaires, is on the first question used to identify employed persons: that is

[4] For a more detailed description of the definitions see cited EU Regulations n. 1575/2000 and n. 1897/2000 and the Eurostat document n. 05/2000 containing the definition of the variables specified in the 1575/2000.

the respondent's perception of his/her own labour status in the first and "having worked for pay or profit, at least one hour" in the last.

6 Evaluation of the Impact on the Employed Estimates

All the changes mentioned above have an impact on the labour status classification. In this paragraph a comparison among the estimates of employed persons produced by the $QLFS$ (January 2004) and by the $CLFS$ (first quarter 2004) will be carried on; the main aim is to evaluate the impact of the changes on the final estimates produced by the labour force survey. In the following we'll indicate:

- $QLFS$: the official estimates produced by the $QLFS$ (old definition, old weighting procedure, pre-census reference population);
- $QLFS_{np}$: the estimates produced by the $QLFS$ data, applying the old definition, the old weighting procedure, the post-census reference population that is used for the $CLFS$ (*np* stays for new population[5]);
- $QLFS_{nw}$: the estimates produced by the $QLFS$ data, applying the old definition, a new weighting procedure as similar as possible to the one adopted by the $CLFS$[6], the post-census reference population that is used for the $CLFS$ (*nw* stays for new weighting procedure with new population);
- $CLFS$: the official estimates produced by the $CLFS$ (new definition, new weighting procedure, post-census reference population);
- $CLFS_{od}$: the estimates produced by the $CLFS$ data, applying a definition of employed persons similar (as much as possible) to the old definition adopted by the $QLFS$ (*od* stays for old definition), the new weighting procedure and the post-census reference population.

We may write:

$$CLFS - QLFS_{np} =$$
$$= CLFS - CLFS_{od} + CLFS_{od} - QLFS_{nw} + QLFS_{nw} - QLFS_{np}$$

and we may define the following figures:

- $CLFS - CLFS_{od}$ is the definition impact;
- $CLFS_{od} - QLFS_{nw}$ is the survey process impact (impact of the new frequency of the survey, the new sample design, the new data collection strategy, the new professional interviewer network, the new questionnaire, etc.);

[5] Due to the availability of the 2001 Population Census results, all the population data previously released (referred to a date following October 2001) have been revised by Istat.

[6] Excluding the new correction factor for total non-response and the constraints 2 and 6.

- $QLFS_{nw}$ - $QLFS_{np}$ is the weighting procedure impact.

Their sum ($CLFS$ - $QLFS_{np}$) is the overall variation. In Tables 1 and 2 all these results have been reported for the main employment characteristics, by geographical areas and gender (estimates and differences are expressed in thousands).

7 Main Results and Concluding Remarks

We may observe that the overall increase on the employed estimate from the $QLFS_{np}$ to the $CLFS$ (+294) is not uniformly distributed among the employment characteristics: the increase is larger for employees with temporary job, self-employed persons, and especially for part-time workers, while employees with permanent job and full-time workers decrease. This trend appears in the three geographical areas, both for men and women.

Stronger differences appear in the Centre-South and for women, where to a considerable increase of part-time workers correspond a decrease of full-time workers.

The more relevant impact is the survey process impact (+214); the definition impact is less relevant (+131) and the new weighting procedure impact goes on the opposite direction (-51); it is interesting to observe that the definition impact and the new weighting procedure impact have a rather uniform distribution among the employment characteristics, the geographical areas and the gender, while the survey process impact shows the same trend as the overall variation.

The definition impact has always a positive sign, for all the employment typologies, in each geographical area, both for men and women, and it produces an overall increase of the employed estimate.

We may synthetically conclude that the $CLFS$, due to the new definition of employed persons and especially due to the new survey process, produces higher estimates for employed people; the $CLFS$ employment distribution is rather different from the one taken by the $QLFS$ and this difference is mainly due to the survey process impact; we may say that the $CLFS$ takes a more relevant part of the "marginal employment" (temporary jobs and part-time workers)

	$QLFS_{np}$	$QLFS_{nw}$	$CLFS_{od}$	$CLFS$
Employed	21771	21720	21934	22065
- Employee	15802	15754	15784	15866
- *with temporary job*	*1427*	*1434*	*1696*	*1714*
- *with permanent job*	*14375*	*14320*	*14088*	*14152*
- Self-employed	5968	5966	6150	6199
- Part-time	1850	1844	2801	2854
- Full-time	19921	19876	19133	19211
North				
Employed	11244	11214	11298	11348
- Employee	8171	8143	8145	8173
- *with temporary job*	*596*	*601*	*704*	*709*
- *with permanent job*	*7575*	*7541*	*7441*	*7463*
- Self-employed	3074	3072	3153	3175
- Part-time	1103	1098	1492	1519
- Full-time	10141	10116	9807	9829
Centre				
Employed	4401	4386	4389	4421
- Employee	3184	3165	3133	3155
- *with temporary job*	*297*	*299*	*336*	*341*
- *with permanent job*	*2886*	*2866*	*2797*	*2814*
- Self-employed	1217	1221	1256	1266
- Part-time	351	353	606	618
- Full-time	4050	4033	3783	3802
South				
Employed	6126	6120	6247	6297
- Employee	4448	4446	4505	4539
- *with temporary job*	*534*	*533*	*656*	*663*
- *with permanent job*	*3914*	*3913*	*3850*	*3875*
- Self-employed	1678	1674	1741	1758
- Part-time	396	393	704	717
- Full-time	5730	5727	5543	5580
Men				
Employed	13470	13462	13331	13390
- Employee	9257	9255	9116	9149
- *with temporary job*	*693*	*704*	*824*	*829*
- *with permanent job*	*8564*	*8552*	*8292*	*8320*
- Self-employed	4213	4207	4215	4241
- Part-time	407	412	625	640
- Full-time	13063	13050	12706	12750
Women				
Employed	8301	8258	8603	8675
- Employee	6545	6498	6668	6717
- *with temporary job*	*734*	*730*	*872*	*885*
- *with permanent job*	*5811*	*5768*	*5795*	*5832*
- Self-employed	1756	1760	1935	1958
- Part-time	1442	1432	2176	2214
- Full-time	6858	6826	6427	6461

Table 1. Comparison among $QLFS$ and $CLFS$ employment estimates

	$CLFS$ - $QLFS_{np}$	$QLFS_{nw}$ - $QLFS_{np}$	$CLFS_{od}$ - $QLFS_{nw}$	$CLFS$ - $CLFS_{od}$
Employed	294 **	-51	214	131
- Employee	64	-49	30	82
- *with temporary job*	*287* **	*7*	*263* **	*18*
- *with permanent job*	*-223* *	*-55*	*-232* *	*64*
- Self-employed	231 **	-2	184 **	49
- Part-time	1004 **	-6	958 **	53
- Full-time	-710 **	-45	-744 **	78
North				
Employed	104	-30	84	49
- Employee	2	-28	3	27
- *with temporary job*	*114* **	*6*	*103* **	*5*
- *with permanent job*	*-112*	*-34*	*-100*	*22*
- Self-employed	102 *	-2	81	22
- Part-time	416 **	-5	394 **	27
- Full-time	-312 **	-25	-309 **	22
Centre				
Employed	20	-15	3	32
- Employee	-29	-18	-32	21
- *with temporary job*	*44* **	*2*	*37* *	*5*
- *with permanent job*	*-73*	*-21*	*-69*	*17*
- Self-employed	49	4	35	10
- Part-time	267 **	2	253 **	13
- Full-time	-248 **	-16	-251 **	19
South				
Employed	171 **	-6	127	50
- Employee	91	-2	60	33
- *with temporary job*	*129* **	*-1*	*123* **	*8*
- *with permanent job*	*-38*	*-1*	*-63*	*25*
- Self-employed	80 *	-4	68	16
- Part-time	321 **	-3	311 **	13
- Full-time	-150 *	-3	-184 **	37
Men				
Employed	-80	-8	-131	58
- Employee	-109	-2	-139	32
- *with temporary job*	*135* **	*11*	*120* **	*5*
- *with permanent job*	*-244* **	*-13*	*-259* **	*28*
- Self-employed	28	-6	8	26
- Part-time	233 **	4	214 **	15
- Full-time	-313 **	-12	-344 **	44
Women				
Employed	375 **	-43	345 **	72
- Employee	172 **	-47	169 **	50
- *with temporary job*	*151* **	*-4*	*142* **	*13*
- *with permanent job*	*21*	*-43*	*27*	*37*
- Self-employed	202 **	4	175 **	23
- Part-time	772 **	-10	744 **	38
- Full-time	-397 **	-33	-399 **	35

* identifies the estimates significant at 90% confidence level;
*** identifies the estimates significant at 95% confidence level.

Table 2. Impact of the changes from $QLFS$ to $CLFS$ on the employment estimates

References

BERGAMASCO, S., GAZZELLONI, S., QUATTROCIOCCHI, L., RANALDI, R., TOMA, A. and TRIOLO, V. (2004): New strategies to improve quality of Istat new capi/cati labour force survey, *European Conference on Quality and Methodology in Official Statistics, Mainz, Germany, 24-26 May.*

COUPER, M.P. (1996): Changes in interview setting under Capi, *Journal of Official Statistics, Statistics Sweden, 12; 3, 301–316.*

DEVILLE, J.C. and SARNDAL, C.E. (1992): Calibration estimator in survey sampling, *Journal of the American Statistical Association, 87, 376–382.*

EUROSTAT (1998): Regulation n. 577/1998, *Official Journal of the European Communities.*

EUROSTAT (2000 a): Regulation n. 1575/2000, *Official Journal of the European Communities.*

EUROSTAT (2000 b): Regulation n. 1897/2000, *Official Journal of the European Communities.*

EUROSTAT (2001): Working group employment statistics - Document n. 05/2001.

GRASSIA, M.G., PINTALDI, F. and QUATTROCIOCCHI, L. (2004): The electronic questionnaire in Istat new capi/cati labour force survey, *European Conference on Quality and Methodology in Official Statistics, Mainz, Germany, 24-26 May.*

ISTAT (2004): La nuova rilevazione sulle forze di lavoro: contenuti, metodologie, organizzazione, Technical report, www.istat.it/Eventi/report_finale.pdf, Roma, 3 giugno 2004.

KELLER, W.J. (1995): Changes in statistical technology, *Journal of Official Statistics, Statistics Sweden, 11.*

NICHOLLS, W.L., BAKER, R.P. and MARTIN, J. (1997): The effects of new data collection technologies on survey data quality, in: *Survey Measurement and Process Quality*, Lyberg L., ed., J. Wiley & Sons, New York.

Using CATPCA to Evaluate Market Regulation*

Giuseppe Coco[1] and Massimo Alfonso Russo[2]

[1] Dipartimento di Scienze Economiche
Università di Bari, Italy
g.coco@dse.uniba.it

[2] Dipartimento di Economia, Matematica e Statistica
Università di Foggia, Italy
m.russo@unifg.it

Abstract. One of the most interesting research area in economics concerns the measurement of relative competiveness of different economic systems. Among the several proposed indicators, a particularly relevant one is the Product Market Regulation (PMR) proposed by the OECD, calculated on the basis of a rich database. This paper uses the same database to compute alternative indicators. The main difference with the OECD indicator is that we propose a less invasive statistical methodology (CATPCA), suitable for the treatment of qualitative data. In addition we remove several arbitrary manipulations of basic data. The calculation delivered a new ranking of the 21 countries analyzed and some new interesting evidence.

1 Introduction

In recent years a considerable amount of effort has gone into the task of comparing different countries' competitiveness through synthetic indices. Within those indices a significant attention is devoted to the dimension of regulatory quality and burden on enterprises. In every existing ranking Italy performs very badly both for the low quality of the regulatory environment and for the excessive role of the Public sector in the marketplace. The most frequently quoted, among those indices are the World Competitiveness Index (WCI), compiled by the International Institute for Management Studies (Garelli (2005)), the Governance Indicators (GI) of the World Bank (Kaufmann et al. (2004, 2005)) and the Product Market Regulation index (PMR) of the OECD (Nicoletti et al. (1999); Conway et al. (2005)).

Only the PMR is built on a data set particularly designed to catch a picture of the regulatory system. This database collects direct information about the regulatory and legal system obtained through a 1998 survey on member countries (updated at 2003) and roughly checked by the OECD team itself.

* The contents of this paper result from joint work of the authors. In particular, Sections 2 and 3 are due to Russo M. while Sections 4 and 5 are due to Coco G.. The autors would like to thank partecipants to the Cladag 2005 meeting and an anonymous referee for useful comments.

Some recent work confirms the relevance of the PMR in explaining diverging economic performance among OECD countries (Gonenc et al. (2001), Nicoletti and Scarpetta (2003)). A reason for dissatisfaction with the PMR lies with the statistical method, that determines quantitative indices from qualitative original information through *subjective quantification*. Moreover, some of the quantified indices are then averaged with *arbitrary weights* in the subsequent aggregation stage. We propose a different, less invasive, statistical methodology for the aggregation of data, tailored for the qualitative nature of most part of the dataset. Furthermore we remove the arbitrary weightings.

2 Data and Aggregation Process Description

We propose a new statistical indicator based on the same database downloadable at the OECD website (www.oecd.org). The information has been preliminarily grouped into a number of areas (variables) defined according to their economic influence. In particular, the 15 defined variables are: 1) Size of public enterprise system - SIZE; 2) Scope of public enterprise system - SCOPE; 3) Special voting rights and control of public enterprise by legislative bodies - SVR-CPEL [1]; 4) Use of command and control regulation - UCCR; 5) Price controls - PC; 6) License and permits system - LPS; 7) Communication and simplification of rule and procedures - CSRP; 8) Administrative burdens for corporations and for sole proprietor firms - ABS [2]; 9) Sector specific administrative burdens - SSAB; 10) Legal barriers - LB; 11) Antitrust exemptions - AE; 12) Ownership barriers - OB; 13) Discriminatory provisions - DP; 14) Tariffs - T; 15) Regulatory barriers - RB.

The 15 variables are measured on different scales, even if the most frequent are qualitative. For this reason we have chosen to measure on an ordinal scale the whole information in the following elaboration, thus not requiring the arbitrary quantification of the qualitative modalities. In the OECD study each single variable defined above has been then univocally attributed to one of three broad regulatory domains: State control over business enterprises (SC), comprising the first 5 variables listed above, that groups the information generally referring to forms of direct intervention of the State in the economy through public ownership or through obligations to a certain particular behaviour (for example binding price regulation); Barriers to entrepreneurship (BE), including variables from 6 to 11, that groups the information concerning the regulation that prevents competition from displaying its full effect; Barriers to international trade and investment (OUT), including the last 4

[1] The two variables SVR and CPEL were originally separated in the OECD index. We grouped them because of the common theme and because the CPEL was built on a single data information.

[2] Also the variable ABS results from the aggregation of two variables in the original elaboration by Nicoletti et al. (1999).

variables, summarizes the information describing the degree of openess of the regulatory system to foreign operators. This structure has been mutuated from the OECD work, with some significant differences (Coco and Russo (2004)).

The main differences concern: quantification of qualitative information collected; statistical methods applied; weightings used. It is important to emphasize that the OECD team quantified qualitative information through a subjective trasformation of ordinal modalities and then aggregated the quantified data through weighted average, using Principal Component Analysis (PCA). Furthermore, in some cases the average was further weighted with the score of another variable. In particular, the score of the variables SCOPE and ABS were used as weights for construction of the variables CPEL, AE, SSAB. This process led to an artificially increased overall weight in the final index of the variable used for weighting. We eliminated the weightings.

3 Statistical Analysis

In order to quantify the amount of regulation in each of the 21 OECD countries surveyed, we have performed twice the CATPCA - Categorical Principal Component Analysis (Gifi (1990)): on the whole (15) set of variables; on each of the three broad regulatory domains (as proposed by OECD team). The second choice, is justified by the well-established *a-priori* knowledge of the economic classification of the examined variables while the first procedure is the best solution in an exploring framework in which the broad regulatory domains follow from the statistical technique. The CATPCA has been performed by the procedure used in SPSS version 11.0 (Meulman and Heiser (2001)). This procedure allows the estimate of the parameters of the model jointly with the optimal quantification (OS-Optimal Scaling) of the categorical modalities. The variables in this work, all measured on an ordinal scale, have been rescaled on a superior order of measurement (numerical). The OS for each variable, aimed at estimating the principal components, has been obtained through an iterative ALS-Alternating Least Squares (Young et al. (1978)). The use of CATPCA, rather than PCA, is even more commendable when the number of elements (countries) of the statistical population to be examined increases (Candel (2001)).

It may be important to establish preliminarly that:

- the examined variables $v(v = 1, 2, ..., V; V = 15)$, to be assigned to each domain $d(d = 1, 2, 3; SC = 1, BE = 2, OUT = 3)$ may be identified as follows: $v^d = 1, 2, ..., V^d$. In each of the domains we then have $V^1 = 5, V^2 = 6, V^3 = 4$ and $V^1 + V^2 + V^3 = V = 15$;
- r_{jv^d} is the measured rank for the variable v^d, belonging to the domain d in country $j(j = 1, 2, ..., J; J = 21)$, while z_{jv^d} identifies its optimal standardized quantification derived form CATPCA;

- a_{iv^d} corresponds to the vth coordinate of the ith estimated dimension $(i = 1, 2, ..., I)$ for the domain d, while with λ_{id} we term the variance explained by the ith principal component in the dth domain.

The CATPCA on the whole set of variables delivered (considering eigenvalues > 1) four principal components that explain 22,68 %, 15,94 %, 12,97 % and 9,9 % of the variance. The three different CATPCA on each broad regulatory domain delivered two principal components for the first (that explain 42,49 % and 21,76 % of the variance) and the third domain (that explain 42,15 % and 23,57 % of the variance) and three components for the second domain [3] (that explain 31,25 % , 24,61 % and 18,69 % of the variance). However, the variance explained with the first procedure (statistically less invasive) is not larger than the one explained with the OECD approach. Morover the economic interpretation of the 4 dimensions delivered by the procedure proves impossible. Taking this into account and considering the high correlation (ρ Spearman = 0,953) of the results obtained with the two approaches, the *a-priori* classification (logical-economic) of the 15 variables in the 3 macro-domains, as summarized in the Figure 1 (modified from Nicoletti et al. (1999) and Conway et al. (2005)), appears sensible.

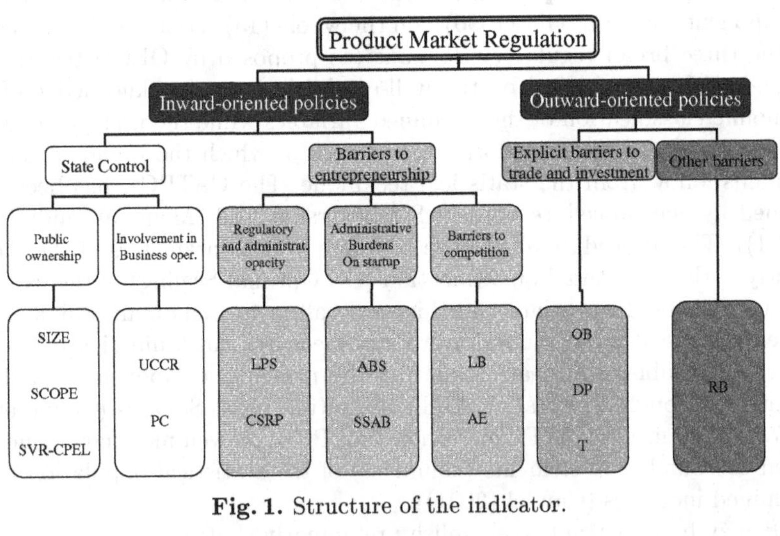

Fig. 1. Structure of the indicator.

The model for the measurement of the regulation index then consists of:

[3] The two sub-domains of SC are respectively: Public ownership and Control of business; the three sub-domains for BE are: Administrative transparency, Burdens on start ups and Barriers to competition; the two sub-domains for OUT are: Barriers to trade and to investment; Other barriers. In the last level one can find all the original 15 variables.

1. *quantification* of the synthetic scoring for each country in each domain through a weighted average,

$$R_{jd} = \frac{\sum_{v^d=1}^{V^d} z_{jv^d} \cdot a_{jv^d}^2}{\lambda_{id}}; \tag{1}$$

2. *further processing* of the three synthetic indicators R_{j1}, R_{j2} and R_{j3}, again through CATPCA. The new procedure delivered two principal components [4]. The general PMR_j indicator for each country is then obtained in the same way as above:

$$PMR_j = \frac{\sum_{d=1}^{3} R_{jd} \cdot a_{iR_d}^2}{\lambda_i}. \tag{2}$$

It must be underlined that using the CATPCA on the whole set of variables, the general PMR_j indicator could be directly obtained in one step.

4 Results

The statistical model with the *a-priori* classification of the variables produces the results reported [5] in Table 1.

Countries	*SC*	*BE*	*OUT*	*PMR*	Countries	*SC*	*BE*	*OUT*	*PMR*
United K.	-1,13	-0,85	-0,53	**-0,81**	Japan	-0,38	0,79	-0,03	**0,11**
Ireland	-0,71	-0,55	-0,53	**-0,59**	Spain	0,50	-0,08	-0,03	**0,12**
Australia	-0,41	-0,58	-0,60	**-0,54**	Canada	-0,52	-0,69	1,28	**0,13**
United States	-0,92	-0,13	-0,13	**-0,37**	Finland	0,37	0,22	-0,03	**0,17**
Netherlands	-0,27	-0,16	-0,40	**-0,29**	Switzerland	-0,38	0,47	0,56	**0,25**
New Zealand	-0,67	-0,40	0,11	**-0,28**	Belgium	0,56	0,61	-0,03	**0,34**
Germany	-0,18	0,19	-0,40	**-0,16**	France	0,49	0,55	0,07	**0,34**
Sweden	-0,49	0,19	-0,02	**-0,10**	Italy	1,08	0,74	-0,53	**0,35**
Denmark	0,25	0,05	-0,40	**-0,07**	Norway	0,87	-0,35	1,23	**0,64**
Austria	0,11	0,25	-0,40	**-0,05**	Greece	1,28	0,21	0,80	**0,77**
Portugal	0,55	-0,49	0,04	**0,04**					

Table 1. Indicators of domain and general PMR.

The results exhibit common patterns with the OECD team study (ρ Spearman $> 0{,}900$ both for the three domains indicators and for the general PMR). Still some country's ranking changes considerably. Summing up, there emerges an Anglo-Saxon group of excellence (United Kingdom, Ireland, Australia) followed by some other English-speaking countries, with a more

[4] The two higher-order domains are: Inward oriented policies (including the domains SC and BE); Outward oriented policies (including only OUT).

[5] In PMR increasing order. Negative (positive) values indicate lower (higher) market regulation.

unequal performance across indicators (United States and New Zealand). It is also rather surprising the mediocre performance of the United States both in the BE indicator and in the OUT indicator, compensated by an excellent SC score. Northern European countries (Germany, Netherlands, Austria and most Nordic countries) follow in the ranking scoring average performance in most indicators. A fourth group is composed by a majority of the European Union countries, among which Italy, characterized by a rather heavy internal regulation. Italy's overall score is significantly different from the one obtained in the OECD study, where it was classified as the worse in the PMR and an outlier, in terms of bad (excessive) internal regulation. On the other side our ranking confirms some other anomalies (Canada, in terms of OUT; Greece and Norway, in terms of general PMR) [6]. We have performed twice a hierarchical cluster analysis: on the whole (15) set of the qualitative variables using the complete linkage Method with city-block measure; on each of the three domain indicators using numerous techniques [7]. The results are particularly steady in both of the procedures and confirm the discussion above. Due to space constraints, we report in Figure 2 only the dendogram obtained by Ward method applied on the three domain indicators [8].

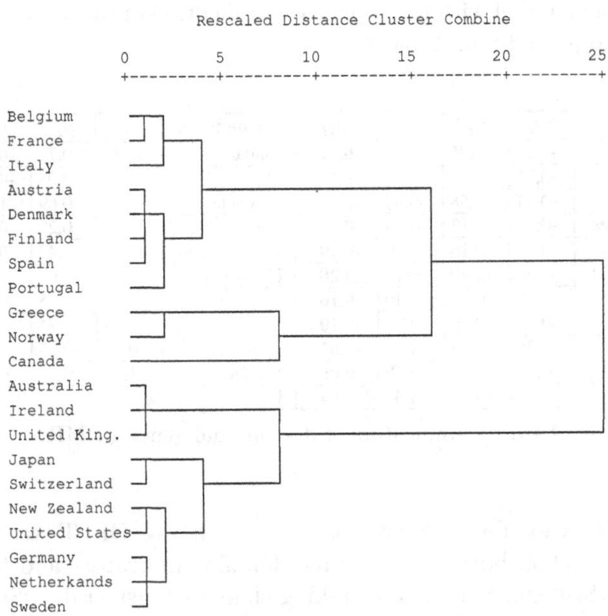

Fig. 2. Dendogram.

[6] In general terms these results hold true also working with the 15 original variables, with the notable exception of Denmark and Finland.

[7] Ward, complete and average linkage (all with city-block or Euclidean distances).

[8] Alternative classifications with different models are available with the authors.

In particular, the United States and New Zealand (depending on the classification method adopted) can be associated to the excellence group or not. The analysis reveals, independently from the method used, a certain degree of similarity among a group of European Union countries (8 out of 14 considered), among which Italy, that is characterized by a negative performance in internal policies (SC and BE), but also by high degree of openness (OUT). In this group Italy stands out as the most heavily regulated in the internal policy dimension but also the most open.

5 Main Conclusion

Evaluating comparatively the competitiveness of different countries is more and more important. There exists a wide consensus that the regulatory environment figures prominently among the main factors that contribute to the competitiveness of each country. A comparative evaluation of the quality of the regulatory environment however poses some difficulties. In particular, qualitative information should be treated with the appropriate technique.

Our indicator, while using the same basic structure as the one in Nicoletti et al. (1999) is preferable to that one. The statistical methodology used here, the CATPCA, is the only one suitable to treat qualitative information. The use of PCA in OECD indicators compels the researchers to performe a subjective quantification of original data, thereby distorting the whole evaluation exercise. Moreover some arbitrary weightings used to calculate the original indicator have been removed. These changes obviously did not deliver a radically different ranking among countries, given the database, but some interesting differences emerge nonetheless.

In particular Italy ranks in any case as one of the worst countries for internal politicies, i.e. taking account jointly of SC and BE, but it is not any more an outlier, independently from the statistical models used. The cluster analysis confirms that, while belonging to the group of worse perfoming countries, it does not represent an absolute anomaly, particularly when compared with a subset of countries within the group (notably France and Belgium).

In the light of these results the policies aimed at reducing regulatory burdens remain a priority for any present and future Italian government. However this evaluation does not confirm the existence of a uniquely negative situation (Nicoletti (2002)), in line with additional subsequent analysis limited to a subset of the data (Coco and Ferrara (2003)). Over-regulation is more likely to be one among many structural weaknesses of our country (as suggested by Blanchard (2002)). Too little attention on the other side has been devoted in our opinion to the particular structure of Italy's indicators. Italy figures as a case of excellence in the OUT indicator, while performing very badly in both internal policy dimensions. An interesting topic for further research may be the investigation of the effects of particular patterns of regulation on economic performance.

References

BLANCHARD, O. (2002): Comment. In: ISAE: *Annual Report on Monitoring Italy.* ISAE, Rome, 190–194.

CANDEL, M. (2001): Recovering the Metric Structure in Ordinal Data: Linear Versus Nonlinear Principal Components Analysis. *Quality & Quantity, 35, 91–105.*

COCO, G. and FERRARA, F. (2003): Sportello Unico, Semplificazioni Amministrative e Riduzione delle Barriere all'Imprenditorialità. In: FORMEZ: *Sportello Unico. Gli Effetti Economici e Amministrativi di una Innovazione.* FORMEZ, Roma, 23–46.

COCO, G. and RUSSO, M.A. (2004): Una Nuova Valutazione Comparativa dei Sistemi di Regolazione Tramite la CATPCA: il Caso Italia. *Quaderno DEMS 11/04, Dipartimento di Economia, Matematica e Statistica, Università di Foggia.*

CONWAY, P. and JANOD, V. and NICOLETTI, G. (2005): Product Market Regulation in OECD Countries: 1998 to 2003. *OECD Working Papers 05/419.*

GARELLI, S. (2005): *IMD World Competitiveness Yearbook: 2005.* International Institute for Management Development, Lausanne.

GIFI, A. (1990): *Nonlinear Multivariate Analysis.* Wiley, Chichester.

GONENC, R. and MAHER, M. and NICOLETTI, G. (2001): The Implementation and the Effects of Regulatory Reform: Past Experience and Current Issues. *OECD Economic Studies, 32.*

KAUFMANN, D. and KRAAY, A. and MASTRUZZI, M. (2004): Governance Matters III: Governance Indicators for 1996, 1998, 2000 and 2002. *World Bank Economic Review, 18, 253–287.*

KAUFMANN, D. and KRAAY, A. and MASTRUZZI, M. (2005): Governance Matters IV: Governance Indicators for 1996–2004. *World Bank Policy Research, Working Paper Series, 3630.*

MEULMAN, J.J. and HEISER, W.J. (2001): *SPSS Categories 11.0.* SPSS Inc., Chicago.

NICOLETTI, G. (2002): Institutions, Economic Structure, and Performance: Is Italy Doomed. In: ISAE: *Annual Report on Monitoring Italy.* ISAE, Rome, 129–189.

NICOLETTI, G. and SCARPETTA, S. and BOYLAUD, O. (1999): Summary Indicators of Product Market Regulation with an Extension to Employment Protection Legislation. *OECD Working Papers 99/18.*

NICOLETTI, G. and SCARPETTA, S. (2003): Regulation, Productivity and Growth - OECD Evidence. *Economic Policy, 36, 9–72.*

YOUNG, F.W. and TAKANE, Y. and DE LEEUW, J. (1978): The principal components of mixed measurement level multivariate data: An alternating least squares method with optimal scaling features. *Psychometrika, 43, 279–281.*

Credit Risk Management Through Robust Generalized Linear Models

Luigi Grossi[1] and Tiziano Bellini[2]

[1] Dipartimento di Economia,
Università di Parma, Italy
luigi.grossi@unipr.it

[2] Ufficio Risk Management,
Banca Monte Parma, Italy
tiziano.bellini@monteparma.it

Abstract. In this work, a robust methodology is developed for the classification of a sample of small and medium firms on the basis of their default probability. The importance of this classification procedure is emphasized by the New Basel Capital Accord (Basel II) for the capital adequacy of internationally active banks. The Basel accord introduces the possibility to adopt models of internal rating for the estimation of the default probability of customers' banks. The reference framework of this paper is the class of generalized linear models which allows to classify units avoiding strict assumptions such those required by the linear discriminant analysis. Another advantage of generalized linear models is the possibility to explore different links between the expected value of the dependent variable and the linear predictor. Parameters are estimated using balance ratios and data coming from Centrale dei Rischi for a set of firms which are customers of a medium sized bank of Northern Italy. Finally, we perform a robust analysis of the model estimates through the forward search in order to monitor the influence of outliers on the final classification.

1 Introduction

A financial institution deciding whether to supply credit assesses if the potential borrower will be able to redeem the credit. According to this goal, financial institutions are engaged in developing rating systems which graduate customers on the basis of their future ability to refund the money supplied and may be applied to classify potential new customers. The main issue of a credit rating system is to identify criteria which separate "good" creditors from "bad" creditors. This issue, apart from its theoretical attractiveness, is gaining importance in financial institutions considering the role of rating systems, not only in day by day lending activity, but also in determining the adequacy of regulatory capital under the Basel Capital Accord (Bank of International Settlements, 2004). Using financial and non-financial risk factors of a sample of more than 600 firms extracted from a financial institution database merged with "Centrale dei Rischi" database, we adopt generalized linear models in order to classify healthy and potentially insolvent firms in classes according with their default probability. One of the most relevant innovation of this paper is the introduction of a robust analysis for distress prediction methods using the forward search methodology (Atkinson and Riani, 2000). The main contribution of the forward search in the framework of rating system is the possibility to improve the classification rule avoiding the influence of outlying firms.

2 Brief Description of the Method: Generalized Linear Models and the Forward Search

In the literature about insolvency prediction, linear discriminant analysis models and multiple logistic regression models have been widely used to discriminate between failed and non-failed firms on the basis of financial ratios. Altmans popular Z-Score and Ohlsons O-Score are, for example, respectively based on linear discriminant analysis and on logistic regression. Neural network models are powerful and have become a popular alternative with the ability to incorporate a very large number of features in an adaptive nonlinear model. For a survey of business failure classification models see, for example, Hand and Henley (1997) and, more recently, Giudici (2003). In the present paper, we adopt generalized linear models (GLM) which are a family of models including logistic regression as a special case. The choice of GLM can be justified in various ways: the first relevant reason is that it is a good compromise between linear discriminant analysis which can be applied only under strict conditions (such as equality of covariance matrices) and nonlinear methods (such as neural networks and genetic algorithms) which are nonparametric models and generally show good forecasting performances, but are black-boxes hard to interpret. Another reason to prefer GLM with respect to discriminant analysis is that GLM methodology allows to use categorical variables. With respect to previous paper based on logistic regression two aspects of the present paper are original: 1) the application of a robust analysis based on the forward search, 2) the introduction of links different from logit which could lead, for some data sets, to better forecasting performances. When handling insolvency data it is natural to label one of the categories as success (healthy) and the other as failure (default) and to assign these the values 0 and 1 respectively. Generally speaking, let Y be the binary response variable which can assume two values according to a particular event which can happen (success) or not (unsuccess) defined as follows:

$$y_i = \begin{cases} 0 & \text{unsuccess} \\ 1 & \text{success} \end{cases} \tag{1}$$

Let $E(Y_i) = \mu_i$ and $\mu_i = x_i'\beta$, where x_i is a vector containing the values of the explanatory variables for the $i-th$ unit and β the corresponding parameter vector, the linear predictor η_i and the mean μ_i are related by the link function

$$g(\mu_i) = \eta_i = x_i'\beta. \tag{2}$$

Binary data, such as defaulting and healthy firms, require a link such that the mean lies between zero and one. The most widely used links for binary data that satisfy this properties are logit, log-log and complementary log-log (cloglog from now on) which are reported as follows:

$$\begin{array}{rl} g(\mu_i) = \log\left(\frac{\mu_i}{1-\mu_i}\right), & \text{logit} \\ g(\mu_i) = \log(\log(\mu_i)), & \text{log-log} \\ g(\mu_i) = \log(-\log(1-\mu_i)), & \text{clog-log} \end{array}$$

As GLM estimates are strongly influenced by outliers, we apply a forward search analysis. The forward search is a general method which has been introduced originally in linear regression models and subsequently extended to other fields of statistics such as multivariate techniques (Atkinson et al. (2004)), structural time series models (Riani, (2004)) and financial time series models (Grossi and Laurini (2004)). The main steps of this procedure are:

1. identification of a basic subset free from outliers. In the case of linear regression models least median of squares estimators are used in order to select the units belonging to the basic subset;
2. ordering of observations according to their degree of accordance to the underlying model using, in the case of linear regression, squared residuals computed on a subset of m observations. The subset size is increased from m to $m+1$ by selecting the least outlying observations from the previous graduation.
3. monitoring of statistics, such as parameter estimates, t-values, and so on along each step of the search.

The output is a complete monitoring of estimates which do not suffer from masking and smearing effects typical of classical backward methods for outlier detection. The forward search for generalized linear models is similar to that for linear regression except that we replace squared least squares residuals with squared deviance residuals. Another point which deserves to be stressed with respect to linear regression models is that, as we are analyzing binary data, we have to avoid including during the search only observations of one kind. This can be done through a balanced search in order to maintain a balance of both kinds of firms (bad and good), that is the ratio of bad and good in the various subsets of the procedure is maintained as close as possible to the ratios in the complete set of n observations (Riani and Atkinson (2001)). For a detailed explanation of the steps of the forward search in generalized linear models, see Atkinson and Riani (2000).

3 Data, Model Application and Main Results

3.1 Brief Description of the Data

The sample is composed of 653 firms from a database of a medium sized financial institution. Sample units have been selected randomly stratifying by industry and by size and excluding both small (turnover under 1 million) and large companies (turnover over 10 million). The remaining firms represent the 10 most important business areas defined by the Bank of Italy. The sample represents roughly 40% of the universe of firms with the former requirements. Because the model we use is based on maximum likelihood and does not work with missing values, we omit from the analysis units which contain a missing value in at least one of the explanatory variables. The number of firms on which we conduct the analysis becomes 523. We merge the December 2004 release of the financial institution database with the December 2004 release of "Centrale dei Rischi" database. The first database contains qualitative variables such as juridical form, economic sector, etc. joined with information on balance sheet for years 2001, 2002, 2003. Into the second we find the monthly history of firms financial exposure with respect to the whole financial system over the period January 2004-November 2004. The total sample of firms has been divided into two groups: healthy and insolvent firms. The first one, excluding firms with missing values, is formed by 456 firms and the second by 67. We define insolvent the firms both defaulted (bankrupted or not) and likely to become defaulting in the near future according with the financial institution qualitative credit rating. Because of the large number of variables found to be sensible indicators of corporate economic capabilities, we decide to use some popular indicators concerning liquidity, profitability, leverage and solvency obtained from balance sheet. We concentrate on year 2003 data. To introduce a dynamic view of the economic performance we calculate the variation of balance ratios in year 2003 with respect to the previous year 2002. In addition, we use some Centrale dei Rischi indicators in order to give evidence of the relationship between the amount of credit the overall banking system offers to the firm (Accorded), the amount of credit used by firms (Usage), the credit usage over the limit fixed by the bank (OverUsage) and warranties supplied (Warranties). In particular, we resume the monthly variables computing the arithmetic mean in the period January 2004-November 2004 and focus the analysis on the following ratios: Usage/Accorded, OverUsage/Accorded, Warranties/Accorded. The final design matrix is formed by 38 variables: juridical form, economic sector, 15 balance ratios, 15 variations of balance ratios, 6 ratios derived from Centrale dei Rischi database. The complete list of variables is not reported for lack of space, but it is available by the authors.

3.2 Construction of Explanatory Variables

Considering the high number of variables, a first challenge of our analysis is to verify whether some variables are more powerful in discriminating healthy from insolvent firms. This selection has been made by means of a forward variable selection method based on the likelihood ratio under the hypothesis of the logit link. That procedure leads to the choice of five variables: i) spread in 2003; ii) banking debts over turnover in 2003; iii) average credit delay in 2003; iv) variation of the duration of monetary cycle in 2003 with respect to 2002; v) usage/accorded with respect to the whole banking system in 2004. Spread in 2003 is a profitability indicator that is obtained deducting from ROA (operational income/total assets) the cost of debt (financial interests/debts). This ratio stresses that firms with higher operational profitability, cleaned from financial interest costs, are usually more likely to be solvent than less profitable firms. The relationship between banking debts at the end of the year 2003 and year turnover gives evidence of firm's financial structure. From an economical point of view, traditionally, firms with a weak financial structure are usually more likely to become insolvent than those with stronger financial structure. The delay accorded to customers to pay their debt is usually expression of the trading power of the firm into the market. When the firm has some marketing difficulties the average delay accorded usually increases. The monetary cycle duration is obtained by difference between the delay in paying suppliers (in days) and the sum of: a) delay accorded to customers (in days); b) stock period (in days). Our analysis emphasizes the role played by the evolution of the duration of monetary cycle from year 2002 till year 2003. The growth of monetary cycle duration shows financial difficulties that will probably lead to insolvency. The ratio usage/accorded stresses the relationship between credit accorded by the whole banking system to the firm and its use of the credit. When a firm reaches high level of credit usage (compared to accorded credit), it is likely that its financial capabilities are not really strong and it will probably fall into insolvency.

3.3 Application of the Forward Search to GLM

Using the selected indicators we estimate three generalized linear models for binary data applying the links cited in section 2.

Figure 1 reports the goodness of link test during the last 60 steps of the forward search. According to the forward search approach, the first values of the lines represented on the figure are computed on the basis of the first 464 observations included in the subset ordered considering the degree of accordance to the model omitting the remaining observations. At each step of the search further observations are included in the subset by respecting the rule of minimising the squared deviance residuals. In the last step, which in the figure corresponds to the extreme right values of the lines, the test is computed using all observations. Horizontal lines indicate 5% asymptotic

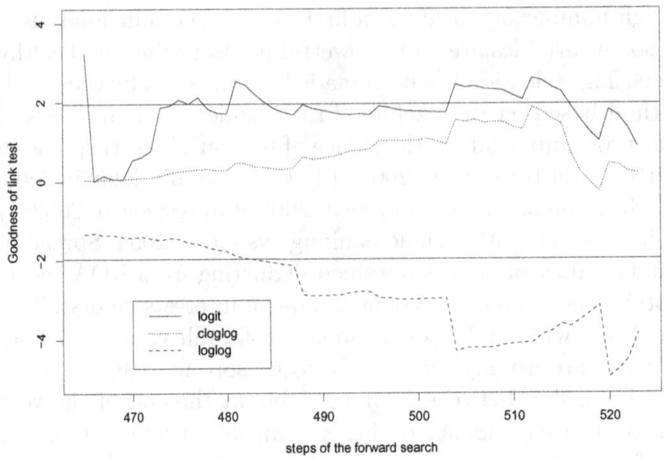

Fig. 1. Goodness of link test along the forward search: logit, cloglog and loglog links

confidence region. As can be noted, the cloglog link (dotted lines) always lies inside the region during the search, the logit link lies across the upper limit, whereas the loglog link lies outside the region in the majority of the steps and must be surely rejected. Thus, the best link looks to be the cloglog even if the logit is very close to the acceptance region. It is worth stressing that the difference of the three links is not so clear observing the values based on the entire sample (last value of the lines).

LOGIT				
		Predicted		
		Good	Sufference	Total
Observed	Good	0.97	0.03	1
	Sufference	0.33	0.67	1
	Total	0.89	0.11	1

CLOGLOG				
		Predicted		
		Good	Sufference	Total
Observed	Good	0.98	0.02	1
	Sufference	0.43	0.57	1
	Total	0.91	0.09	1

Table 1. Classification error of logit and cloglog links at the end of the forward search

Logit and cloglog models have been both estimated and lead to very similar results (see Table 1): the total classification error is about 7% with both models when all units are included in the estimation procedure. Nevertheless, the logit link returns a classification error of insolvent firms (insolvent firms classified as healthy) substantially lower than that given by the cloglog link: 33% vs 43% given all observations. Thus, the cloglog better classifies healthy firms (classification error of healthy firms: 1.7% vs 3.1%) and worse the insolvent. From the financial institution point of view, it is more serious to misclassify an insolvent firm as healthy than the opposite and the logit link should be preferred. The analysis of classification errors given till now does not consider the presence of outliers. To this purpose the robust analysis of the forward search could be very useful.

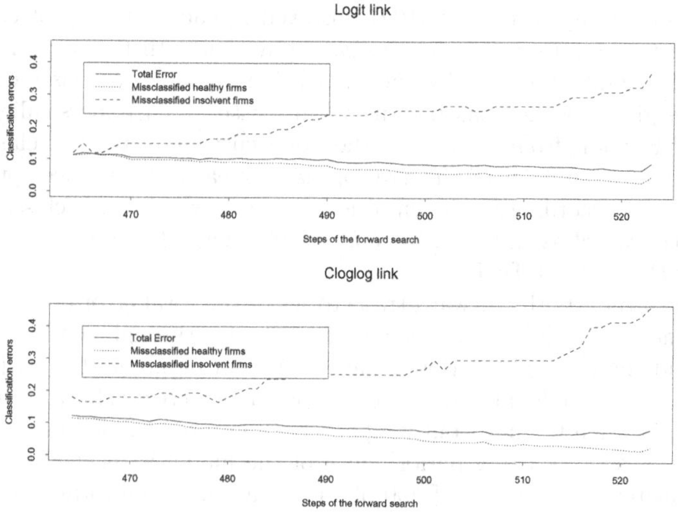

Fig. 2. Classification errors during the forward search: logit (upper panel) and cloglog (lower panel) link

Figure 2 shows trajectories of classification errors during the forward search for logit and cloglog links. It is very interesting to note that adding observations to the initial subset causes a slow decrease of total classification error and error in classifying healthy firms (lower lines). On the opposite the proportion of misclassified insolvent firms increases very quickly: in the case of the logit link this error goes from about 10% when the subset is composed by 460 observations to 33% corresponding to estimates on the whole sample. This particular behavior can be explained considering that observations which are included in the last steps are healthy firms which are financially similar to insolvent firms and wrongly influence the classification rule. For example, the last observation included by the forward search is a healthy firm with debts to banks three times greater than total sales: this ratio is clearly greater than the maximum values among insolvent firms. Finally, observing Figure 2, we gain that the cloglog link is more influenced by outliers with respect to logit: the trajectory of misclassified insolvent firms presents a neat jump around step 515 and, around step 500, classification error of insolvent firms with the two links is equivalent.

3.4 The Forward Search on Rating System

The Basel Agreement states that banks can use internal rating systems to determine the regulatory capital according to the ability of customers to refund their debts (see for example Altman and Saunders (2001)). In this section we evaluate the ability of the model estimated in the previous section to forecast the right rating class of borrowers. This goal can be reached comparing the average probability of default (PD) observed in each category of customer according to the internal qualitative rating system with the posterior probability of default predicted by the model. The qualitative classification of customers given by the bank identifies four classes of customers called A, B, C and DEF, going from the best to the worst customers, the last class being composed by defaulted firms. The rating class for each firm has been predicted by the model according to the empirical PD observed in each class, that is: $(PD < 0.0023)$: class A, $(0.0023 < PD < 0.02)$: class B, $(0.02 < PD < 0.28)$: class C, $(PD > 0.28)$: DEF.

Figure 3 reports the forward trajectories of the total error obtained forecasting the qualitative rating system of the banks through the logit (upper panel) and the cloglog link (lower panel). As can be noted, the trajectories remain roughly stable until step 500 and begin to increase sharply after that step, that is when the most outlying observations are included in the subset. Thus, the best forecasting performance of the model is reached estimating the parameters on a subset of 500 observation out of the total sample size. Note that the forecasting error of the model is computed at each step of the procedure by excluding the observations less in accordance with the underlying model, which can be considered as out of sample. Therefore, the forward search gives in-sample forecasts for observations included in the main subset and out-of-sample forecasts for the remaining units.

4 Final Remarks and Extensions for Further Research

Starting from the statements of the Basel Agreement, we have analyzed a method to classify bank customers according to their future ability to refund money. In the literature about rating system the problem of influential observations has not been deeply analyzed. Notwithstanding outliers can strongly bias model estimates and the forecasting performance of the procedure. In this paper a robust analysis of generalized linear models for the classification of firms has been presented. The robustification of the models has been

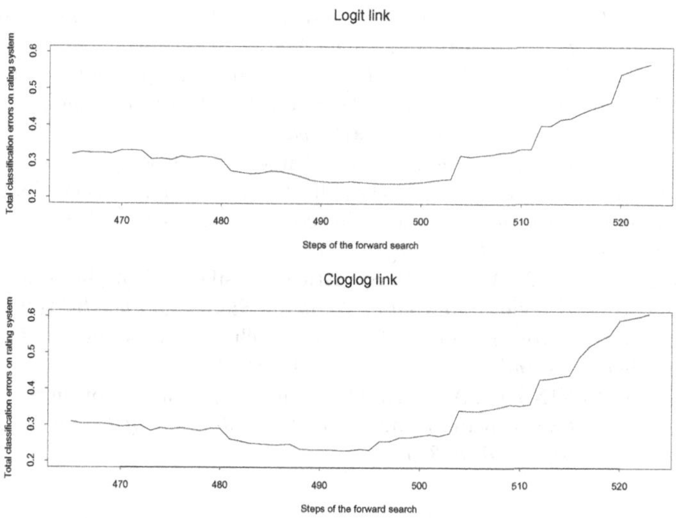

Fig. 3. Total classification errors using the categories of the internal rating system along the last 60 steps the forward search: logit (upper panel) and cloglog (lower panel) link

made through the forward search which is an iterative procedure which allows to monitor the influence of single o group of observations on the model estimates. Among all possible link functions which can be used in the GLM framework, the application of a goodness of link test suggested by Atkinson and Riani (2000) has restricted the attention to logit and cloglog links. The forward search analysis has shown that the presence of outlying firms can dramatically influence the classification errors, particularly that which leads to misclassify insolvent firms. In this paper five variables have been selected on the basis of the likelihood ratio under the hypothesis of a logit link and without considering the influence of outliers. Greater attention deserve in the next future the selection of variables which should be integrated in the forward search. Another point which should be deepened in future research regards the development of calibratory tools to evaluate how significant is the influence of observations on parameters. This will be done by means of simulated envelopes.

References

ALTMAN, E.I. and SAUNDERS, A. (2001): An analysis and critique of the BIS proposal on capital adequacy and ratings. *Journal of Banking and Finance, 25, 25–46.*

ATKINSON, A.C. and RIANI, M. (2000): *Robust Diagnostic Regression Analysis.* Springer Verlag, New York.

ATKINSON, A.C., RIANI, M. and CERIOLI, A. (2004): *Exploring Multivariate Data with the Forward Search.* Springer Verlag, New York.

BANK for INTERNATIONAL SETTLEMENTS (2004): International Convergence of Capital Measurement and Capital Standards: a Revised Framework. *Report, Basel, http://www.bis.org/publ/bcbs107.pdf.*

GIUDICI, P. (2003): *Applied Data Mining.* Wiley, West Sussex, U.K.

GROSSI, L. and LAURINI, F. (2004): Effects of extremal observations on the test for heteroscedastic components in economic time series. *Applied Stochastic Models in Business and Industry, 20, 115–130.*

HAND, D. and HENLEY, W. (1997): Statistical classification methods in consumer credit scoring: A review. *Journal of the Royal Statistical Society, 160, 523–541.*

RIANI, M. (2004): Extension of the forward search to time series. *Studies in non Linear Dynamics and Econometrics, 8, Article 1.*

RIANI, M. and ATKINSON, A.C. (2001): A unified approach to outliers, influence, and transformations in discriminant analysis. *Journal of Computational and Graphical Statistics, 10, 513–544.*

Classification of Financial Returns According to Thresholds Exceedances

Fabrizio Laurini

Department of Economics
University of Parma, Italy
flaurini@stat.unipd.it

Abstract. Properties of a panel of financial time series are explored, aiming at classifying market shares according to their extremal returns behaviour. Existing methods for optimal portfolio selection involve estimation of correlation coefficient, whose properties for measuring dependence in financial time series are questionable. Alternatively, for stationary processes of financial returns, the mean size of cluster of thresholds exceedances leads to define a measure of extremal dependence more accurate than correlation. Further functionals that might help to optimal portfolio selection, are, for instance, the total loss occurred to a stock during an extreme event or the time-length duration of a loss in a stress period. Combining functionals of financial returns it is possible to clustering shares properly and setting up a tool for portfolio selection. The performance of this method is assessed, through an application to real financial time series, by means of standard Markowitz theory of optimal selection of shares.

1 Introduction and Motivation

Financial institutions face problems of optimal portfolio selection, its optimization and its hedging over time. Classification methods based on cluster analysis might be useful in portfolio selection as they provide a set of additional tools, yet somehow qualitative, for exploring common features among market shares.

For stock prices P_t, $t = 1, \dots, T$ and the return process $X_t = \log(P_t) - \log(P_{t-1})$ a panel of portfolio selection is usually carried adopting an optimization algorithm which involves estimation of mean, variance and covariance over time of X_t (typically stationary). According to optimal portfolio selection, share allocation is chosen in order to minimize risk and maximize profits. In practice, shares belong to a portfolio according to their *average* historical behaviour, and future profits and losses are estimated with sample mean (expected return) and sample variance/covariance (riskiness). The pairwise analysis of correlations between returns gives suggestions for optimal selection (or asset allocation) and hedging of a portfolio, by suitably choosing proportions of shares having (somehow) opposite behaviour.

An empirical finding typical of financial time series returns is that extremes tend to occur in small clusters, which commonly arise independently each other. These clusters convey peculiar information about the extremal pattern of a share, that cannot be consistently estimated by any moments of the marginal distribution of X_t. Thus, as long as extreme behaviours of portfolio shares are observed, there is evidence suggesting to switch attention from "average" to "extreme" features of financial returns. Therefore, there may be interest in classifying financial data by grouping series having similar extreme pattern.

In order to classify returns according to their extreme behaviour, we only consider those returns that are observed during an extreme event (stress period). An extreme event is of kind $\{X_t > u\}$ where u is a suitably high threshold, typically a high quantile of the marginal distribution of X_t. Every share faces several stress periods over time and such periods may be very different from share to share. However, for a given share, stress periods are assumed to arise independently each other, and the target is to quantify properly what happens during all the stress periods of a return process. Shares with similar stress periods will belong to same group.

Formally, for a stationary time series, like financial returns, a relevant quantity that summarizes short-term extremal behaviour during a stress period is the extremal index $\theta_X \in (0, 1]$ (Leadbetter et al., 1983). Loosely, for a given stationary time series, θ_X^{-1} represents the average cluster size of extreme values.

However, the extremal index θ_X of a series is just a cluster functional, but coarser information is available by exploring further properties of clusters of threshold exceedances. Therefore, the approach followed here is to analyze functionals of clusters of threshold exceedances. For instance, θ_X^{-1} is the functional that takes independent clusters and averages their size. For a series of financial returns, functionals of practical interest are the average size and average duration of losses during a stress period. For each time series we provide estimates of these functionals of threshold exceedances.

In practice, the estimate of the extremal index θ_X of a stationary process X_t, $t \in \mathbb{Z}$, requires clusters of exceedances of a suitable high threshold u to be identified. Traditional methods, asymptotically equivalent (Smith and Weissman, 1994), seem to suffer from very high bias at levels of u adopted in practice, like 95-th or 99-th quantile of the marginal distribution of X_t. A generalization of such methods, which accounts for the volatility process of returns, is proposed by Laurini and Tawn (2003) where bias is considerably reduced.

Other approaches to classify clusters of thresholds exceedances can be used, since results from Hsing et al. (1988) allow to approximate the process of threshold exceedances with clustered non-homogeneous Poisson processes, where the intensity of the process depends on the value of the extremal index θ_X. Here, we do not focus on such point process approach.

The paper is organized as follows. Section 2 describes how clusters of exceedances can be identified in financial returns and their functionals evaluated. Section 3 shows the application of the proposed method for a panel of shares and includes some subsections where performances are explored using standard tools of portfolio selection. Finally, a last section closes with some comments and final remarks.

2 Clusters Identification and Functional Evaluation

We recall that for a given financial series of stock prices P_t, $t = 1, \ldots, T$ we can define returns as $X_t = \log P_t - \log P_{t-1}$. It is widely recognized that returns are stationary, uncorrelated but dependent. Discrete-time models for the volatility process σ_t take multiplicative form $X_t = \sigma_t \epsilon_t$, where ϵ_t is *iid* with finite mean and variance; for fixed t, σ_t and ϵ_t are assumed to be independent.

Extremes in returns series tend to occur in clusters and we identify clusters according to the algorithm of Laurini and Tawn (2003). Let $F_X(x) = \Pr\{X \leq x\}$ and $F_X^{\leftarrow}$ its inverse. Consider two thresholds u and c, and define $u^\star := F_X^{\leftarrow}(u)$ and $c^\star := F_\sigma^{\leftarrow}(c)$, with $u^\star \geq c^\star$ and define for $i < j$

$$L_{i,j}^{\sigma} = \min(\sigma_i, \ldots, \sigma_j) \text{ and } M_{i,j}^{X} = \max(X_i, \ldots, X_j)$$

with the interpretation that for $j < i$ the variables are not defined. Independent clusters C_g, $g = 1, \ldots, G$, are identified by sets satisfying the constraints

$$\{\{M_{2,m}^X \leq u\} \cup \{\bigcup_{i=2}^{m} V_{i,c,u}\}\} \bigcap \{X_1 > u\}$$

where the set

$$V_{i,c,u} = \{c < L_{2,i-1}^{\sigma}, M_{2,i}^{X} \leq u, \sigma_i \leq c\} \quad \text{for } i = 2, \ldots, m,$$

corresponds to the process remaining below u for variables $X_2, \ldots, X_{i-1}$ but dropping below c for the first time for σ_i. Thus clusters terminate either if there are $m - 1$ consecutive value of $X_j < u$ or if the conditional variance of the process becomes small enough between two exceedances. At this stage σ_t is assumed to follow a stationary GARCH(1,1) process, i.e., for some positive parameters a, b_1, b_2 the volatility process has form $\sigma_t^2 = a + b_1 X_{t-1}^2 + b_2 \sigma_{t-1}^2$.

For each identified cluster C_g of length N_g, cluster functionals we are interested in can be written as follows.

Cluster size $S := \sum_{g=1}^{G} (\sum_{j=1}^{N_g} I_{(X_j > u)})/G$
Cluster loss $L := \sum_{g=1}^{G} (\sum_{j=1}^{N_g} (X_j - u) I_{(X_j > u)})/G$
Cluster duration $D := \sum_{g=1}^{G} \left(\max_{j=1,\ldots,N_g} j I_{(X_j > u)} - \min_{j=1,\ldots,N_g} j I_{(X_j > u)} \right) /G$

where $I_{(X_j > u)}$ is the indicator function for a threshold exceedance within a cluster. Examples of such cluster functionals are given in Figure 1, where on top left we sketch two independent clusters of exceedances and, clockwise, we show cluster size, cluster loss and cluster duration respectively.

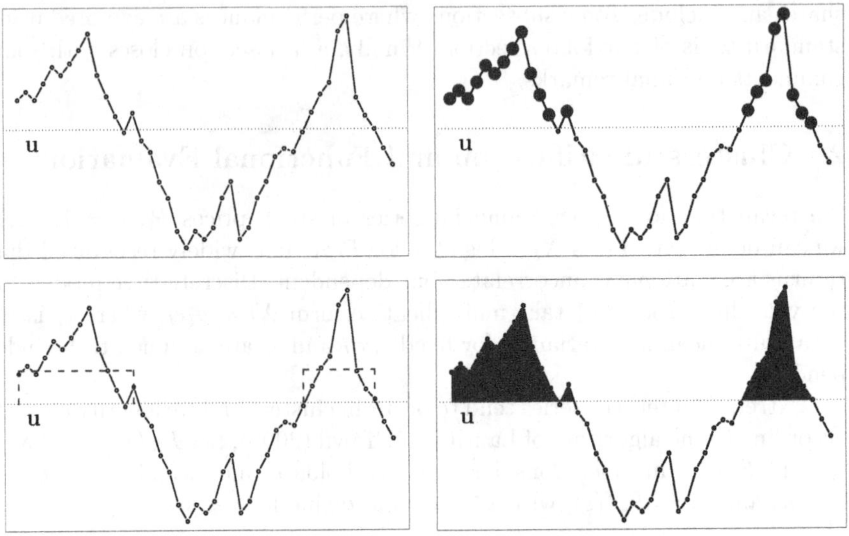

Fig. 1. Example: given two identified clusters (top left), consider the cluster functionals "cluster size" (big dots in top right panel), "cluster loss" (shaded area in bottom right panel) and "cluster duration" (dotted are in bottom left panel)

3 Application to Panel of Financial Returns

We use data from the MIB30 Italian stock index with daily observations from Jan 2000 to Dec 2002, and we link this analysis to that of Cerioli et al. (2005). Rather than the original 30 series analyzed by Cerioli et al. (2005) we focus on the 24 series that do not have missing values. For the analysis we are carrying in this work missing values rise numerical (and methodological) problems that we decided to avoid. Moreover, instead of considering separately FINECO and BIPOP, we decided to merge those shares, which nowadays are called FINECOGROUP.

3.1 Analysis of Average Features

In this section we explore the analysis that can be carried with standard Markowitz approach of portfolio selection (see Markowitz, 1952). Markowitz portfolio selection assumes to choose shares that have minimum risk among those with highest expected return. Minimum risk has to be intended as smallest sample variance. This approach, which is not suited to hedging portfolios, has the drawback of using the variance as a measure of risk, and recent research in financial econometrics has highlighted that small (historical) sample variance does not imply low riskiness. An introduction of Markowitz theory and Capital Asset Price Model (CAPM) can be found in Amenc and Le Sourd (2003).

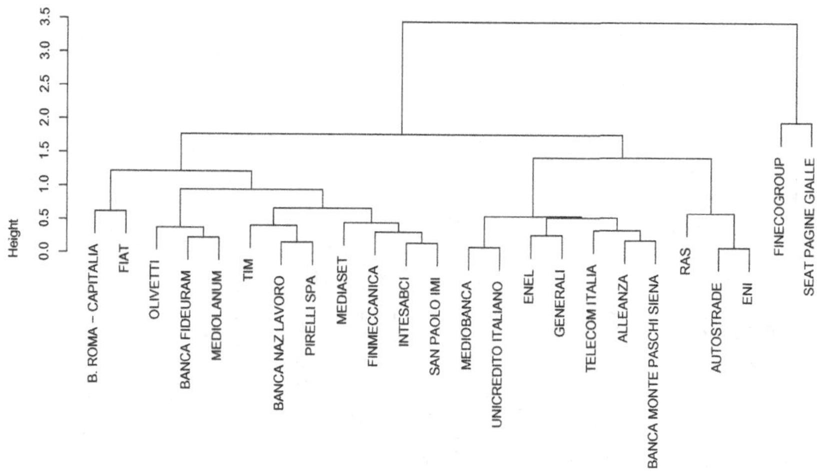

Fig. 2. Dendrogram with classification of average features. This dendrogram is obtained by choosing the Euclidean distance for sample mean and sample variance (standardized) and average method of aggregation

The motivation of our approach to classify shares differently than what suggested by Markowitz and CAPM theory is linked to standard analysis that can be performed on portfolio by (among others) the `R` package `fPortfolio` (R Development Core Team, 2005), which compute a k-means and hierarchical clustering on the portfolio of shares.

With our data the best allocation achieved with k-means method is when the number of clusters is $N_g = 3$, according to Kalinski index. Results of the hierarchical method of clustering are broadly consistent and we discuss this issue as follow. In order to obtain the dendrogram we consider each share as a single unit, recording sample mean and sample variance and setting up a distance based on such sample measures (suitably standardized). The hierarchical clustering, is derived by choosing the Euclidean distance and by adopting the "average method" for aggregation.

Figure 2 shows the dendrogram obtained by hierarchical clustering, and it is broadly consistent with results obtained by the `R` package `fPortfolio`. Therefore, by using either hierarchical or non-hierarchical methods with average features we cannot conclude whether we have a group of "risky" share and a group of "non-risky" assets.

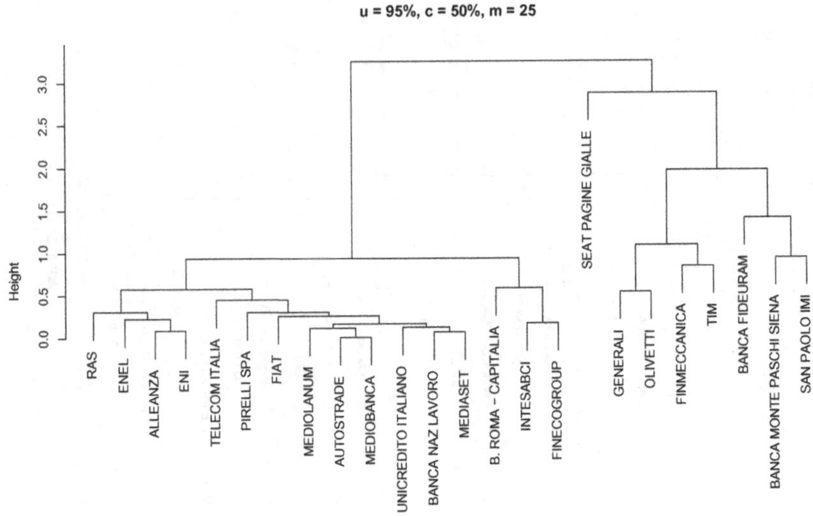

Fig. 3. Dendrogram with classification of extreme features. The dendrogram is obtained by choosing the Euclidean distance for sample extremes functionals introduced in Section 2, suitably standardized, and the average method of aggregation. Choice of thresholds are $u = F_X(0.95)$, $c = F_\sigma(0.5)$ and runs length is $m = 25$

3.2 Analysis of Extreme Features

In financial econometrics literature it is widely recognized that sample variance is not suited to measure risk, and many researchers assume the variance is not constant over time. Therefore, the Markowitz and CAPM methods can dramatically fail at providing an optimal solution for portfolio selection. We propose to investigate properties of portfolio of shares by studying local extreme values for each financial time series as follows.

For each observed series of returns, we identify clusters of extremes observations and evaluate cluster functionals according to what introduced in Section 2. The choice of parameters for cluster identification is $u = F_X(0.95)$, $c = F_\sigma(0.5)$ and $m = 25$, but results are robust to other parameters' choice. For each returns series we recall that the volatility series $\hat{\sigma}_t$ is obtained by fitting a GARCH(1,1) model.

Figure 3 shows the dendrogram which is obtained after standardization of cluster functionals, following exactly the same approach of dendrogram in Figure 2.

From both figures it seems that homogenous groups can be identified. It is less obvious whether we have two or three groups (but we assume to have identified the three groups GR.A.1: B.ROMA–SAN PAOLO; GR.A.2: MEDIOBANCA–ENI; GR.A.3: FINECOGROUP & SEAT) in Figure 2.

There is evidence for identification of two different groups, from dendrogram of Figure 3, which is obtained using thresholds exceedances and their functional evaluation. From dendrogram in Figure 3 we observe a low-risk group including RAS–FINECOGROUP (GR.E.1) and a much more risky group SEAT–SAN PAOLO (GR.E.2). The main difference from the two methods, which come by comparing Figure 2 and Figure 3, is related to the split of FINECOGROUP & SEAT, and the inclusion of several risky shares in GR.E.2 when extreme features are considered.

3.3 Comparing Market Efficiency

Differences among the two methods studied so far ("average" and "extreme") can be assessed by comparing the market efficiency of each identified group through the equilibrium estimate of Markowitz theory. Such equilibrium can be assessed by the tangency of the Capital Market Line with curve of efficient frontier. We have explored the market equilibrium considering:

- all data of portfolio composed by the 24 time series (where an equilibrium exists);
- 3 groups identified by estimation of average features (equilibrium is found in one group only);
- 2 groups identified by functionals of thresholds exceedances (equilibrium is found in one group).

Results are compared in Figure 4 and Figure 5, which are the standard outputs routinely analyzed using in the R package `fPortfolio`. In all plots, the x-axis an y-axis are, respectively, the expected return and expected (marginal) variability. The point of equilibrium (when exists) is denoted with the symbol $\oplus$. For each panel we report the scatter of all shares, labeled with numbers, and the line of the efficient frontier (gray dotted line). The point of equilibrium is reached by the best allocation of shares, that might involve a weight set equal to zero for some shares. We also report the point of equal weights portfolio with the symbol $\blacktriangle$.

By exploring Figure 4 we can draw some interesting conclusions. To start with, equilibrium is not always obtainable, suggesting that CAPM theory can fail in finding an optimal set of weights. However, when the equilibrium is found (all shares, GR.A.2 and GR.E.1) differences of performance seem negligible. Notice that in some cases (GR.A.3) the equilibrium does not exist, but we can plot the efficient frontier and assess the performance when we consider the equal weights portfolio.

It seems that by identifying shares according to thresholds exceedances we can provide an alternative method to CAPM. When CAPM perform well, the approach that we have introduced performs equally well. Therefore, yet

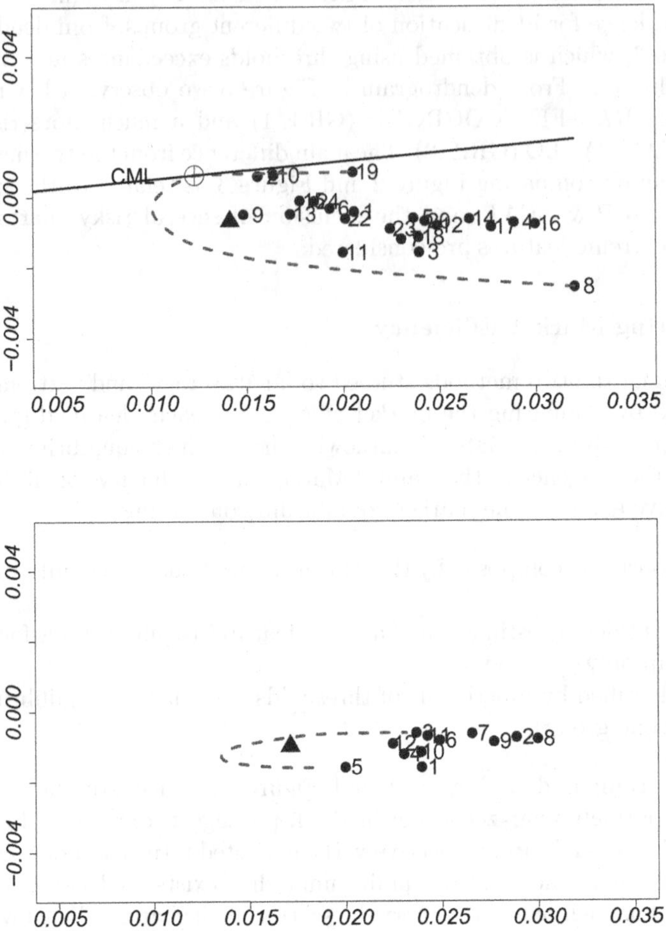

Fig. 4. Efficient frontier for portfolio. Top panel refers to the set of all shares. Bottom panel refers to the combination of shares belonging to group GR.A.3, where equilibrium cannot be found

not mathematically justified, the method analyzed here has the advantage of measuring risks suitably, by exploring local changes in the volatility process. Setting up the mathematical background for a proper optimization tool seems a promising avenue for future research. In principle this can be achieved by the maximization of expected returns under restriction that a functional of threshold exceedance is fixed at some suitable level (e.g. the extremal index $\theta_X = \max$ or cluster loss $L = \min$).

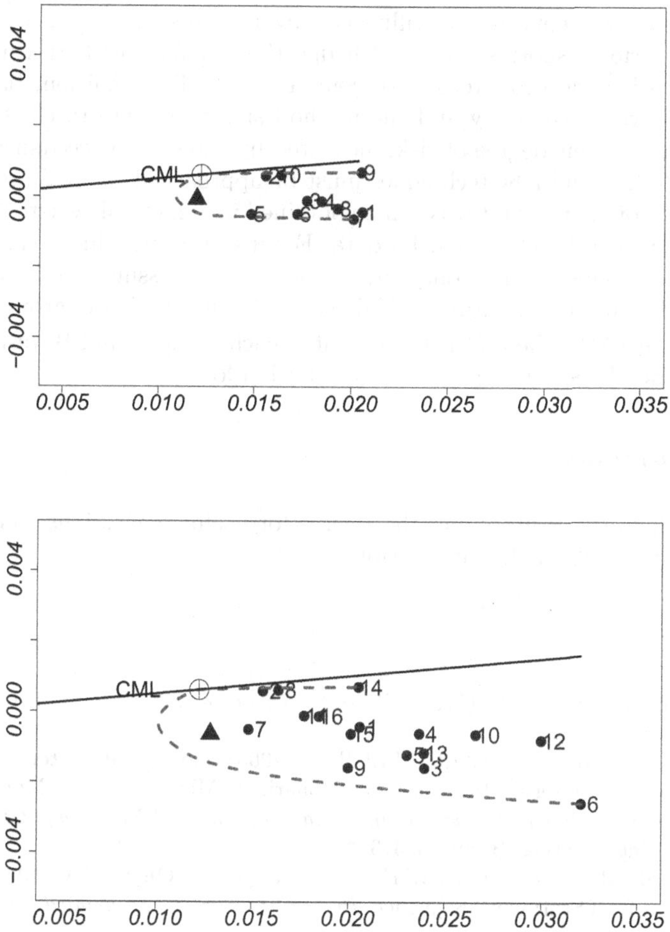

Fig. 5. Efficient frontier for portfolio with combination of shares belonging to group GR.A.2 (top panel), and efficient frontier for portfolio with shares belonging to group GR.E.1 (bottom panel)

4 Some Final Remarks

A point raised by comparing performances of two methods of classifying financial returns, is that a portfolio selected according to average features suffers from high uncertainty to market conditions, but further insights can be provided by exploring within-group properties and by simulating scenarios using Monte Carlo. However, the suggestion we make to reduce risks, is to select shares in a portfolio within the group GR.E.1.

Methods for optimal share allocation and portfolio selection might be strongly influenced by business cycles in stock markets (e.g. "bull" or "bear" periods). Markowitz theory is not influenced by business cycles, since investors might perform short selling and hedge the portfolio with risk-free securities, which include short-term zero coupon bonds. By definition, risk-free securities have no volatility and the method suggested here to classify shares according to their degree of risk, measured by threshold exceedances, would eventually fail and new techniques must be approached.

Classification of financial returns can be obtained with methods where the number of groups can be estimated by data. For instance, in a hierarchical approach, define J cluster functionals $\vartheta_1, \ldots, \vartheta_J$; we can assume that each cluster feature belongs to a mixture of K densities f_l (the number of groups), i.e. $\vartheta_1 \sim \sum_{l=1}^{K} f_l(\mu_i) P_i$, where P_i is the weight of each group. A full Bayesian setting is provided by specifying a prior distribution for μ_i.

Acknowledgments

A special thank to the Editors and the referee for useful suggestions which improved the first draft of the manuscript.

References

AMENC, N. and LE SOURD, V. (2003): *Portfolio Theory and Performance Analysis.* Wiley, New York.

CERIOLI, A., LAURINI, F. and CORBELLINI, A. (2005): Functional cluster analysis of financial time series. In: Vichi, M., Monari, P., Mignani, S. and Montanari, A. (Eds.): *Studies in Classification, Data Analysis, and Knowledge Organization.* Springer-Verlag, Berlin, 333–342.

HSING, T., HÜSLER, J. and LEADBETTER, M.R. (1988): On the Exceedance Point Process for a Stationary Sequence. *Probability Theory and Related Fields, 78, 97–112.*

LAURINI, F. and TAWN, J.A. (2003): New estimators for the extremal index and other cluster characteristics. *Extremes, 6, 189–211.*

LEADBETTER, M.R., LINDGREN, G. and ROOTZÉN, H. (1983): *Extremes and Related Properties of Random Sequences and Processes.* Springer-Verlag, New York.

MARKOWITZ, H. (1952): Portfolio selection. *Journal of Finance, 7, 77–91.*

R DEVELOPMENT CORE TEAM (2005) R: A language and environment for statistical computing. *R Foundation for Statistical Computing, Wien, Austria. URL http://www.Rproject.org, ISBN 3900051070.*

SMITH, R.L. and WEISSMAN, I. (1994): Estimating the Extremal Index. *Journal of the Royal Statistical Society, Series B, 56, 515–528.*

Nonparametric Clustering of Seismic Events

Giada Adelfio[1], Marcello Chiodi[1], Luciana De Luca[2], and Dario Luzio[2]

[1] Dipartimento di Scienze Statistiche e Matematiche "Silvio Vianelli" (DSSM), Università di Palermo, Italy
adelfio@dssm.unipa.it, chiodi@unipa.it

[2] Dipartimento di Chimica e Fisica della Terra (CFTA)
Università di Palermo, Italy
deluca@unipa.it, luzio@unipa.it

Abstract. In this paper we propose a clustering technique, based on the maximization of the likelihood function defined from the generalization of a model for seismic activity (ETAS model, (Ogata (1988))), iteratively changing the partitioning of the events. In this context it is useful to apply models requiring the distinction between independent events (i.e. the background seismicity) and strongly correlated ones. This technique develops nonparametric estimation methods of the point process intensity function. To evaluate the goodness of fit of the model, from which the clustering method is implemented, residuals process analysis is used.

1 Introduction

Clustering methods, largely used in many applicative fields, are also a statistical tool useful to analyze the complexity of the seismic process, which can be considered as the superposition of two different components: the background seismicity, and the space-time clustered one, induced by main earthquakes. A cluster of earthquakes is formed by the main event of the sequence, foreshocks and aftershocks, that could, respectively, occur before and after the mainshock. On the other hand isolated events are spontaneous earthquakes that don't trigger a sequence of aftershocks. Space-time characteristics of principal earthquakes are close to those of a Poisson process that is stationary in time, since the probability of occurrence of future events is constant in time irrespectively of the past activity, even if nonhomogeneous in space.

Because of the different seismogenic features controlling the kind of seismic release of clustered and background seismicity (Adelfio et al. (2005)) several mathematical models, describing correlated and uncorrelated component of seismicity, are defined.

For example, to describe the magnitude frequency of principal events for the given time and space area, the Gutenberg-Richter law is mainly used (Console (2001)):

$$\log_{10} N(M) = a - b(M - M_0) \tag{1}$$

where $N(M)$ is the number of earthquakes with magnitude equal or larger than M in a given time interval, a is a measure of the level of seismicity, and

b, typically close to 1, describes the relative number of small events in a given interval of time, and M_0 is a threshold value for the magnitude.

On the other hand the time aftershock activity is well described by the modified Omori formula (Utsu (1961)), for which the occurrence rate of aftershocks at time t following the earthquake of time τ is described by:

$$g(t) = \frac{K}{(t - \tau + c)^p}, \quad \text{with } t > \tau \tag{2}$$

with K constant, c and p characteristic parameters of seismic activity of region. Like the parameter b in (1), p is useful for characterizing the pattern of seismicity, indicating the decay rate of aftershocks in time.

For the short-term (or real-term) prediction of seismicity and to estimate parameters of phenomenological laws we need a good definition of the earthquake clusters, that are superimposed to the background seismicity and shade its principal characteristics.

For these purposes, in this paper, after a brief description of the peculiarities of seismic data and the definition of aftershock seismicity model, a clustering technique is proposed. Considering the seismic activity as the realization of a particular point process, it iteratively finds out a partitioning of the events, for which the contributions to the likelihood function are maximized. Finally, to check the validity of the assumptions, a sketch of the residual analysis is reported.

2 Sources of Data and Peculiarities

To study the seismic history of an area seismic catalogs are used; they contain information about each earthquakes for which instrumental registrations are available, reporting the main variables identifying recorded events: the size of an earthquake in units of magnitude, a logarithmic measure of earthquake strength; the origin time, that is the date and time when earthquake occurs; the hypocenter of the earthquake that is the position on the surface of the earth (latitude and longitude) and a depth below this point (focal depth), which however is seldom used because of the uncertainty of its measure.

Because of uncertainty about the values of these quantities, especially for the space coordinates, catalogs often report also information about the quality of the location of events, useful for evaluating the reliability of an earthquake location. To improve the catalog estimates of locations sophisticated seismological methods of *relocation* are often used.

Depending on the magnitude of the earthquake, additional information is sometimes available, as waveforms or focal mechanisms. No catalog however can be retained complete, in the sense that some events of low magnitude may not appear. In fact events of low magnitude far from the seismological network are less likely to be detected. So any catalog should have a *threshold of completeness* (M_0), which is the value of magnitude for which all events with magnitude greater than M_0, occurred in the study space-time region, are observed.

3 Definition of ETAS Model

Models for space-time earthquakes occurrences can be distinguished in two widely class (Schoenberg and Bolt (2000)): probability models describing earthquakes catalogs as a realization of a branching or epidemic-type point process and models belonging to the wider class of Markov point processes (Ripley and Kelly (1977)), that assume previous events have an inhibiting effect to the following ones. The first type models could be identified with self-exciting point processes, while the second are represented by self-correcting processes as the strain-release model.

ETAS (epidemic type aftershocks-sequences) model (Ogata (1988), is a self-exciting point process, representing the activity of earthquakes in a region during a period of time, following a branching structure. In particular it could be considered as an extension of the Hawkes model (Hawkes (1971), which is a generalized Poisson cluster process associating to cluster centers a branching process of descendants. ETAS model hypotheses can be outlined as follows:

1. the background events are the immigrants in the branching process and each ancestor produces offspring independently, depending on its magnitude (triggering ability);
2. the time and space probability distribution for an offspring are function respectively of the time and space lag from its ancestor;
3. magnitude distribution of clustered events is independent from that of background seismicity.

To provide a quantitative evaluation of future seismic activity the conditional intensity function is crucial. It is p. In general, the conditional intensity function of a space-time point process, proportional to the probability that an event with magnitude M will take place at time t, in a point in space of coordinates (x, y), can be defined as:

$$\lambda(t, x, y|H_t) = \lim_{\Delta t, \Delta x, \Delta y \to 0} \frac{Pr_{\Delta t \Delta x \Delta y}(t, x, y|H_t)}{\Delta t \Delta x \Delta y} \tag{3}$$

where H_t is the space-time occurrence history of the process up to time t; $\Delta t, \Delta x, \Delta y$ are time and space infinitesimal increments; $Pr_{\Delta t \Delta x \Delta y}(t, x, y|H_t)$ represents the history-dependent probability that an events occurs in the volume $\{[t, t + \Delta t) \times [x, x + \Delta x) \times [y, y + \Delta y)\}$.

On the basis of ETAS model assumptions, the intensity function of this non-stationary Poisson Process, conditioned to the history H_t is:

$$\lambda(t, x, y \mid H_t) = f(x, y) + \sum_{k: t_k < t} f(x - x_k, y - y_k) \frac{K_k \exp[\alpha(M_k - M_0)]}{(t - t_k + c)^p}. \tag{4}$$

It describes seismic activity as the sum of the activity rate $f(x,y)$ of background activity, constant in time, and the triggering density, referred to the clustered component, following a branching structure. In (4) the triggered density represents the conditional distribution of events within the clusters, describing the occurrence of offspring shocks following the main event. It is obtained as the product of space, time and size functions: time aftershock activity is represented by a non stationary Poisson process according to the modified Omori formula, such that the occurrence rate of aftershocks is correlated with the mainshock's magnitude.

4 Clustering of Seismic Events

In literature several methods are proposed to decluster a catalog. Gardner and Knopoff (1974) identify and remove aftershocks (with the purpose of analyzing the features of the background seismicity) defining a rectangular time-space window around each mainshock. The size of the windows depends on the mainshocks magnitude. Reasenberg (1985) identifies aftershocks by modelling an interaction zone about each earthquake assuming that any earthquake that occurs within the interaction zone of a prior earthquake is an aftershock and should be considered statistically dependent on it

Zhuang et al. (2002), propose a stochastic method associating to each event a probability to be either a background event or an offspring generated by other events, based on the ETAS model for clustering patterns described in the previous section; a random assignment of events generates a thinned catalog, where events with a bigger probability of being mainshock are more likely included and a nonhomogeneous Poisson process is used to model their spatial intensity. This procedure identify two complementary subprocess of seismic process: the background subprocess and the cluster subprocess or the offspring process.

5 The Proposed Clustering Technique

The method of clustering here introduced is based on the local maximization of the likelihood function of ETAS model describing clustering phenomena in seismic activity. Differently by ETAS model, in our approach the shape of intensity function components in time and space doesn't depend on any parametric assumption. Since for a spatial-temporal point process the likelihood function is:

$$\log L = \sum_{i=1}^{n} [\log \lambda(x_i, y_i, t_i | H_t)] - \int_X \int_Y \int_0^T \lambda(x, y, t | H_t) dx dy dt$$

our iterative procedure clusters seismic events assigning each induced event of coordinates (x_i, y_i, t_i, M_i) to that mainshock of coordinates (x_k, y_k, t_k, M_k)

that maximizes the product

$$\hat{f}(x_i - x_k, y_i - y_k)\hat{g}(t_i - t_k)\hat{\kappa}(M_k) \qquad (5)$$

without making any parametric assumption about the shape of the intensity function $\hat{f}(\cdot)$ and $\hat{g}(\cdot)$. Even if inside each cluster the space and time component are separable, in the overall intensity function relative to the whole space and time domain there is not such separability, since the point process is assumed to be clustered on the space-time location of k mainshocks.

We estimate functions $\hat{f}(\cdot)$ and $\hat{g}(\cdot)$ inside the j-th cluster by a bivariate and univariate kernel estimator respectively:

$$\hat{f}(x, y) = \sum_{i=1}^{n_j} \frac{1}{2\pi h_x h_y} \exp\left[-\frac{(x - x_i)^2}{2h_x^2} - \frac{(y - y_i)^2}{2h_y^2}\right]$$

where m_j is the number of points of the j-th cluster and the summation is extended to the points belonging to that cluster.

The windows width h_x, h_y are found using a formula (Silverman, 1986, pag.45) which approximately minimizes the mean integrated squared error: $h_x = A\hat{\sigma}_x n^{-1/5}$, $h_y = A\hat{\sigma}_y n^{-1/5}$, where $\hat{\sigma}_x$ and $\hat{\sigma}_y$ are the estimated dispersions for each coordinate in each cluster, and A is a constant depending on the type of kernel (1.06 for normal kernel). We used a simple bivariate normal kernel with independent components, since the kind of smoothing allows however the estimation of complex geographical patterns, not only of ellyptical shape.

For the time density estimation, the univariate version of the above kernel estimator is used, and for the window width a multiple of the Silverman' s value is used, in order to obtain smoother estimates, since the decaying time should theoretically be smooth inside each cluster, starting from the time of the mainshock event.

For the $\kappa(M)$ triggering function, that describes the expected number of dependent shocks generated by a main event with size M_k, the functional form $\kappa(M_k) = \exp\{\beta(M_k - M_0)\}$ is used; β measures the magnitude efficiency of an earthquake in generating its offsprings and is estimated from data, and M_0 is the completeness threshold of the whole catalog.

5.1 Steps of the Algorithm

The algorithm has been implemented trough software R (R Development Core Team (2005)); it is an iterative procedure which consists of the following steps:

1. consider a starting classification of the events in the catalog; $N_a \Rightarrow$ number of events which in the starting classification have been identified as aftershocks; $N_s \Rightarrow$ number of events which in the starting classification have been identified as mainshocks and isolated events;

2. Estimate the space and time densities ($f(x_i - x_k, y_i - y_k)$ and $g(t_i - t_k)$, with $i = 1, \ldots, N_a$; $k = 1, \ldots, N_s$) and the magnitude function $\kappa(M_k)$ for each ith event, considering all N_s events of the background seismicity, ordered in time; write their product in the (i, k) cell of a $N_a \times N_s$ matrix, identified from the ith secondary event that we want to assign and from the kth principal event;
3. Locate the mainshock k_i that maximizes the product (5) for each i;
4. At the end of each iteration isolated events could be assigned to found clusters if their contribution to the likelihood function is greater than a fixed threshold l_1. Similarly, events assigned to clusters could become isolated events if their contribution to the likelihood function is lower than a given threshold l_2. l_1 and l_2 are chosen on the basis of the empirical distribution of the internal likelihood of each cluster and on the basis of the percentiles of a χ^2 distribution, the approximate distristribution of log-likelihood quantities.
5. At the end of each iteration this procedure returns the events classified in mainshocks (here identified with the maximum magnitude events in the clusters), aftershocks or foreshocks (corresponding to each mainshock) and isolated events.

   The iterative procedure stops when the current classification does not change after a whole loop of reassignment.

6 Analysis of the Results and the Residuals Process

This clustering technique is applied to a catalog containing 1756 seismic events occurred in the Southern Tyrrhenian Sea between January 1988 and October 2002. The seismicity of this zone consist mainly of aftershock sequences and more seldom of isolated events.

We did not compare our method with the result of the Zhuang's stocastic declustering based on the ETAS model, since parameters have different meanings and also because Zhuang's technique is more properly used on single aftershocks sequences, while the aim of our method is to deal with catalogues with multiple aftershocks sequences.

The starting classification found out 185 clusters. The iterative procedure stops at the 9th iteration, finding out 150 clusters, of which 7 have more than 10 earthquakes (Fig. 1). We observe that clusters with few events tend to be disaggregated and their events moved to the nearest clusters or to the set of isolated events.

Starting from this result, residuals analysis techniques are used to justify the model as the basis of the clustering algorithm and to test the goodness of fit of the model to the specific space and time patterns of data.

In a time point process, residuals, obtained by an integral transformation of times occurrence, consist of an homogeneous Poisson process with unit intensity (Cox and Isham (1980)). Since the transformed inter-arrival times are i.i.d. exponential random variables with unit mean, to assess the assumption of time independence of the principal events, the inter-times distribution of observed residuals is analyzed.

For spatial point process $\boldsymbol{X}$ in a bounded region $W \subset \Re^d$, $d \geq 2$, the lack of ordering in two dimensional space does not allow to generalize conditional intensity of a spatial-temporal or simply temporal process given the past of process up to time t. In this case other techniques, based on *Papangelou conditional intensity* $\lambda(u, \boldsymbol{x})$, but not shown here for the sake of brevity, are used.

On the basis of statistical and graphical results (see Fig. 2) we could conclude that principal events found out through our algorithm likely come from a Poisson process that is homogeneous in time even if nonhomogeneous in space.

7 Conclusive Remarks

The clustering technique here presented has been defined in a specific context: to solve the problem of clustering of a seismic catalog, starting from the likelihood function of a model chosen to describe the underlying clustering process. Although some aspects are not definitively solved, it is an advantageous method of classification. It finds out the classification of seismic events that maximizes the likelihood function of the point process modelling the seismic phenomena. It returns clusters of earthquake that have a good seismic interpretation and the estimation of the intensity function of the point process which has generated seismic events in space, time and magnitude domains.

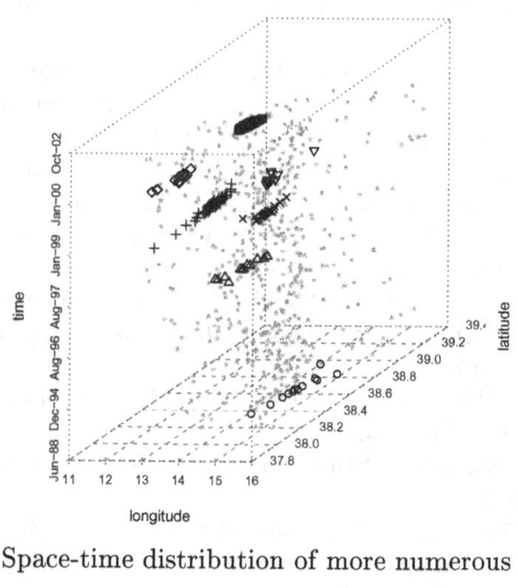

Fig. 1. Space-time distribution of more numerous clusters

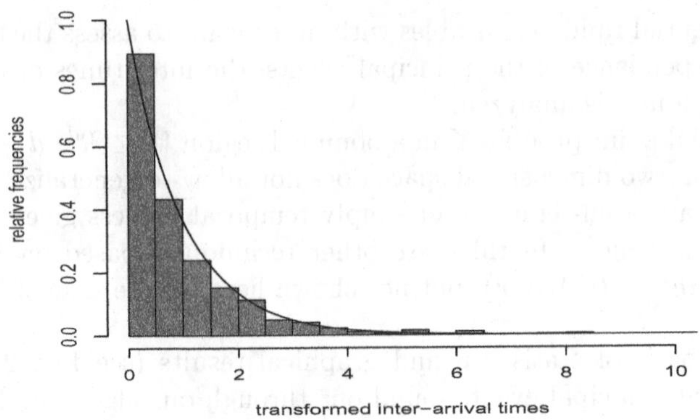

Fig. 2. Inter-times of temporal residuals: empirical and theoretical distribution

References

ADELFIO, G., CHIODI, M., DE LUCA, L., LUZIO, D. and VITALE, M. (2005): Southern-Tyrrhenian seismicity in space-time-magnitude domain. Submitted for publication.

CONSOLE, R. and LOMBARDI, A.M. (2001): *Computer algorithms for testing earthquake forecasting hypotheses.* Istituto Nazionale di Geofisica e Vulcanologia, Roma.

COX, D.R. and ISHAM, V. (1980): *Point Processes.* Chapman and Hall.

GARDNER, J. and KNOPOFF, L. (1974): Is the sequence of Earthquakes in Outhern California, With Aftershock Removed, Poissonian?. *Bulletin of the seimological Society of America, 64, 1363-1367.*

HAWKES, A. and ADAMOPOULOS, L. (1973): Cluster models for erthquakes-regional comparison. *Bull. Int. Statist. Inst.,45 (3), 454-461.*

OGATA, Y. (1988): Statistical Models for Earthquake Occurrences and Residual Analysis for Point Processes. *Journal of the American Statistical Association, 83, 401, 9-27.*

R DEVELOPMENT CORE TEAM (2005) R: A language and environment for statistical computing. *R Foundation for Statistical Computing, Wien, Austria. URL http://www.Rproject.org, ISBN 3900051070.*

RESEMBERG, P. (1985): Second-order moment of central california seismicity. *Journal of Geophysical Research, 90, B7, 5479-5495.*

RIPLEY, B.D. and KELLY, F.P. (1977): Markov point processes. *Journal of the London Mathematical Society, 15, 188-192.*

SCHOENBERG, F. and BOLT, B. (2000): Short-term exciting, long-term correcting models for earthquake catalogs. *Bulletin of the Seismological Society of America, 90(4), 849–858.*

SILVERMAN, B.W. (1986): *Density Estimation for Statistics and Data Analysis.* Chapman and Hall, London.

UTSU, T. (1961): A statistical Study on the Occurrence of aftershocks. *Geophys. Mag., 30, 521-605.*

ZHUANG, J., OGATA, Y. and VERE-JONES, D. (2002): Stochastic Declustering of Space-Time Earthquake Occurrences. *Journal of the American Statistical Association, 97, 458, 369-379.*

A Non-Homogeneous Poisson Based Model for Daily Rainfall Data

Alberto Lombardo[1] and Antonina Pirrotta[2]

[1] Dipartimento di Tecnologia Meccanica, Produzione e Ingegneria Gestionale Università di Palermo, lombardo@dtpm.unipa.it
[2] Dipartimento Ingegneria Strutturale e Geotecnica Università di Palermo, pirrotta@stru.diseg.unipa.it

Abstract. In this paper we report some results of the application of a new stochastic model applied to rainfall daily data. The Poisson models, characterized only by the expected rate of events (impulse occurrences, that is the mean number of impulses per unit time) and the assigned probability distribution of the phenomenon magnitude, do not take into consideration the datum regarding the duration of the occurrences, that is fundamental from a hydrological point of view. In order to describe the phenomenon in a way more adherent to its physical nature, we propose a new model simple and manageable. This model takes into account another random variable, representing the duration of the rainfall due to the same occurrence. Estimated parameters of both models and related confidence regions are obtained.

1 Introduction

One of the most popular stochastic model describing rainfall is the Neyman-Scott's model (Calenda and Napolitano, 1999). If a time correlation induced by the transaction matrix exists (transaction from rainy to non-rainy days and vice-versa), this model tries to eliminate such correlation, opportunely increasing the sampling time interval. In fact, as such interval increases, the correlation coefficients usually decay. This model seems more appropriate in regions with high rainfall and for data collected with a high sampling rate (say 10 minutes). In such cases it is possible to have interesting information about the modality characterizing the phenomenon, such as driving rain. Unfortunately, for the data object of our analysis – rainfall daily data from January 1, 1960 to December 31, 1998 recorded at a Sicilian hydrological site – such procedure has demonstrated inadequate. In fact, the transaction matrices remained non-homogeneous for all tried time interval (2 days, 3 days, ... , 7 days). This fact has leaded to believe that the Neymann-Scott's model is unfit for the data under study. In order to overcome this drawback, in this paper a new model is introduced; it is based on interpreting the rainfall phenomenon by properly filtering a Poisson white noise.

2 The Poisson Stochastic Process

In many cases of engineering interest, the random nature of particular loads – as wind, earthquake, rainfall – requires a stochastic representation model, consequently the response of the system needs to be adequately described in probabilistic sense by evaluating the response statistics (Stratonovich, 1963).

It is well known that, in several practical problems, the white noise could highly simulate such loads, in particular the process can be considered normal or non-normal.

The normal white noise is fully characterized by statistics up to the second order, while the non-normal process requires higher order statistics to be defined. The simple structure of normal white noise, by a mathematical point of view, explains the high popularity in using this process and sometimes the goodness of global statistic response encourages this use further on. Although the aforementioned considerations, the normal white noise is not proper in modelling point rainfall occurrences, because it does not reproduce fundamental physical characteristics of this event. Normal white noise is continuous and then it does not take in due consideration the intermittent characteristic of rainfall, in consequence of the alternate dry-wet structure of the rainfall itself. On the contrary, the Poisson white noise process, denoted as $W_p(t)$, is defined in the form

$$W_p(t) = \sum_{k=1}^{N(t)} Y_k \, \delta\left(t - T_k\right) \tag{1}$$

and consists of a train of Dirac's delta impulses $\delta(t - T_k)$ occurring at Poisson distributed random times T_k. $N(t)$ is the number of impulses in the time interval $[0, t)$ with initial condition $N(0) = 0$ with probability one. The random variable Y_k, that constitutes the spike amplitude at the corresponding time T_k, are assumed to be mutually independent and independent of the random instants T_k. This process is completely characterized by the expected rate $\lambda(t)$ of events (impulse occurrences), that is the mean number of impulses per unit time, and the assigned probability distribution of Y. The number of impulses $N(dt)$ in the time interval $t \div t + dt$ in mean is given by $E\left[N\left(dt\right)\right] = \lambda\left(t\right) dt$. The entire probabilistic description of the process $W_p(t)$ is given by its *cumulants*

$$K_{W_p}(t_1, t_2, \ldots, t_s) = \lambda\left(t\right) E\left[Y^s\right] \delta\left(t_1 - t_2\right) \ldots \delta\left(t_1 - t_s\right) \tag{2}$$

This model is simple and can be used for simulating rainfall events, assuming T_k as their initial point and Y_k as their magnitude, in order to estimate the statistics of hydrological response variables. However it does not contain the fundamental datum regarding duration and behaviour during the occurrence of the phenomenon. In fact, by an hydrological point of view, it is different if a same amount of rainfall occurs only in one day or in several consecutive days. We assume that the rainfall distribution within the day is negligible in any case.

In order to describe the phenomenon in a more exhaustive way, we propose a new model that is simple and manageable. This model takes into account another random variable R, representing the duration of the rainfall due to the same occurrence. Let us assume that the daily rainfall follows a law having a starting point (the first day of a cluster of rainy days) and an endpoint. A cluster is so constituted by at least one rainy day. The proposed model is a *linear stochastic differential equation driven by a non-homogeneous Poisson Process with stochastic parameters*

$$\dot{H}(t) = A(t)\,H(t) + B(t)\,W_p(t) \tag{3}$$

with $A(t)$ and $B(t)$ random variables ruling the decaying rate of the rainfall, i.e. the number of the days in the cluster and the cumulated amount of rainfall in that cluster, respectively.

Contrarily to model (1), this model:

i) allows to model intermittent rainfall processes with random duration of each cluster;
ii) admits that, if today it is rainy, tomorrow the probability to have a rainy day is higher;
iii) allows to use typical tools of the stochastic differential calculus for forecast models.

The differential equation can be studied shifting the time until to identify T_k with the origin, without taking into account of the previous clusters. That is, assuming the stationary condition, the response can be written as:

$$\dot{H}_k(t) = A_k H_k(t) + B_k Y_k \delta(t) \tag{4}$$

with A_k and B_k related to the kth cluster, $H_k(t)$ the relative volume of rainfall and $\dot{H}(t)$ its first derivative. The solution is known in closed form:

$$\begin{cases} H_k(t) = B_k Y_k e^{A_k t} & ; \ \forall t \geq 0 \\ H_k(t) = 0 & ; \ \forall t < 0 \end{cases} \tag{5}$$

The random parameters must guarantee that the whole rainfall in the cluster is Y_k and the duration of the kth cluster is R_k. For the first constrain we impose:

$$\int_0^\infty H_k(t)\,dt = Y_k = B_k Y_k \int_0^\infty e^{A_k t} dt \tag{6}$$

and, integrating, it results $A_k = -B_k$, so obtaining the following differential equation:

$$\dot{H}_k(t) = A_k H_k(t) - A_k Y_k \delta(t) \tag{7}$$

For the second constrain it is sufficient to impose that the slope of $H_k(t)$ intersects the temporal axis in R_k. It follows that $A_k = -1/R_k$. Therefore it is possible to write:

$$\dot{H}_k(t) = -\frac{1}{R_k} H_k(t) + \frac{1}{R_k} Y_k \delta(t) \tag{8}$$

and than the whole process is given by the stochastic differential equation:

$$\dot{H}(t) = -\frac{1}{R} H(t) + \frac{1}{R} \sum_{k=1}^{N(t)} Y_k \delta(t - T_k) = -\frac{1}{R} H(t) + \frac{1}{R} W_p(t) \tag{9}$$

where R is the length of clusters, that is a random variable with assigned distribution. This equation represents a differential equation excited by an external Poisson Process.

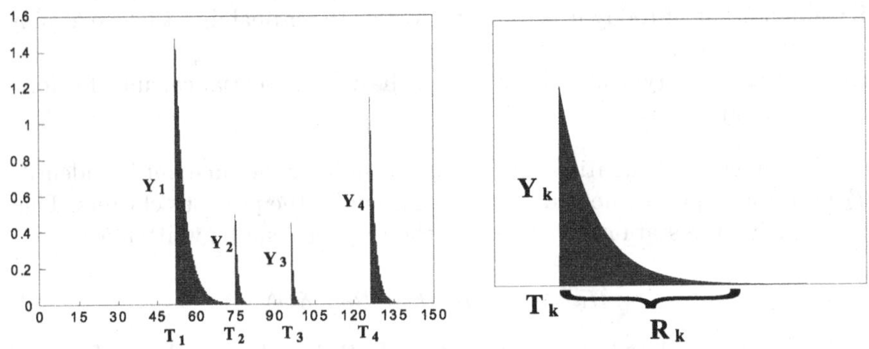

Fig. 1. A sampling of rainfall process modeled by (9)

For this equation the statistics of any order can be calculated (see Caddemi and Di Paola (1995), Di Paola and Falsone (1993 a, b), Di Paola and Vasta (1997), Di Paola (1997), Pirrotta (1998)). The moment of order j for each realization of the random variable R, say R_s, is given by:

$$E\left[H^j\right] = -\frac{j}{R_s} E\left[H^j\right] + \lambda \sum_{q=1}^{j} \frac{j! E\left[Y^q\right]}{q! R_s^q (j-q)!} E\left[\tilde{H}^{j-q}\right] \tag{10}$$

The previous formula describes a system of linear differential equations which solution is immediate for any j. The characteristic function and the density function are also easily obtainable. By looking at (10) it is apparent that any order of statistics depends on the expected rate $\lambda(t)$ of events (impulse occurrences), on $p_Y(y,t)$ and $p_R(r,t)$ (probability density function of whole rainfall in the cluster and the cluster duration, respectively).

The moments of R are obtainable by integrating the realization of the random variable R by its domain r and by integrating the differential equation (10), multiplied by the probability density function of R.

The proposed model is robust in the sense that is auto-adaptive, i.e. the further records of rainfall will improve the estimation of the model parameters making it more realistic.

3 Modelling Daily Rainfall Data in Sicily

When modelling rainfall in Sicily, the following considerations have to be taken into account:

- the arrivals of storm origin constitute a random variable with distribution according to a Poisson process, with expected rate of arrivals different month by month, peculiarity of a non-homogeneous Poisson process;
- the magnitudes of daily rainfall are characterized by a distribution that is independent of the arrival random variable;
- when a storm arrives it follows that, if today it is rainy, it will be very probable that tomorrow it will rain.

3.1 The Non-Homogeneous Poisson Model

The first model we tried to use, see (1), describes the arrivals by a non-homogenous Poisson model with rate varying month by month (Grunwald and Richard, 2000, Sitaraman, 1991). In order to estimate such rates, it is necessary to state the long period stationary condition, assuming that data belonging to the same day in different years are replicates of the same random variable. However, from some analyses it derives that the non-homogeneous Poisson model, even within each month, cannot be fit, because some conditions do not hold:

i) the occurrence must be a without memory process, i.e. the rainfall probability must be equal for both cases in which the day before was rainy or not; it implicates that:
ii) the waiting time for the next occurrence must follow the negative exponential distribution.

These conditions do not hold in the data under study, even within each month.

Looking for a way for overcoming this lack, we have split the data in two groups: days preceded by a rainy day {1} and days preceded by a non-rainy day {0}, and have carried out again the previous analysis, obtaining the results in Figure 3.1.

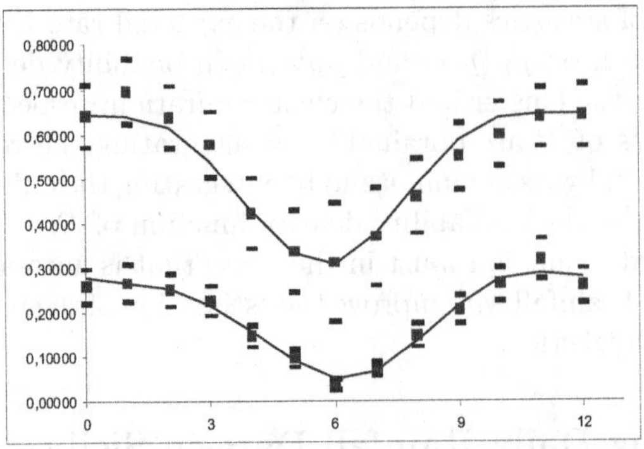

Fig. 2. Behaviour of the estimated probability of having a rainy day, and related confidence interval (95% confidence level), for different months, interpolated by a periodical function (with 2 harmonics), for days preceded by a non-rainy day (lower curve) and days preceded by a rainy day (upper curve)

In such way the model reverts to a *non-homogeneous double stochastic* (or semi-Markovian) model, because the rainfall probability depends, month by month, only on the condition of the previous state: the probability that tomorrow we will have a rainy day depends not only on the month, but also on the fact that today it is rainy or not. Now the tests we carried out (no memory model) have leaded to accept this model as admissible.

At this point it is necessary to model the rainfall quantity distribution for rainy days. Among the several distributions we tried (Gamma, Gumbel, Normal, Lognormal, Weibull), the one that constantly fit better has resulted the Weibull distribution, that, as it is well known, is characterized by two parameters and has distribution and density function

$$F(x) = 1 - \exp\left\{-(x/\alpha)^{\beta}\right\} \quad \text{and} \quad f(x) = \frac{\beta}{\alpha}\left(\frac{x}{\alpha}\right)^{\beta-1} exp\{-(x/\alpha)^{\beta}\} \quad (11)$$

respectively. It is possible to calculate the likelihood function and its log transformation in order to obtain the maximum likelihood estimates a and b of the parameters α and β, respectively. Anyway it is necessary to verify if it is opportune to estimate the pairs of parameters not only month by month, as it seems obvious, but also splitting the two groups of data $\{0\}$ and $\{1\}$ before mentioned. At this aim it is not sufficient simply to watch the "distance" of several pairs of parameters, but it is necessary to base oneself on inferential considerations, remembering that we handle *estimated* parameters.

A graphical procedure results straightforward. Instead of constructing explicit tests, it is possible to depict a conjoint confidence region of a pair of parameters and superimpose the pairs of regions one upon another. If two regions have overlapping areas – and larger these areas are – higher is the likelihood that two populations, which two samples come from, have same parameters; on the contrary, if two regions have no overlapping areas, then it is reasonable to reject the hypothesis that two populations have same parameters, safe a I-type error risk that can be approximately fixed. As the parameters are estimated by the maximum likelihood method, their asymptotic sampling distribution can be assumed as normal. However, as the estimates of the two parameters are not independent, their conjoint distribution has to be calculated. The point estimation of such parameters is available in many statistical softwares (we used Minitab); on the contrary, for constructing and graphically representing the confidence regions with assigned confidence level, it is necessary to refer to the likelihood function of parameter vector θ, given the estimate vector t and its covariance matrix Σ:

$$L(\theta \mid t) = \frac{\exp\left\{-\frac{1}{2}(\theta - t)^T \Sigma^{-1} (\theta - t)\right\}}{(2\pi \left|\Sigma\right|)^{n/2}} \tag{12}$$

The confidence region can be obtained as that region on which the hypervolume equals the required confidence level and such as all inner points have likelihood higher than outer points. This constraint, thanks to the unimodality of the distribution, leads to focus the attention on the isolevel curves, as depicted by common contour plots for which $L(\theta|t) = cost$. Fortunately this multivariate problem can be reduced to an univariate one, by remembering that the argument of the Gaussian exponential is distributed as a χ^2 random variable. Therefore the problem simplify to calculate the curve that assures that:

$$(\theta - t)^T \Sigma^{-1} (\theta - t) = \chi^2_{(1-\alpha,\, n)} \tag{13}$$

where $\chi^2_{(1-\alpha,\, n)}$ is the $100(1-\alpha)$ percentile of the χ^2 distribution, $1-a$ is the required confidence level and the degrees of freedom n are equal to the parameters object of inference. In this case $\theta = \{\alpha, \beta\}$, $t = \{a, b\}$ and $n = 2$. The problem of computing the covariance matrix remains. Here again the asymptotic properties of maximum likelihood estimators come to help, assuring that the inverse of such matrix Σ^{-1} equals the opposite of the expectation of the Hessian H, i.e. the matrix of the second derivative of the log likelihood $l(\theta \mid t)$ in respect of the pair of the parameters. Remembering that the matrix Σ is positive definite, the previous second order function is an ellipse, with positive or negative orientation, depending on the sign of the mixed derivative (covariance). The explicit form of such function is:

$$\begin{aligned}&(\theta - t)^T E[H](\theta - t) = \\ &= (\alpha_1 - a_1, \beta_2 - b_2)^T E\begin{bmatrix} -\frac{\partial^2 l(\alpha,\beta)}{\partial\alpha^2} & -\frac{\partial^2 l(\alpha,\beta)}{\partial\alpha\partial\beta} \\ -\frac{\partial^2 l(\alpha,\beta)}{\partial\alpha\partial\beta} & -\frac{\partial^2 l(\alpha,\beta)}{\partial\beta^2} \end{bmatrix} (\alpha_1 - a_1, \beta_2 - b_2)\end{aligned} \tag{14}$$

where:

$$\frac{\partial^2 l}{\partial \alpha^2} = -\frac{\beta}{\alpha^2}\left\{(\beta+1)\sum_{i=1}^{n}(x_i/\alpha)^\beta - n\right\}$$

$$\frac{\partial^2 l}{\partial \beta^2} = -\frac{n}{\beta^2} - \sum_{i=1}^{n}(x_i/\alpha)^\beta log^2(x_i/\alpha) \tag{15}$$

$$\frac{\partial^2 l}{\partial \alpha \partial \beta} = -\frac{n}{\alpha} + \frac{\beta}{\alpha}\sum_{i=1}^{n}(x_i/\alpha)^\beta log(x_i/\alpha) + \frac{1}{\alpha}\sum_{i=1}^{n}(x_i/\alpha)^\beta$$

The expectations of the second derivatives can be estimated through the sampling data, substituting them by the sampling means: the very large sample sizes allow to state that such estimates are surely very accurate.

At this point it is possible to draw a line representing an area where an estimated rate of the likelihood volume is contained, say the 90%. Comparing the regions related to two groups (rainy days preceded by a non-rainy day and rainy days preceded by a rainy day, the graphs are here not reported for saving space) the hypothesis that the parameters α and β month by month are equal for both groups is acceptable, with the exception of April. (This case can be reasonably regarded as an admissible I type error). Therefore we can conjointly estimate the pair of parameters of the Weibull distribution per each month. The confidence regions for pairs of parameters are reported in Figure 3.1.

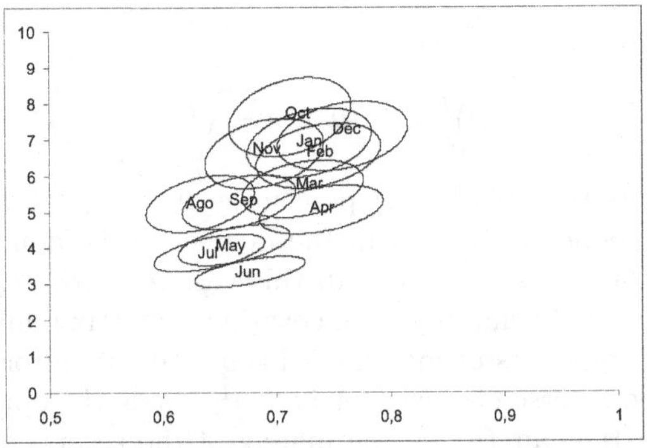

Fig. 3. Confidence region (90% confidence level) for the pairs of monthly parameters of the Weibull distribution representing rainfall

We can see the presence of perhaps four typical behaviour: the first one (autumn-winter), characterizing the months from October to February, with α within 6 and 8; the second one (spring) involving the month of March and April, with α within 5 and 6 and β within 0.7 and 0.8; the third one (late summer and early autumn) for months of August and September, with α again within 5 and 6, but β lower than before, within 0.6 and 0.7; last one is the typical southern high summer from May to July, with α within 3 and 4 and β within 0.6 and 0.7.

3.2 The New Proposed Model

Let we pass to the differential model proposed (9). It is based on the distribution of two quantity: *duration* and *global magnitude* of the recorded phenomenon at each occurrence (cluster). Therefore it is necessary to record, per each cluster of consecutive rainy days, the number of such days and the cumulated rain fallen in those days. Because the analysis must be carried out month by month, when a cluster extends on two consecutive months, the datum will be assigned to the first month, by convention. At this point, to simplify the analysis one can sum all the rainfall of consecutive rainy days and assign this quantity as the first day rainfall. This means to consider a Poisson process with variable expected rate of rain origin arrivals. This rate varies (approximately) according to a sinusoidal function (see Figure 3.2).

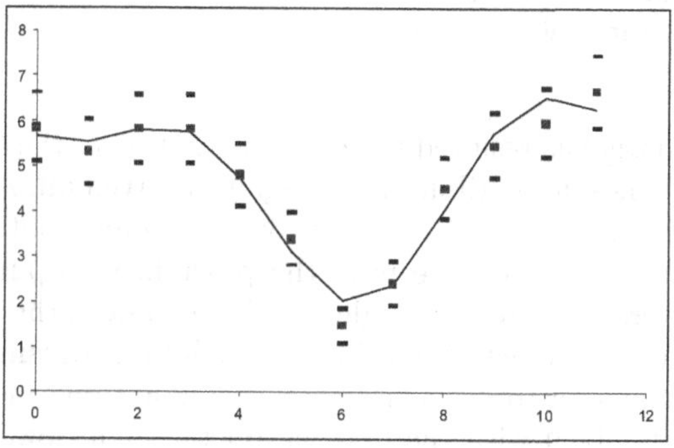

Fig. 4. Monthly behaviour of the expected number of clusters of rainy days, and relative confidence interval (95% confidence level), interpolated by a periodical function (with 2 harmonics)

For characterizing the distribution of the duration some difficulties have been encountered. At the beginning, it seemed natural to turn the attention towards discrete random variables, in particular to those coming from the Bernoullian scheme: Binomial, Poisson and Negative Binomial, characterized by the fact that expectation is greater than, equal to, or lesser than variance, respectively. Unfortunately the comparison of the first two sampling moments does not showed a constant relation, month by month: for some months was greater the mean, for others the variance, with deviates not attributable only to random effects. At this point we turned out attention towards continuous distribution. In particular it seemed that the Lognormal distribution assures a good fitting of the data. In Figure 3.2 the confidence intervals of the pairs of parameters, obtained month by month, of the Lognormal distributions are reported. In this case we can neglect the covariance between the parameter pairs, because they are the sampling mean and the standard deviation, that are independent in normal case. As it can be seen from this figure, here it is again possible to hypothesize periodical functions, describing the trend of these two parameters.

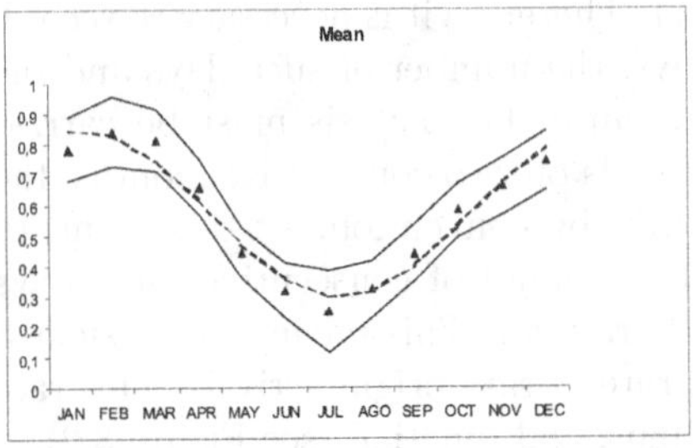

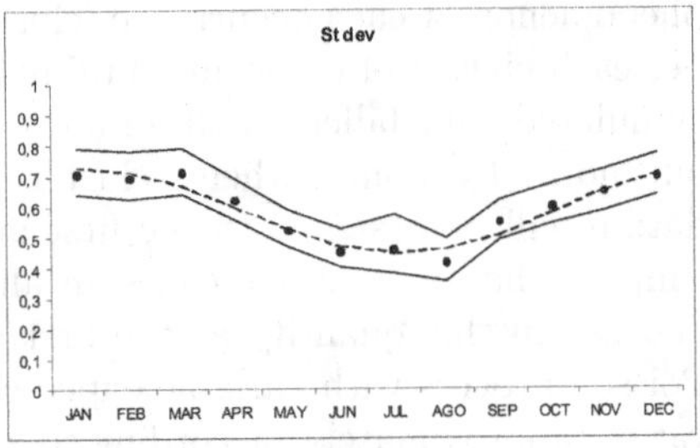

Fig. 5. Confidence intervals (95% confidence level) for the pairs of monthly parameters of the Lognormal distribution, interpolated by a periodical function with 1 harmonic, representing rainfall cumulated duration

Less problematic has resulted the characterization of the cumulated magnitude. Here the chosen distribution was again the Weibull. As it can be seen from the confidence regions of the pairs of parameters, obtained month by month (see Figure 3.2), here we have the possibility to put together some pairs of parameters too. We can single out three models: the first one related to autumn-winter (from October to March) with form parameter from 0.7 to 0.9 and scale parameter from 15 and 25; a spring-autumn model (April-May and August-September) with the first parameter slight lower (from 0,6 and 0,8) and scale parameter very lower (nearly 10); and finally a summer model (June-July) with still lower parameters (the first one 0.65 to 0.75 and the second nearly 5).

4 Conclusions

In Sicilian region, on the base of the examined rainfall data, it has been possible to observe that:

- The probability of having a rainy day depends on the fact that the previous day has been a rainy or non-rainy day. This behaviour is easily understandable by considering the mechanism generating rainfall, related to atmospheric disturbances passing on the region territory and the extension of the cloud masses.

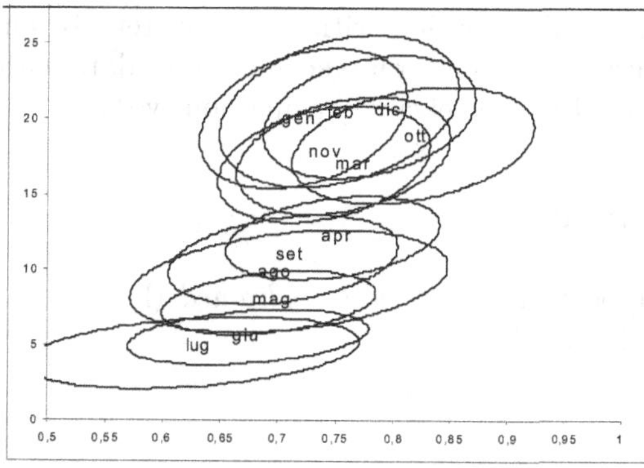

Fig. 6. Confidence region (90% confidence level) for the pairs of monthly parameters of the Weibull distribution representing rainfall cumulated magnitude

- The occurrence of a rainy day can be modeled as a Poisson random phenomenon, with rate varying month by month and depending on the previous day status (rainy, non-rainy).
- In order to represent the phenomenon in a way that is more adherent to its physical characteristics, a new model has been proposed, that takes into account not only the rainfall magnitude, but also the duration of each cluster of rainy days.
- The two models cannot be numerically compared, because in the first one the magnitude represents the rainfall of one rainy day, in the second one the rainfall of a cluster of rainy days. Anyway, for these data in both models it has been possible to represent this quantity by the same random variable.
- The main advantage of the second model lies in its probabilistic representation, that makes it more useful for hydrological aims, rather then short term forecasting. In fact it can be used for generating "simulated rains", or its estimated parameters can be introduced directly in more complex hydrological models.
- Moreover the new proposed model can be extended to data coming from other hydrological sites, in order to develop a model valid for a larger area (say, region). At the moment we do not know if changing site or time period the representation of amplitude and duration will keep constant. However, the model seems quite flexible to fit well intermittent data, as those dealt with in this paper.
- Because this study is in a prototypal phase, until now the computational procedure is not optimized, in fact data handling and calculations have been implemented by means of different softwares; but it can be easily improved.
- Eventually, the proposed model can be made closer to reality by adding another parameter describing the average behaviour of rainfall in a single cluster; this leads to solve another differential equation that models the

height of rainfall distribution within the cluster. To aim at this, however, it is necessary to have a larger number of data within each cluster, therefore it could be conceivable having hourly data.

Acknowledgments

The financial support of the *Project Hydromed* (http://www.hydromed.it/hydromed/) is acknowledged.

References

CADDEMI, S. and DI PAOLA, M. (1995): Ideal and physical white noise in stochastic analysis, *Int. Journal Nonlinear Mechanics, 31,(5), 581–590.*

CALENDA, G. and NAPOLITANO, F. (1999): Parameter estimation of neyman-scott processes for temporal point rainfall simulation, *Journal of Hydrology, 225, 45–66.*

DI PAOLA, M. (1997): Linear systems excited by polynomials of filtered poisson pulses, *Journal of Applied Mechanics, 64, 712–717.*

DI PAOLA, M. and FALSONE, G. (1993 a): Itô and stratonovich integral for delta-correlated processes, *Probabilistic Engineering Mechanics, 8, 197–208.*

DI PAOLA, M. and FALSONE, G. (1993 b): Stochastic dynamics of nonlinear systems driven by non normal delta-correlated process, *Journal of Applied Mechanics, 60, 141–148.*

DI PAOLA, M. and VASTA, M. (1997): Stochastic integro-differential equation and differential equations of nonlinear systems excited by parametric poisson pulses, *International Journal of Nonlinear Mechanics, 32(5), 855–862.*

GRUNWALD, G.K. and RICHARD, H.J. (2000): Markov models for time series with mixed distribution, *Environmetrics, 11, 327–339.*

PIRROTTA, A. (1998): Non-linear systems under delta-correlated processes handled by perturbation theory, *Probabilistic Engineering Mechanics, 13(4), 283–290.*

SITARAMAN, H. (1991): Approximation of some markov-modulated poisson processes, *ORSA Journal on Computing, 3(1), 12–22.*

STRATONOVICH, R.L. (1963): *Topics in the Theory of Random Noise*, Gordon and Breach, New York.

A Comparison of Data Mining Methods and Logistic Regression to Determine Factors Associated with Death Following Injury

Kay Penny[1] and Thomas Chesney[2]

[1] Centre for Mathematics and Statistics, Napier University, Craiglockhart Campus, Edinburgh, EH14 1DJ, Scotland, U.K.
k.penny@napier.ac.uk

[2] Nottingham University Business School, Jubilee Campus, Wollaton Road, Nottingham, NG8 1BB, U.K.
Thomas.Chesney@nottingham.ac.uk

Abstract. A comparison of techniques for analysing trauma injury data collected over ten years at a hospital trauma unit in the U.K. is reported. The analysis includes a comparison of four data mining techniques to determine factors associated with death following injury. The techniques include a classification and regression tree algorithm, a classification algorithm, a neural network and logistic regression. As well as techniques within the data mining framework, conventional logistic regression modelling is also included for comparison. Results are compared in terms of sensitivity, specificity, positive predictive value and negative predictive value.

1 Introduction

Trauma is the most common cause of loss of life to those under forty (Trauma Audit and Research Network (TARN), 2004), and many of these deaths are preventable. In 1991, in response to a report from the Royal College of Surgeons of England, the government supported a pilot trauma system at the North Staffordshire Hospital in Stoke-on-Trent in the U.K. (Oakley et al., 2004). The Trauma Audit and Research Network (TARN, 2004) records injury details such as the patient's sex and age, the mechanism of injury, various measures of the severity of the injury, initial management and subsequent management interventions, and the outcome of the treatment, including whether the patient lived or died. The system is intended to develop effective care for injured patients through process and outcome analysis and dissemination of results. TARN data are collected throughout England although the current study concerns only that data collected in the University Hospital of North Staffordshire. The University Hospital of North Staffordshire is the major trauma centre in the area, and receives referrals from surrounding hospitals.

The aim of this research is to determine factors associated with death for all levels of injury severity for patients admitted between 1992 and 2003. The results of the data mining techniques and a conventional logistic regression approach are reported, and evaluated according to their predictive ability.

2 Methods

The data were de-identified by the data owner prior to analysis for reasons of confidentiality and privacy, and the data were held securely. The data were collapsed into a flat file (single table), as many data mining techniques require the data to be in this format (Breault, 2002). Two different approaches to the statistical analysis of these data were considered; data mining techniques and a conventional logistic regression modelling approach. The data mining techniques considered include a classification algorithm (C5.0) a classification and regression tree (CART), logistic regression (LR), and a neural network (NN). The analysis was carried out using the statistical packages SPSS 11.0.1 and Clementine 7.0.

Factors considered for inclusion in the models are summarised in Table 1 and include patient's age and sex, year of admission to hospital, primary receiving hospital, and whether the patient was referred from another hospital, mechanism of injury and type of trauma. Various injury severity scores were also considered including Injury Severity Score (ISS) (Baker et al., 1974), Abbreviated Injury Scores (AIS) (Baker et al., 1990), a new injury severity score (NISS) (Osler et al., 1997), and the Glasgow Coma Score (GCS) (Teasdale and Jennet, 1974).

Decision tree algorithms such as C5.0 and CART are classifiers that create a set of rules to arrive at a result. C5.0 has been successfully used previously in medical data mining (see for example: Aguilar-Ruiz et al., 2004; Jensen, 2001). A decision tree has attributes (or to use the data mining terminology, "factors") from the dataset as its nodes. Each branch off its nodes are the values that each factor can take. For example, the node for the factor, "patient referred from another hospital" would have two branches, "yes" and "no". To construct the tree, the factors in the dataset must be split according to how much information they contribute to the classification task. The more information they contribute, the higher up they are in the tree. In this fashion, the tree is organised with as few nodes as possible between the top of the tree and the final branch where classification is made. One important point to note is that in a decision tree, the rules are not meant to be stand alone and the order of the rules is vital.

Neural networks on the other hand, attempt to model human intelligence using the neurons in a human brain as an analogy. Input is fed through the neurons in the network which transform them to output a probability, in this case, the probability that a patient will die. An exhaustive prune was used to create the ANN. In the ANN, all the neurons are fully connected and each is a feed-forward multi layer perceptron which uses the sigmoid transfer function (Watkins, 1997). The learning technique used was back propagation. This means that, starting with the given topology, the network was trained, then a sensitivity analysis is performed on the hidden units and the weakest were removed. This training/removing was repeated for a set length of time.

Age in years
Sex (Male or Female)
Age group (years): 0-15; 16-25; 26-35; 36-50; 51-70; over 70
Year of admission (1992-2003)
Primary receiving hospital
Referred from another hospital (yes or no)

Mechanism of injury group:
Motor vehicle crash; Fall greater than 2m; Fall less than 2m; Assault; Other

Type of trauma (blunt or penetrating)
Injury Severity Score (**ISS**)
New Injury Severity Score (**NISS**)

Abbreviated injury scores (AIS):
Head, Face, Lower limb, Neck, Chest, External, Abdomen, Cervical-spine, Upper limb, Thoracic-spine, Lumbar-spine

Glasgow coma score (GCS) group:
Mild brain injury, total GCS: 13-15
Moderate brain injury, total GCS: 9-12
Severe brain injury, total GCS: 3-8

Glasgow coma scores (GCS):
Eye response; Motor response; Verbal response

Table 1. Factors considered in the analyses

As well as within the data mining framework, conventional LR modelling was included for comparison. In medical applications it is usual practice to develop a logistic regression model using the complete data set, and the model is then tested on the same set of data used to build the model. However, to allow comparison with the data mining methods presented in this paper, the conventional LR model has been developed using the same training data set as that used for the data mining methods, and tested on the validation data set. The main difference between the conventional LR model and the LR method within the data mining framework, is that the conventional LR method was developed in SPSS with user intervention to determine a parsimonious model with good predictive ability, yet as simple a model as possible. Hence the conventional approach is more subjective than the data mining approach which includes the main effects for all variables in the model, however little they add to the predictive ability of the model.

For training and testing the data mining methods, the data were randomly split into a training data set and a validation data set. A two thirds/one third split was used, keeping the same proportions of live/die outcomes in each. There are 7787 records in the training data set and 3896 in the validation data set. The data are however, very imbalanced as more patients lived (94%) than died (6%), therefore boosting, which is available within the Clementine software, was used in an effort to improve the modelling for the C5.0 method.

In many data mining efforts the evaluation criterion is the accuracy i.e. the percentage of correct classifications made by an algorithm, however, in medical data mining consideration must also be given to the percentage of false positives and false negatives made. The evaluation criteria included for testing the classification algorithms are sensitivity, specificity, positive predictive value (PPV) and negative predictive value (NPV). A receiver operator curve (ROC) analysis is used to compare the areas under the curve for the two logistic regression models.

Factor		**OR (95% CI)**	**p-value**
Mechanism of injury:	Motor vehicle crash	1.00	<0.001
	Fall < 2 metres	0.73 (0.51, 1.04)	0.078
	Fall > 2m	0.52 (0.33, 0.82)	0.005
	Assault	0.58 (0.25, 1.35)	0.203
	Other	1.42 (0.95, 2.12)	0.090
Age in years:	0 - 15	1.00	<0.001
	16 -25	0.94 (0.52, 1.71)	0.850
	26 - 35	1.47 (0.79, 2.73)	0.222
	36 - 50	1.41 (0.77, 2.57)	0.270
	51 - 70	3.89 (2.16, 7.02)	<0.001
	70 or over	24.21 (13.34, 43.97)	<0.001
Referred from another hospital		0.39 (0.27, 0.57)	<0.001
New Injury Severity Score (NISS)		1.09 (1.08, 1.10)	<0.001
GCS group:	Mild	1.00	<0.001
	Moderate	1.96 (1.14, 3.37)	0.015
	Severe	12.57 (8.66, 18.23)	<0.001
Year of admission:	1992	1.00	<0.001
	1993	1.18 (0.62, 2.23)	0.616
	1994	0.83 (0.42, 1.65)	0.592
	1995	0.59 (0.30, 1.15)	0.123
	1996	0.43 (0.22, 0.84)	0.013
	1997	0.97 (0.51, 1.84)	0.916
	1998	0.63 (0.33, 1.20)	0.160
	1999	0.79 (0.41, 1.53)	0.490
	2000	0.66 (0.29, 1.51)	0.325
	2001	1.32 (0.65, 2.69)	0.440
	2002	0.88 (0.44, 1.79)	0.728
	2003	2.56 (1.27, 5.15)	0.008

Table 2. Conventional Logistic Regression Model

3 Results

The results of the conventional LR model show that patients with severe injuries according to the NISS and GCS, involvement in motor vehicle accidents, older age and not being referred from another hospital, are all associated with increased odds of death (see Table 2). Correcting for the other factors included in the model, the odds of death vary according to year of admission and appear to be decreasing over the years until 2003. The variables which are significant in the LR model within the data mining framework include most of the injury severity scores, type of trauma, mechanism of injury, age group and year of admission. However, this model is much more complex and includes many more variables, many of which are not statistically significant in the model. The LR model produced within the data mining framework is more accurate than the conventional LR model, as confirmed by the receiver operator curve (ROC) analysis which gave areas under the curve of 0.93 for the conventional LR and 0.96 for the LR validation data within the data mining context.

The NN model output gives details of the relative importance of the variables, which are: age group, 12 AIS scores, NISS, year of admission, GCS and mechanism of injury (see Table 3). Both the C5.0 and CART models produce a set of rules comprising combinations of injury severity measures and age of patients.

Relative importance of most prominent factors:

Age group	0.38
11 Abbreviated injury scores between	0.23-0.41
New Injury Severity Score (NISS)	0.31
Year of admission	0.26
GCS - eye response	0.25
Mechanism of injury	0.23
GCS - motor response	0.21
GCS - verbal response	0.20

Table 3. Neural Network Results

The evaluation of the data mining methods with the validation data is shown in Table 4, alongside the evaluation of the conventional LR model. Results are compared in terms of sensitivity, specificity, PPV and NPV in correctly determining patient death following injury. A cut-off value of 0.5 has been used in the LR evaluations. The NN model has a high sensitivity of 96.0% as almost all of the true deaths have been correctly predicted. The sensitivity is comparatively poor for the C5.0 and conventional LR models (65.2% and 54.3% respectively). However, the PPV is highest for these two models, showing that a higher proportion of patients, who are predicted to die, did actually die.

4 Conclusions

The data mining methods considered here predict death with good accuracy, in particular the NN method. Neural networks are known to produce good results across a number of applications; however, a major disadvantage is that the results cannot be explained easily. Although the relative importance of the explanatory variables is given, the effect of each explanatory variable

	Data Mining Methods				
Evaluation Criteria	C5.0	CART	Neural Network	Logistic Regression	Conventional Logistic Regression
Sensitivity	65.2%	78.0%	96.0%	91.5%	54.3%
Specificity	94.3%	88.0%	87.7%	86.9%	97.9%
PPV	0.422	0.293	0.333	0.407	0.623
NPV	0.977	0.984	0.959	0.994	0.971

Table 4. Predictive Evaluations of Methods

on the outcome variable is unknown. Hence neural networks are less useful if the aim is to explain which factors are likely to lead to death or survival.

The other three data mining methods provide rules (C5.0 and CART) or odds ratios (LR), which give an insight into the characteristics associated with death. The conventional LR model appears to provide less accurate results; this is due to the fact that only six independent variables were included in the conventional LR model, whereas the LR model within the data mining approach includes all variables available for consideration. The LR model from the conventional approach is less complex and easier to interpret than the LR model derived by data mining. Therefore, using the conventional model, the odds of death could be calculated quite quickly for a trauma patient if the values of the variables included in the model were available.

Data mining can be thought of as an in-depth search to find information that previously went unnoticed in the mass of data collected (Giudici, 2003), and has been used in medical applications in recent years (Cios and Moore, 2002; Breault et al., 2002). Lavrac (1999) summarises and discusses some techniques specifically for medical data mining and finds that decision trees often give an appropriate explanation of decisions although they are not favoured by physicians. Physicians may feel that, after pruning, too few parameters are taken into account. Bratko (1986) found medical experts preferred trees that had redundant rules, which had been removed during pruning then re-inserted, despite the fact that this led to a reduction in accuracy.

As well as providing reassurance that the trauma injury measures in current use are valid and reliable measures, data mining a medical database such as this one has the potential to discover factors or combinations of factors which can predict death. Although the models are very complex, they can be pruned to simpler models, but at a loss of accuracy. If the model is to be used in the field i.e. to calculate a patient's risk of death following injury, then a simpler, but less accurate, model would be needed. However, if the aim is to explore a database for previously unknown risk factors, then it is reasonable to consider a more accurate but complex model.

Survival analysis of these data was also considered. However, the dataset contains 33% of missing values for date of death or discharge, of which, 70%

are missing for patients who died. Hence the results may be biased due to the high proportion of missing data for this variable.

Acknowledgements

The authors wish to thank David Chesney, Simon Davies, Nicola Maffuli and Peter Oakley for their advice and comments on this paper.

References

AGUILAR-RUIZ, J.S., COSTA, R, DIVINA, F. (2004): Knowledge discovery from doctor-patient relationship. *Proceedings of the 2004 ACM Symposium on Applied Computing (Nicosia, Cyprus, March 14 - 17, 2004).* ACM Press, New York.

ASSOCIATION FOR THE ADVANCEMENT OF AUTOMOTIVE MEDICINE (1990): *The abbreviated injury scale, 1990 revision.* Des Pleines, IL., Association for the Advancement of Automotive Medicine.

BAKER, S.P., O'NEILL, B., HADDON Jr., W. and LONG, W.B. (1974): The injury severity score: a Method for describing patients with multiple injuries and evaluating patient care. *Journal of Trauma, 14, 187–196.*

BRATKO, I. (1986): Machine Learning: Between accuracy and interpretability. In: G. Della Riccia, H.J. Lenz, R. Kruse (Eds.): *Learning, Networks and Statistics.* Springer, CISM Courses and Lectures No 382, 163–177.

BREAULT, J.L., GOODALL, C.R. and FOS, P.J. (2002): Data mining a diabetic data warehouse. *Artificial Intelligence in Medicine, 26 (1,2) 37–54.*

CIOS, K.J and MOORE, G.W. (2002): Uniqueness of medical data mining. *Artificial Intelligence in Medicine, 26 (1, 2) 1–25.*

GIUDICI, P. (2003): *Applied Data Mining.* Wiley, Chichester.

JENSEN, S. (2001): Mining medical data for predictive and sequential pattern. *PKDD Challenge on Thrombosis Data.* http://lisp.vse.cz/challenge/pkdd2001/jensen.pdf (Accessed 12/01/06).

LAVRAC, N. (1999): Selected techniques for data mining in medicine. *Artificial Intelligence in Medicine, 16, 3–23.*

OAKLEY, P.A., MACKENZIE, G., TEMPLETON, J., COOK, A.L. and KIRBY, R.M. (2004): Longitudinal trends in trauma mortality and survival in Stoke-on-Trent 1992-1998. *Injury, 35, 379–385.*

OSLER, T., BAKER, S. and LONG, W. (1997): A modification of the injury severity score that both improves accuracy and simplifies scoring. *J Trauma, 41, 922–926.*

TARN (2004): The Trauma Audit and Research Network. http://www.tarn.ac.uk/introduction/FirstDecade.pdf. (Accessed 3/11/04).

TEASDALE, G. and JENNETT, B. (1974): Assessment of coma and impaired consciousness. A practical scale. *Lancet (ii), 81–83.*

WATKINS, D. (1997): Clementine's Neural Networks Technical Overview. http://www.cs.bris.ac.uk/~ cgc/METAL/Consortium/secure/neural_overview.doc. (Accessed 12/01/06).

Author Index